Reflexion and Control
Mathematical Models

Communications in Cybernetics, Systems Science and Engineering

ISSN: 2164-9693

Book Series Editor:

Jeffrey Yi-Lin Forrest
International Institute for General Systems Studies, Grove City, USA
Slippery Rock University, Slippery Rock, USA

Volume 5

Reflexion and Control
Mathematical Models

Dmitry A. Novikov &
Alexander G. Chkhartishvili

Trapeznikov Institute of Control Sciences, Russian Academy of Sciences, Moscow, Russia

CRC Press
Taylor & Francis Group
Boca Raton   London   New York   Leiden

CRC Press is an imprint of the
Taylor & Francis Group, an **informa** business

A BALKEMA BOOK

CRC Press/Balkema is an imprint of the Taylor & Francis Group, an informa business

© 2014 Taylor & Francis Group, London, UK

Typeset by MPS Limited, Chennai, India
Printed and Bound by CPI Group (UK) Ltd, Croydon, CR0 4YY

Library of Congress Cataloging-in-Publication Data

Novikov, D. A. (Dmitry Aleksandrovich), 1970– author.
    Reflexion and control : mathematical models / Dmitry A. Novikov & Alexander G. Chkhartishvili,
Trapeznikov Institute of Control Sciences, Russian Academy of Sciences, Moscow, Russia.
        pages cm. — (Communications in cybernetics, systems science and engineering,
ISSN 2164-9693 ; volume 5)
        Summary: "The book considers the basics of mathematical modeling of reflexive processes
in control. The models of informational and strategic reflexion allow describing and studying the
behavior of reflexing subjects, investigating the relationship between payoffs gained by agents and
their reflexion ranks, solving problems of informational and reflexive control in organizational,
economic, social and other systems, and military applications among others (the book contains
over 30 examples of possible applications in these fields), describing uniformly many phenomena
connected with reflexion, viz., implicit control, informational control via the mass media and
reflexion in psychology"—Provided by publisher.
    Includes bibliographical references and index.
    ISBN 978-1-138-02473-1 (hardback) — ISBN 978-1-315-77554-8 (eBook PDF)   1. Game theory.
2. Control theory.   I. Chkhartishvili, A. G. (Aleksandr Gedevanovich), author.   II. Title.
    QA269.N68   2014
    515'.642—dc23

                                                                              2014000188

Published by: CRC Press/Balkema
                 P.O. Box 11320, 2301 EH Leiden, The Netherlands
                 e-mail: Pub.NL@taylorandfrancis.com
                 www.crcpress.com – www.taylorandfrancis.com

ISBN: 978-1-138-02473-1 (Hbk)
ISBN: 978-1-315-77554-8 (eBook PDF)

- "See how the minnows come out and dart around where they please! That's what fish really enjoy!"
- "You're not a fish – how do you know what fish enjoy?"
- "You're not me, so how do you know I don't know what fish enjoy?"

From a Taoist parable

- "Well, of course, Archbishop, the point is that you believe what you believe because of the way you were brought up."
- "That is as it may be. But the fact remains that you believe I believe what I believe because of the way I was brought up, because of the way you were brought up."

From D. Myers' book "Social Psychology"

Table of contents

Editorial board   ix
About the authors   xi

Introduction   1

1   Reflexion in decision-making   17
   1.1   Individual decision-making   17
   1.2   Interactive decision-making: Games and equilibria   19
   1.3   General approaches to the description of informational and
         strategic reflexion   24

2   Informational reflexion and control   29
   2.1   Informational reflexion in two-player games   29
   2.2   Awareness structure of games   32
   2.3   Informational equilibrium   37
   2.4   Graph of a reflexive game   40
   2.5   Regular awareness structures   45
   2.6   Reflexion rank and informational equilibrium   50
   2.7   Stable informational equilibria   59
   2.8   True and false equilibria   62
   2.9   The case of observable actions of agents   64
   2.10 Reflexive games and Bayesian games   68
   2.11 Informational control   73
   2.12 Modeling of informational impact   80
   2.13 Set-type awareness structures   89
   2.14 Transformation of awareness structure   95
   2.15 Concordant informational control   98
   2.16 Reflexion in planning mechanisms   106

3   Strategic reflexion and control   113
   3.1   Strategic reflexion in two-player games   113
   3.2   Reflexion in bimatrix games and games of ranks   118
   3.3   Boundedness of reflexion ranks   133
   3.4   Reflexive structures and reflexive control   134

4   Applied models of informational and reflexive control   **145**
  4.1   Implicit control   145
  4.2   The mass media and informational control   153
  4.3   Reflexion in psychology   155
    4.3.1   Playing chess   155
    4.3.2   Transactional analysis   157
    4.3.3   The Johari window   159
    4.3.4   Ethical choice   160
  4.4   Reflexion in bélles-léttres   161
  4.5   Reflexive search games   167
  4.6   Manufacturers and intermediate sellers   171
  4.7   The scarcity principle   174
  4.8   Joint production   176
  4.9   Market competition   181
  4.10 Lump sum payments   183
  4.11 Sellers and buyers   189
  4.12 Customers and executors   192
  4.13 Corruption   194
  4.14 Bipolar choice   195
  4.15 Active expertise: Informational reflexion   199
  4.16 The cournot oligopoly: Informational reflexion   205
  4.17 Resource allocation   207
  4.18 Insurance   210
  4.19 Product advertizing   216
  4.20 The hustings   218
  4.21 Rank-order tournaments   219
  4.22 Explicit and implicit coalitions in reflexive games   221
  4.23 Active forecast   231
  4.24 Social networks   234
  4.25 Mob control   240
  4.26 The reflexive partitions method   243
    4.26.1 Diffuse bomb   244
    4.26.2 The colonel Blotto game   253
    4.26.3 The Cournot oligopoly: strategic reflexion   259
    4.26.4 The consensus problem   263
    4.26.5 Active expertise: strategic reflexion   264
    4.26.6 Transport flows and evacuation   266
    4.26.7 A stock exchange   268

Conclusion   **273**

References   275
Subject index   283

Editorial board

About the authors

Dmitry A. Novikov was born in 1970. He is doctor of science in engineering, Professor and corresponding member of the Russian Academy of Sciences. At present, he is the deputy director of the Trapeznikov Institute of Control Sciences of the Russian Academy of Sciences and head of the Control Sciences Department of the Moscow Institute of Physics and Technology. He has authored more than 400 scientific works on the theory of control in interdisciplinary systems, including works on methodology, system analysis, game theory, decision-making and mechanisms of control in social and economic systems. He is scientific adviser of more than 30 doctors of science and PhDs.

Alexander Chkhartishvili was born in 1970. He is doctor of science in mathematics. At present, he is the head of the Research Group at the Institute of Control Sciences of the Russian Academy of Sciences. He has authored more than 150 scientific works on game theory, decision-making and mechanisms of control of social and economic systems.

Introduction

This book is dedicated to modern approaches to mathematical modeling of *reflexion in control* (including an important class of game-theoretic models – *reflexive games* describing the interaction of subjects making decisions based on an hierarchy of beliefs about essential parameters, beliefs about beliefs, etc.).

Reflexion. A fundamental property of human entity lies in the following. In addition to *natural* ("objective") *reality*, there exists its image in human minds. Furthermore, an inevitable gap (mismatch) takes place between the latter and the former. In the sequel, the described image will be called a part of *reflexive reality*.

Traditionally, purposeful study of this phenomenon relates to the term "reflexion." The term reflexion (from Latin *reflex* "bent back") means:

- a principle of human thinking, guiding humans towards comprehension and perception of one's own forms and premises;
- subjective consideration of a knowledge, critical analysis of its content and cognition methods;
- the activity of self-actualization, revealing the internal structure and specifics of the spiritual world of a human.

The term *"reflexion"* was first suggested by John Locke. However, in different philosophical systems (those of Locke, Leibniz, Hume, Hegel, etc.), reflexion has various interpretations. In psychology, systematical treatment of reflexion dates back to the 1960s (Lefebvre's scientific school). Note one may view reflexion in another interpretation connected with a *reflex* (a reaction of a living organism to excited receptors). In the present book, we employ the first (philosophical) definition of reflexion.

To elucidate the whole essence of reflexion, let us consider the case of a single subject. He/she possesses certain beliefs about natural reality; however, a subject may perform reflexion (construct images) with respect to these beliefs (thus, generating new beliefs). Generally, this process is infinite and results in formation of reflexive reality. The reflexion of a subject with respect to his/her own beliefs of reality, principles of his/her activity, etc., is said to be *self-reflexion* or *reflexion of the first kind*. We emphasize that most social research works concentrate on self-reflexion. In philosophy, self-reflexion represents the process of the individual's thinking about beliefs in his/her own mind [127].

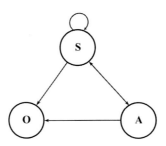

Figure I Ways of estimating.

Reflexion of the second kind takes place with respect to other subjects; it covers the beliefs of a subject about possible beliefs, decision principles and self-reflexion of other subjects.

Reflexion ranks. To provide a common description of reflexive imaging, psychology involves the following approach [127]. Consider interrelations among three elements (see Fig. 1), *viz.*, the subject of activity (S), the object of activity (O) and another subject (A). The arrows designate separate acts of "thinking" ("constructing images").

We use various sequences of characters ("S," "O," and "A") to characterize relations among the elements. The order of characters corresponds to (a) who assesses (constructs images of) what or (b) who performs reflexion with respect to what. The object of activity is assumed "passive" (thus, performing no reflexion).

First order relations (zero-rank reflexion) include the following *estimates*:

SO – the estimate of the results of the subject's activity by himself/herself (*self-appraisal of the results*);
SS – the estimate of the subject by himself/herself (individual *self-appraisal*);
SA – the estimate of another subject by the subject of activity (as an individual);
AO – the estimate of the results of the subject's activity by another subject;
AS – the estimate of the subject by another subject (as an individual).

Being passive, the object appears unable to estimate; moreover, we do not consider self-appraisal of another subject (AA). Therefore, the above five relations exhaust feasible combinations of the relations of the first order.

The subject of activity and another subject may think about the relations shown in Fig. 1. This yields first-rank reflexion.

Second order relations (first-rank reflexion). Here one should distinguish between:

– self-reflexion (reflexion of the first kind), which corresponds to SS-type sequences, i.e., subject's thoughts about his/her self-appraisal and self-appraisal of his/her results:
SSO – the subject's thoughts about his/her appraisal of his/her results;
SSS – the subject's thoughts about his/her self-appraisal;

and

– reflexion of the second kind (the remaining sequences):

SAO – the subject's thoughts about the estimate of his/her activity results by another subject ("what others think about the results of my activity");

SAS – the subject's thoughts about the estimate given to him/her by another subject ("what others think about me");

ASS – another subject's thoughts about the subject's self-appraisal;

ASO – another subject's thoughts about the subject's self-appraisal of his/her activity results;

ASA – another subject's thoughts about the estimate given to him/her by the subject of activity.

Third order relations (second-rank reflexion). Naturally, in this case we find numerous combinations. Some are provided below: SASO – the subject's thoughts about another subject's thoughts about his/her self-appraisal of the subject's results ("what others think about my estimates of my results"); ASAO – another subject's thoughts about the subject's thoughts about the estimate given to his/her activity results by another subject, and so on.

Similarly, this framework serves to describe relations of higher orders (higher reflexion ranks).

Examples. Below we choose several examples of second-order reflexion that illustrate the following. In many cases, making correct conclusions is possible only by taking the position of other subjects and analyzing their feasible reasoning.

The first example concerns the classical *dirty face game*) [65], also known as the three wise men puzzle or the problem of husbands and unfaithful wives [121].

Imagine the United Kingdom of the Victorian period; two passengers, Bob and his niece Alice, sit in a compartment of a railway carriage. They both have dirty faces. However, neither of them blushes with shame (Victorians would definitely blush if somebody observed them with a dirty face). And so, we conclude that neither passenger knows anything about his/her dirty face (even though observing the dirty face of the companion).

Suddenly, the Conductor enters the compartment and takes notice of a sitting passenger with a dirty face. Subsequently, Alice blushed. Actually, she understood that her face was dirty. However, how could she realize that? Hadn't the Conductor reported what she knew before?

Let us follow Alice's line of reasoning. Alice: "Suppose that my face is dirty. Being aware of that one of us is dirty, Bob should then have concluded that he is dirty and should have blushed. Meanwhile, Bob is not ashamed; this implies the premise of my clean face is false – my face is actually dirty, and I should have blushed."

As a matter of fact, the Conductor added information on Bob's knowledge to Alice's awareness. Previously, she knew nothing about Bob's awareness that one of them has dirty face. In other words, the Conductor's message made the knowledge of a sitting passenger with a dirty face common knowledge.

Another example involves the *coordinated attack problem*) [70]. There also exist close problems, *viz.*, the electronic mail game [151] and others (see the reviews in [56, 76, 166]).

Consider the following situation. Two army divisions are located on two hills, whereas their enemy is in the valley. Gaining a victory over the enemy is possible only through a coordinated attack by both divisions. The commander of division 1 (*General A*) sends to the commander of division 2 (*General B*) a herald with the message "We attack at dawn." The herald can be intercepted by the enemy; and so, *General A* has to wait for message 2 from *General B* (which confirms that message 1 has been received). But message 2 can be intercepted by the enemy, as well. Hence, *General B* has to wait for the confirmation that *General A* has received his confirmation. And so on – *ad infinitum*. The problem lies in defining the maximal number of messages (confirmations) required for attacking. Still, the conclusion is as follows. Within the stated conditions, a coordinated attack is impossible, and the way out consists in adopting probabilistic models [116, 117].

The third example deals with the classical *problem of two brokers* [138]. Assume that two brokers gambling on a stock exchange apply different decision support systems. It happens that a network administrator illegally duplicates these systems and sells the opponent's system to each broker. Afterwards, the administrator tries to sell the following information to each broker: "Your opponent has your decision support system." The next initiative of the administrator is attempting to sell the following information: "Your opponent knows that you have his decision support system." And so forth. The problem is how brokers should use information acquired from the administrator. What information is important at different iterations?

Thus, we have briefly studied the examples of second-rank reflexion. Now, let us discuss when reflexion is essential. Suppose that the only subject performing reflexion is an economic agent striving for maximization of his/her goal function by choosing one of several ethically admissible actions. In this case, natural reality is incorporated in the goal function as a certain parameter, whereas the results of reflexion (beliefs about beliefs, etc.) do not represent arguments of the goal function. Consequently, one may claim that self-reflexion is useless as not modifying the agent's choice.

Note that the subject's actions can depend on reflexion when actions appear nonequivalent ethically. That is, the *utilitarian* aspect runs parallel to the deontological (*ethical*) aspect, see [95, 96]. However, generally economic decisions are ethically neutral. Thus, we analyze the interaction among several subjects.

Consider a situation of decision-making with several participating subjects (such decision situations are called *interactive*). The goal function of each subject includes the actions of other subjects. In other words, these actions represent a part of natural reality (they are conditioned by reflexive reality though). Reflexion and analysis of reflexive reality become necessary. Prior to exploring the basic approaches to mathematical modeling of reflexion effects, we describe the correlation between two major categories of this book – "reflexion" and "control."

Reflexion and control. Let us start with defining the essence of control. *Control* is an element, a function of organized systems of different nature (biological, social, technical, etc.), preserving their definite structure, sustaining their mode of activity and implementing the program or goal of their activity; *control* is a purposeful impact exerted on a controlled system to ensure its required behavior [135].

In what follows, we discuss the general statement of a control problem. Assume there is a *control subject* (a principal) and a *controlled system* (*control object* – in terminology of technical systems – or a *controlled subject*). The state of a controlled system depends on external disturbances, control actions applied by a principal and possibly on actions performed by the controlled system (if the latter represents an

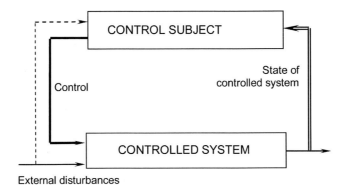

Figure 2 The structure of a control system.

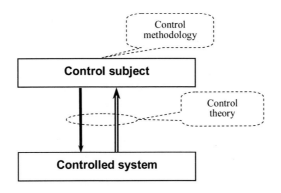

Figure 3 Control methodology and control theory.

active subject), see Fig. 2. The principal's problem lies in choosing control actions (see the thick line in Fig. 2) to ensure the required behavior of a controlled system taking into account information on external disturbances (see the dashed line in Fig. 2).

The so-called input-output structure of a control system (Fig. 2) is typical for control theory dealing with control problems in systems of different nature. The presence of *feedback* (see the double line in Fig. 2) which provides a principal with information on the state of a controlled system is the key (but not compulsory!) property of a control system. Some researchers **interpret feedback as reflexion** (as an image of the controlled system's state in the "mind" of a control subject). This forms the first aspect of interrelation between control and reflexion.

A series of scientific directions investigate the interaction and activity of a control subject and controlled system. Control science (or *control theory* in the terminology of corresponding experts) mostly focuses on the interaction between a control subject and controlled system – see Fig. 3. *Control methodology* [130] is the theory of organizing of control activity, i.e., the activity performed by a control subject. We emphasize that *activity* can be mentioned only with respect to active subjects (e.g., a human being, a group, a collective). In the case of passive (e.g., technical) systems, the term "functioning" is used instead. In the sequel, we believe that a control subject and controlled system appear active (otherwise, there is a clear provision for the opposite). Hence, **each of them may perform (at least) self-reflexion**, constructing "images" of the process, organization principles and results of his/her own activity. This is the second aspect of interrelation between control and reflexion.

Searching for *optimal control* (i.e., the most efficient admissible control) requires the control subject's ability to predict the controlled system's response to certain control actions. One of the prerequisites is a model of the controlled system. Generally speaking, a *model* is an image of a certain system; an analog (a scheme, a structure or a sign system) of a certain fragment of the natural or social reality, a "substitute" for the original in cognition process and practice. A model can be considered as an image of a controlled system in the mind of a control subject. **Modeling** (as a process of "reflecting," i.e., constructing this image) **can be viewed as reflexion.** Furthermore, a controlled system may predict and assess the activity performed by a control subject. And so, we obtain the third aspect of interrelation between control and reflexion.

The fourth aspect lies in the following. **A control subject or controlled system performs reflexion with respect to external subjects and objects,** phenomena or processes, their properties and laws of activity/functioning. For instance, the matter concerns an external environment (for a control subject), an external environment and/or other elements of a controlled system (for a fixed element of a controlled system). Indeed, suppose that a controlled system includes several active agents; each of them may perform reflexion with respect to the others. It is exactly this aspect – mutual reflexion of controlled subjects – which is discussed in detail below (see Chapters 2–3).

The listed quartet of aspects corresponds to zero reflexion rank ("estimating," see above). By analogy to Fig. 1, one can give a uniform description to reflexion of higher ranks. First-rank reflexion covers the control subject's beliefs about the estimate of other controlled subjects (agents) by a given agent. Second-rank reflexion touches the estimate of these beliefs by a controlled system. And so on.

Of crucial importance here is that **the process and/or result of reflexion can be controlled**, i.e., can represent a component of the controlled system's activity, being modified by a control subject for a definite goal. Precisely this relationship between control and reflexion enables informational control and reflexive control studied in this book! Actually, we present theoretical results and applications in the field of controlling reflexion.

In this context, let us make a digression as follows. The results of modeling and informational/reflexive control derived for social, economic, organizational and other systems (including human beings) have recently been extended to artificial technical systems. For instance, consider the so-called *multi-agent systems* (MAS) [157]. Such systems consist of numerous interacting autonomous agents having technical or informational nature (a classical example is a group of mobile robots). Multi-agent systems are remarkable for interaction decentralization and agents' multiplicity; these features lead to fundamentally new and important emergence properties (autonomy, lower vulnerability to adverse effects, etc.).

MAS have a complex (hierarchical) internal structure. The typical functional structure of an agent includes several hierarchical levels – see Fig. 4. *Operational level* (execution level) serves for implementing certain actions (e.g., stabilization of motion along a given trajectory). *Tactical level* is intended for choosing actions (e.g., planning

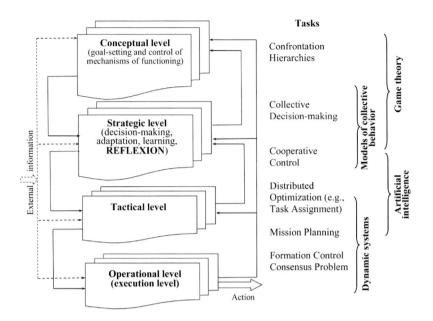

Figure 4 MAS: The hierarchical architecture of an agent.

of actions – trajectories selection or solution of distributed optimization problems). Actions can be chosen taking into account interaction with other agents. *Strategic level* answers for decision-making, learning and adaptivity of agents, as well as for control cooperativity (coordinated solution of a common task by a set of agents). An agent should have the capacity for strategic decision-making, adaptation, learning and **reflexion**. Finally, *conceptual level* corresponds to goal-setting principles. Each level employs a certain framework (as a rule, methods being applicable at a certain level can be used at higher hierarchical levels – see Fig. 4).

One modern tendency of the theory of multi-agent systems, game theory (see below) and artificial intelligence theory lies in that researchers strive to integrate these scientific directions. Yet, game theory and artificial intelligence theory aim at higher levels of agents' architecture. Within algorithmic, computational and evolutionary game theories [3, 106, 175]), one would observe "transition downwards," i.e., from the uniform description of a game to its decentralization and analysis of the feasibility of implementing autonomously the mechanisms of equilibrium behavior and realization. In fact, similar "decentralization" trends can be found in operations research [168]. On the other hand, the theory of MAS moves "upwards" in a parallel way due to the local character of scientific communities. The theory of MAS aspires after better consideration of strategic behavior and "intellectuality of agents" (including their capacity for reflexion). The behavior and interaction of active subjects is described by game theory. Today, game theory is a major tool for studying systems with incorporated human beings.

Game theory. Formal (mathematical) models of human behavior have been constructed and studied for over 150 years. Gradually, these models find wider application in control theory, economics, psychology, sociology, etc., as well as in practical problems. Most intensive development dates from the 1940s, the appearance of *game theory* often connected with J. von Neumann and O. Morgenstern's famous book *Theory of Games and Economic Behavior* [125] published in 1944.

In the sequel, we will understand a *game* as the interaction of subjects with non-coinciding interests. Still, an alternative interpretation treats a game as a type of unproductive activity whose motive consists not in the corresponding results, but in the process of activity itself. In addition, we refer to [88, 127], where the notion of a game is assigned a broader sense.

Game theory represents a branch of applied mathematics, which analyzes models of decision-making in the conditions of noncoinciding interests of opponents (*players*); each player strives to influence the situation in his/her favor [67, 75, 121]. In what follows, a decision-maker (a player) is called an *agent*. In the present book, we focus *par excellence* on noncooperative static normal-form games, where agents choose their *actions* one-time, simultaneously and independently. The only exception lies in dynamic models of collective decision-making discussed in Section 3.4.

Therefore, the major task of game theory is describing the interaction among several agents with noncoinciding interests, where the results of agent's activity (payoff, utility, etc.) generally depend on actions of all agents [75, 121]. Such description yields a forecast of a rational and "stable" outcome of the game – the so-called *game solution* (*equilibrium*).

Describing a *game* means specifying the following parameters:

– *a set of agents*;
– *preferences of agents* (relationships between payoffs and actions). Each agent is supposed to strive to maximize his/her payoff (and so, the behavior of each agent appears purposeful);
– *a set of feasible actions of agents*;
– *awareness of agents* (information on essential parameters, being available to agents at the moment of their choice);
– *sequence of moves* (the sequence of choosing actions).

Roughly speaking, a set of agents determines *who* participates in a game. Next, preferences reflect what agents *want*, sets of feasible actions describe what agents *can*. Finally, awareness and sequence of moves correspond to what agents *know* and *when* they choose actions, respectively.

The above parameters define a game; unfortunately, they are insufficient for forecasting its outcome, i.e., a solution (or an equilibrium) of the game – the set of rational and stable actions of agents. Nowadays, game theory suggests no universal concept of equilibria. By adopting different assumptions regarding principles of the agent's decision-making, one can construct different solutions. Thus, designing an equilibrium concept forms a basic problem for any game-theoretic research; this book does not represent an exception, either. Reflexive games are defined as a direct interaction among agents, where they make decisions based on hierarchies of their beliefs. In other words, awareness of agents is extremely important. And so, let us discuss this component in greater detail.

The role of awareness. Common knowledge. In game theory, psychology, distributed systems and other fields of science (see the overviews in [66, 117]), one should consider not only agents' *beliefs* about essential parameters, but also their beliefs about the beliefs of other agents, etc. The set of such beliefs is called the *hierarchy of beliefs*. We will model it using the tree of awareness structure of a reflexive game (see below). In other words, situations of interactive decision-making (modeled in game theory) require that each agent "forecasts" opponents' behavior prior to his/her choice. And so, each agent should possess definite beliefs about the view of the game by his/her opponents. On the other hand, opponents should do the same. Consequently, the uncertainty regarding the game to-be-played generates an infinite hierarchy of beliefs of game participants.

Consider an example of such an hierarchy. Suppose there exist two agents, namely, A and B. Each agent can have individual nonreflexive beliefs about an uncertain parameter θ (the *state of nature*). Denote these beliefs by θ_A and θ_B, respectively. Yet, performing the *first-rank reflexion*, each agent may think about their opponent's beliefs. The described beliefs (known as *beliefs of the second order*) will be designated by θ_{AB} and θ_{BA}, where θ_{AB} are the beliefs of agent A about the beliefs of agent B, θ_{BA} are the beliefs of agent B about the beliefs of agent A. Moreover, the process continues – within the framework of further reflexion (the *second-rank reflexion*) each agent may think about the opponent's beliefs about his/her beliefs. This yields beliefs of the *third order*, θ_{ABA} and θ_{BAB}. The process of generating beliefs of higher orders can be infinite (indeed, there are no logic restrictions to further increase of reflexion rank). The whole set of all beliefs – θ_A, θ_B, θ_{AB}, θ_{BA}, θ_{ABA}, θ_{BAB}, etc. – forms the hierarchy of beliefs.

A special case of awareness concerns *common knowledge*, when beliefs of all orders coincide. A rigorous definition of common knowledge was introduced in [100]. Notably, common knowledge is a fact with the following properties:

1) all agents know it;
2) all agents know 1;
3) all agents know 2 and so on – *ad infinitum*.

The formal model of common knowledge was originally proposed in [8]. Later on, many investigators refined and redeveloped it [9, 11, 57, 58, 59, 76, 83, 84, 102, 116, 159].

The present book is almost completely dedicated to models of agents' awareness in *game theory* (*viz.*, hierarchies of beliefs and common knowledge). Thus, we give several examples demonstrating the role of common knowledge in different fields of science – philosophy, psychology, etc. (see also the overview in [56]).

In *philosophy*, common knowledge has been studied in *convention* analysis [100, 172, 173]. For instance, consider Road Regulations in a certain country; they state that each participant of road traffic must follow these regulations and has the right to expect the same behavior from other participants. But other participants of road traffic must be also sure that the rest observe the Road Regulations, and so on. Hence, the convention "Follow the Road Regulations" must be a common knowledge.

In *psychology*, one would face the notion of *discourse* (from Latin *discursus* 'argument'). It means human thinking in words, being mediated by past experience; discourse acts as the process of connected logical reasoning, where a next idea stems from the previous one. The importance of common knowledge in discourse comprehension has the following illustration in [48, 56].

Two persons leave a movie theater. One asks another, "What did you think of the movie?" The second person understands the question only in the following case.

He/she understands the matter concerns the movie they have just seen. In addition, the second person must understand that this fact is understood by the first person. On the other hand, the asking person must be sure that the responding person understands that the matter concerns the movie they have just seen, and so on. Notably, adequate interaction (communication) between these persons requires that the movie forms their common knowledge (people must agree about language usage [100]).

Mutual awareness of agents turns out to be significant in *distributed computer systems* [57, 59, 76], *artificial intelligence* [70, 110] and other fields.

Game theory often assumes that all[1] parameters of a game are a *common knowledge*. In other words, each agent knows (a) all parameters of the game, (b) the fact that the rest of the agents know (a), and so on – *ad infinitum*. Such assumption corresponds to the *objective description of a game* and enables addressing the *Nash equilibrium*[2] concept [124] as a forecasted outcome of a noncooperative game (a game, where agents do not agree about coalitions, data exchange, joint actions, redistribution of payoffs, etc.). Thus, the assumption regarding common knowledge allows claiming that all agents know which game they play and that their beliefs about the game coincide.

Instead of agents' actions, one may consider something more complicated – agents' *strategies*. A strategy represents a mapping of all information available to an agent into a set of his/her feasible actions. For instance, we mention strategies in a multi-step game, mixed strategies, strategies in Howard's metagames [86, 87] (see also informational extensions of games in [67]). However, in these cases the rules of play are a common knowledge. Finally, it seems possible to believe that a game is chosen randomly according to a certain probability distribution making up a common knowledge – the so-called *Bayesian games* [63, 78, 121].

Generally, each agent may possess individual beliefs about parameters of a game. And so, each belief corresponds to a *subjective description of the game* [67] (see also modern models of awareness in [50, 61, 81, 149]). Consequently, agents participate in the game, having no objective views of it or interpreting this game in different ways (rules, goals, the roles and awareness of opponents, etc.). Unfortunately, still no universal approaches have been proposed for equilibria design under insufficient common knowledge.

On the other part, within the "reflexive tradition" of the humanities, the surrounding world of each agent includes the rest of the agents; moreover, beliefs about other agents get reflected during the process of reflexion (in particular, variations of beliefs may result from nonidentical awareness). However, researchers have not succeeded in deriving constructive formal outcomes in this field to date.

Hence, an urgent problem lies in designing and analyzing mathematical models of games, where agents' awareness is not a common knowledge and agents make decisions based on hierarchies of their beliefs. Such a class of games is called **reflexive games** [42, 137, 138]. We will provide a formal definition later.

---

[1] If the initial model incorporates uncertain factors, specific procedures of uncertainty elimination are involved to obtain a deterministic model.

[2] An agents' action vector is a Nash equilibrium if none of them benefits by unilateral deviation from it (provided that the rest of the agents choose the corresponding components of the Nash equilibrium). A more rigorous definition can be found below.

The term "reflexive games" was introduced by V. Lefebvre in 1965, see [99]. However, the cited work and his other publications [96-99] represented qualitative discussions of reflexion effects in interaction among subjects (actually, no general concept of solution was suggested for this class of games). Similar remarks apply to [55, 69, 161, 169], where a series of special cases of players' awareness was studied. The research work [138] concentrated on systematical treatment of reflexive games and an endeavor of constructing *a uniform equilibrium concept* for these games.

Prior to outlining the major content of this book, let us describe the basic approaches used below.

The basic approaches and structure of this book. In fact, the monograph [138] is our first work dedicated to models of reflexion in the game-theoretic context. Many years have elapsed since that time, and this line of investigations gained further development (e.g., see [39, 74, 132]). The present book reflects recent advances in the corresponding field. It includes the primary results derived by us and our colleagues, as well as reviewing the approaches adopted by other researchers.

Chapter 1 (Reflexion in Decision-making) possesses an introductory character; notably, models of individual and interactive decision-making are considered, the awareness required for implementing well-known equilibrium concepts is analyzed, and famous models of common knowledge and hierarchy of beliefs are described.

Recall that a reflexive game is a game, where agents' awareness does not form a common knowledge[3] and agents make decisions based on hierarchies of their beliefs. According to game theory and reflexive models of decision-making, it seems reasonable to distinguish between strategic reflexion and informational reflexion.

Informational reflexion is the process and result of agent's thinking about (a) the values of uncertain parameters and (b) what his/her opponents (other agents) know about these values. Here the "game" component actually disappears – an agent makes no decisions.

Strategic reflexion is the process and result of agent's thinking about which decision-making principles his/her opponents (other agents) employ under the awareness assigned by him/her via informational reflexion.

Therefore, informational reflexion often relates to insufficient mutual awareness, and its result serves for decision-making (including informational reflexion). Strategic reflexion takes place even in the case of complete awareness, preceding an agent's choice of action. In other words, informational and strategic reflexion can be studied independently, but both occur in the case of incomplete or insufficient awareness.

Chapter 2 (Informational Reflexion and Control) deals with formal models of informational reflexion and informational control. A key factor in reflexive games consists in agents' awareness (hierarchy of beliefs). Hence, its formal description involves the notion of an *awareness structure* – a (generally, infinite) tree whose nodes correspond to information (beliefs) of agents about essential parameters, beliefs of other agents, etc. An example of such an hierarchy can be found below.

The concept of an awareness structure enables giving a formal definition to certain intuitively apprehensible notions such as adequate awareness of agent 1 about agent 2, mutual awareness, identical awareness, etc.

---

[3]Naturally enough, research results in the field of reflexive games turn into corresponding results in the field of classic games if awareness is a common knowledge (see below).

The notion of a *phantom agent* is a key to reflexive games analysis in this book. Let us discuss it at the qualitative level (omitting the technicalities, see Chapter 2).

Suppose that two agents, namely, A and B, interact in a certain situation. Of course, each agent possesses an image of the other; agent A has an image of agent B (denoted by AB), while agent B has an image of agent A (denoted by BA). These images coincide with or differ from the reality. For instance, agent A may possess an adequate belief about agent B (the identity $AB = B$ holds true) or may not.

The following question rises immediately. Is the identity $AB = B$ possible in principle? You know, B represents a real agent, whereas AB is merely his/her image! This philosophical question requires going into subtleties; let us merely emphasize a couple of important facts. First, the matter concerns modeling of individual behavior in a specific situation rather than complete understanding of an individual. During everyday communication with different people, we often face situations of adequate and inadequate perception of an individual by other individuals.

Second, within the framework of formal (game-theoretic) modeling of human behavior, an agent as a participant of a certain situation is described by a (relatively) small set of characteristics. The latter can be completely known to another agent (exactly as to a researcher).

Consider the case when B and AB differ (formally, due to incomplete information on B available to A or due to trusting false information). Choosing certain actions, A takes into account not B, but the latter's image, i.e., AB. Reformulating this statement, one may say that A subjectively interacts with AB. And so, AB can be called a *phantom agent*. Really, this agent does not exist, but appears in the mind of *real agent A*. Consequently, the phantom agent affects actions of agent A, i.e., it affects the reality.

We give an elementary example. A believes that B is his/her friend. At the same time, B knows this fact and is a foe of A (the so-called "betrayal"). Obviously, such a situation includes phantom agent AB described by "B is a friend of A"; actually, this subject is missed. On the other hand, B is adequately informed of A, notably, $BA = A$.

Thus, the idea consists in studying phantom agents (existing in the minds of real and other agents) in addition to real agents (actually participating in a game). Real and phantom agents perform reflexion, enduing phantom agents with some awareness reflected in an awareness structure.

There may be infinitely many (real and phantom) agents participating in a game. This means a potentially infinite number of acts of reflexive imaging (infinite depth of the tree of an awareness structure). In any common situations, one can construct infinitely large assertions such as "I know...," "I know that you know...," "I know that you know that I know ...," "I know that you know that I know that you know...," and so on. Yet, in practice such a "stupid infinity" seems pointless – starting from a definite moment, beliefs get stabilized and further increase in reflexion ranks yields nothing new. Therefore, in real situations an awareness structure possesses finite *complexity*, i.e., the corresponding tree has a finite number of pairwise-different subtrees. In other words, a game comprises a finite number of real and phantom agents[4].

---

[4]In the limiting case of common knowledge, a phantom agent of level 1 coincides with its prototype (real agent) and the corresponding tree has the depth of 1. More specifically, the rest of the subtrees duplicate higher-level trees.

The notion of a phantom agent enables the following. First, determining a reflexive game as a game of real and phantom agents. Second, defining an *informational equilibrium* as the generalization of a Nash equilibrium to the case of a reflexive game; here each (real and phantom) agent evaluates his/her *subjective equilibrium* (an equilibrium in the game he/she believes they are playing using the available hierarchy of beliefs about objective and reflexive reality [40].

A convenient tool of informational equilibrium analysis lies in the *graph of a reflexive game*. In this graph, nodes answer to real and phantom agents and each node-agent has incoming arcs from nodes-agents whose actions affect the payoff of the given agent (in his/her subjective equilibrium). The number of incoming arcs equals the number of real agents minus unity. The graph of a reflexive game can be constructed without specifying the goal functions of agents. In this case, the graph reflects the qualitative interrelation of awareness of reflexing agents (instead of the quantitative ratio of their interests). Moreover, this graph provides a comfortable and expressive means of describing reflexion effects (see Section 2.4).

Let us get back to the example above. The graph of the reflexive game between two agents acquires the form $B \leftarrow A \leftrightarrow AB$. Real agent B (the betrayer) is adequately aware of agent A, which interacts with phantom agent AB (B representing a friend of A).

Strategic reflexion is considered in *Chapter 3* of the present book. The following observation can be made. Suppose that an agent models opponents' behavior by assigning definite reflexion ranks to them and him/her. Then the initial game turns into a new game, where the agent's strategy consists in choosing reflexion ranks.

Studying the process of reflexion in the new game leads to another new game, and so on. Furthermore, even if the former game incorporates a finite set of feasible actions, the latter game would have an infinite set of feasible actions (the number of different reflexion ranks). Hence, the primary problem of strategic reflexion analysis is evaluating the maximal reasonable rank of reflexion. In Chapter 3, this problem is solved for bimatrix games (Section 3.2) and models accounting for the bounded abilities of a human being in the field of data processing (Section 3.3).

We provide an example of strategic reflexion – *Penalty kick* in soccer (we also refer to *Hide-and-seek* and *Misère in Preferans*, see Section 3.2). Agents represent a kicker and a goal-keeper. For simplicity, suppose that the kicker chooses between two actions, *viz.*, "shooting in the left corner of goal" and "shooting in the right corner of goal". Accordingly, the goal-keeper has two actions, "catching the ball in the left corner" and "catching the ball in the right corner". If the goal-keeper guesses right, he/she catches the ball.

Let us model the reasoning of agents. Assume that the goal-keeper knows the kicker often chooses the right corner of the goal. Hence, he/she should catch the ball in the right corner. Yet, if the kicker knows that the goal-keeper knows his/her common way of shooting, the goal-keeper should model the reasoning of the kicker. He/she may think as follows, "The kicker knows that I know his/her common way of shooting. And so, the kicker expects me to catch the ball in the right corner and may shoot in the left corner. In this case, I should catch the ball in the left corner." If the kicker possesses sufficient reflexion depth, he/she may guess the reasoning of the goal-keeper and try outwitting the opponent by shooting in the right corner. The same line of reasoning can be followed by the goal-keeper; as the result, he/she would catch the ball in the right corner.

Both the kicker and goal-keeper may infinitely increase reflexion depths by thinking in the place of each other. Furthermore, none of them have rational grounds to stop at a certain step. Hence, in modeling of mutual reasoning, one would not a priori define the outcome of this game. The game, where agents choose between two actions, can be substituted by another game, where agents choose reflexion ranks assigned to an opponent. But this game also admits no rational solution, since each agent may model an opponent's behavior by considering a "twice reflexive game" and so on – *ad infinitum*.

In such a situation, the only aspect assisting agents lies in bounding the depth of their reflexion. The initial set of feasible actions is finite; consequently, the situation repeats itself starting from reflexion rank 2. Indeed, the kicker shoots in the right corner, being at zero and second (any even) level of reflexion. Thus, the goal-keeper has to guess whether the kicker's reflexion rank is even or not.

The maximal reflexion rank to-be-possessed by an agent for embracing the whole variety of game outcomes is called the *maximal rational rank of reflexion*. By failing to bear certain opponent's strategies in mind, an agent runs the risk of decreasing his/her payoff. As it turns out, the maximal rational rank of reflexion is finite in many cases; the corresponding formal results are presented in Sections 2.6 and 3.2. In the example *Penalty kick*, the maximal rational rank of reflexion performed by agents constitutes 2.

Suppose that the goal-keeper has no information on the common way of shooting by the kicker. Hence, the latter's actions are symmetric (i.e., the left and right corners appear "equivalent"). Still, it seems possible to introduce asymmetry for pursuing one's own goals. For instance, the goal-keeper may twitch to a certain corner of goal, "inviting" the kicker to shoot in another corner (subsequently, the goal-keeper jumps exactly in the "distant" corner). A more sophisticated strategy consists in the following. A goal-keeper's team-mate shows him/her the corner the kicker would shoot in (such that the kicker notices the hint). Subsequently, the goal-keeper jumps in the opposite corner. Finally, we emphasize that both techniques have been successfully adopted in football many times.

The concepts of an awareness structure, informational equilibrium and the graph of a reflexive game form the model of a reflexive game, which enables the following. First, it provides the uniform methodology and mathematical framework to describe and analyze various situations of collective decision-making by agents possessing different awareness, to study the impact of reflexion ranks on agents' payoffs, to obtain conditions of existence and implementability of informational equilibria, etc. Many examples of possible applications are discussed below.

Second, the suggested model of a reflexive game allows investigating the influence of reflexion ranks (the depth of an awareness structure) on agents' payoffs. The results derived in Sections 2.5, 2.6 and 3.2 indicate that (under slight assumptions) the maximal rational rank of reflexion is bounded. In other words, in many cases infinite increase in reflexion rank seems unreasonable in the sense of agents' payoffs.

Third, the suggested model of a reflexive game makes it possible to establish existence conditions and properties of an informational equilibrium, as well as to pose constructively and correctly the ***problem of informational control***. In this problem, a principal has to find an awareness structure such that the informational equilibrium implemented in it appears most beneficial to him/her. The problem of informational control is stated and solved in Section 2.11 for several special cases. The corresponding theoretical results are adopted in applied models discussed in *Chapter 4*.

In *Chapter 3* we consider models of strategic reflexion. This is done according to the logic of describing informational reflexion used in Chapter 2. Similarly to informational control for informational reflexion, in Section 3.4 we formulate the **problem of reflexive control** (for strategic reflexion). In addition, *Chapter 4* presents some models of reflexive control.

Finally – in the fourth place – the language of reflexive games (awareness structures, graphs of a reflexive game, etc.) is convenient to describe reflexion effects in psychology (see *Playing chess, Transactional analysis, Ethical choice*, etc.), to analyze art works, as well as to model organizational, economic, social and other systems. Details can be found in Chapter 4.

Alternatively, the structure of this book can be viewed from decision theory positions (see Fig. 5 and Fig. 1.1 below). The elementary (basic) model of *decision-making* lies in the choice problem solved by an individual (a decision-maker, DM) under complete awareness. Possible extensions of this model are the cases of natural or/and game uncertainty. The latter comprises uncertainty (incomplete awareness of a DM) regarding opponents' awareness (informational reflexion) or their decision-making principles

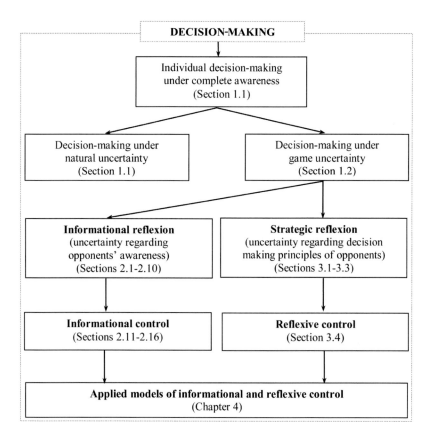

Figure 5 The logic and structure of this book.

(strategic reflexion). Purposeful impacts on DM's beliefs about opponents' awareness or decision-making principles are the essence of informational and reflexive control.

Thus, we have presented the structure and content of this book. Actually, several approaches to reading can be proposed. The first one is linear (successive reading of all chapters). The second approach is intended for a reader mostly interested in formal models (observational reading of Chapters 2–3 and glancing over the examples in Chapter 4). The third approach aims at a reader concerned with practical interpretations rather than mathematical subtleties (observational reading of the Introduction, the examples in Chapter 4 and the Conclusion).

The authors are deeply grateful to A. Yu. Mazurov, Cand. Sci. (Phys.-Math.) for careful translation of this book into English, as well as for helpful remarks and feedback.

Chapter 1

Reflexion in decision-making

Chapter 1 presents the model of individual decision-making (Section 1.1), overviews some major solution concepts of noncooperative games, discusses necessary assumptions imposed on awareness and mutual awareness of agents according to these solution concepts (Section 1.2), as well as analyzes conventional models of awareness and common knowledge (Section 1.3).

1.1 INDIVIDUAL DECISION-MAKING

Following [135, 136], let us state the model of agent's decision-making. Suppose that an agent can choose an *action x* from a set of *feasible actions X*. By choosing the action $x \in X$, the agent obtains the payoff $f(x)$, where $f: X \to \Re^1$ represents a real-valued *goal function* reflecting agent's preferences.

Accept *the hypothesis of rational behavior* which states the following. Under all available information, an agent chooses actions leading to the most beneficial values of his/her goal function. This hypothesis is not the only possible one – for instance, see the concept of bounded rationality [158]. According to the hypothesis of rational behavior, an agent chooses an alternative from the set of "best" alternatives. In the present case, this is a set of alternatives, where the goal function attains its maximum.

Hence, the agent's choice is determined by the *rule of individual rational choice* $P(f, X) \subseteq X$, which separates a set of the most beneficial actions[1] (from the agent's view):

$$P(f, X) = \text{Arg} \max_{x \in X} f(x).$$

Now, complicate the model by assuming the following. In addition to his/her actions, the agent's payoff depends on the value of an uncertain parameter $\theta \in \Theta$ – *the state of nature*. Notably, choosing the action $x \in X$ and realizing the state of nature $\theta \in \Theta$ lead to the agent's payoff $f(\theta, x)$, where $f: \Theta \times X \to \Re^1$.

In this general case (under an uncertain parameter – the state of nature), there exists no unambiguously "best" action. Choosing an action, an agent should "predict" or guess the state of nature.

---

[1] Appropriate maxima or minima are supposed to exist.

Therefore, introduce *the hypothesis of determinism*: an agent strives for eliminating the existing uncertainty (based on all available information), so as to make decisions under complete information [93, 135]. Equivalently, the final criterion used by a *decision-maker* (DM) must not involve uncertain parameters. According to the hypothesis of determinism, an agent has to eliminate the existing uncertainty in external (agent-independent) parameters. A possible approach lies in introducing definite suppositions on their values.

Depending on available *information I* on uncertain parameters, one can distinguish among [135]

–   *interval uncertainty* (we know merely the set Θ of feasible values of uncertain parameters);
–   *probabilistic uncertainty* (in addition to the set Θ, we know the probabilistic distribution $p(\theta)$ of uncertain parameters);
–   *fuzzy uncertainty* (in addition to the set Θ, we know the membership function for the values of uncertain parameters) [140].

This book *par excellence* addresses the elementary (*point-type*) case – agents possess beliefs about a certain value of the state of nature.

Make the following *assumption* regarding uncertainty elimination procedures adopted by an agent. Interval uncertainty is eliminated by evaluating the *maximal guaranteed result* (MGR). Probabilistic uncertainty is eliminated by computing the expected value of the goal function. Fuzzy uncertainty is eliminated by establishing the set of maximally undominated alternatives[2].

Denote by $f \underset{I}{\Rightarrow} \widehat{f}$ the uncertainty elimination procedure, i.e., the process of passing from the goal function $f(\theta, x)$ to the goal function $\widehat{f}(x)$ incorporating no uncertain parameters. Consequently, we have $\widehat{f}(x) = \min\limits_{\theta \in \Theta} f(\theta, x)$ in the case of interval uncertainty, $\widehat{f}(x) = \int\limits_{\Theta} f(x, \theta) p(\theta) d\theta$ in the case of probabilistic uncertainty, etc.

Uncertainty elimination yields a deterministic model, i.e., the rule of individual rational choice takes the form

$$P(f, X, I) = \text{Arg} \max\limits_{x \in X} \widehat{f}(x),$$

where I means information used by an agent in uncertainty elimination $f \underset{I}{\Rightarrow} \widehat{f}$.

Up to this moment, we have studied individual decision-making; to proceed, consider *game uncertainty*. Here essential aspects concern agent's beliefs about the set of feasible values of *opponents' action profile* (actions chosen by other agents according to certain or uncertain behavioral principles).

---

[2]These assumptions are not the only possible ones. Employing other approaches (e.g., the hypothesis of optimism or the hypothesis "weighted optimism-pessimism" (the Hurwitz criterion) instead of the MGR) would generate alternative solution concepts. Still, the analysis scheme would be similar to the one below.

1.2   INTERACTIVE DECISION-MAKING: GAMES AND EQUILIBRIA

Model of a game. To describe *collective behavior* of agents, it is not sufficient to define the preferences of the agents and the rules of rational individual choice. The following aspect has been underlined earlier. In a single-agent system, the hypothesis of rational (individual) behavior implies that the agent seeks to maximize his/her goal function by a proper choice of the action. In the case of several agents, we should account for their mutual influence. Consequently, a *game* arises as an interaction among the agents such that the payoff of each agent depends both on his/her action and on the actions of the rest agents. Suppose that (due to the hypothesis of rational behavior) each agent strives to maximize his/her function by choosing a proper action; evidently, in the case of several agents, a rational action of each agent depends on the actions of other agents[3].

Consider the following game-theoretic model of noncooperative interaction among n agents. Each agent chooses an *action* x_i belonging to a *feasible set* X_i, $i \in N = \{1, 2, \ldots, n\}$ (the *set of agents*). Moreover, the agents choose their actions *one-time, simultaneously and independently.*

The gain of agent i depends on his/her action $x_i \in X_i$, on the *opponents'* action profile $x_{-i} = (x_1, x_2, \ldots, x_{i-1}, x_{i+1}, \ldots, x_n) \in X_{-i} = \prod_{j \in N \setminus \{i\}} X_j$ (for agent i, opponents are agents belonging to the set $N \setminus \{i\}$) and on the state of nature[4] $\theta \in \Theta$. Actually, it represents a real-valued *gain function* $f_i = f_i(\theta, x)$, where $x = (x_i, x_{-i}) = (x_1, x_2, \ldots, x_n) \in X' = \prod_{j \in N} X_j$ stands for the action vector of all agents (also known as the *action profile*). Under a fixed state of nature, the set $\Gamma_0 = (N, \{X_i\}_{i \in N}, \{f_i(\cdot)\}_{i \in N})$ composed of the set of agents, their feasible actions and goal functions is said to be a *normal-form game*. A solution to the game (an *equilibrium*) is a set of stable action vectors of the agents. For additional information, see monographs and textbooks on game theory and collective decision-making [63, 105, 108, 118, 119, 121, 141, 146].

According to the hypothesis of rational behavior, each agent strives for choosing the best action (in the sense of his/her goal function) for a given action profile. Fix agent i; for him/her, the *environment* is the set of states of nature $\theta \in \Theta$ and the *opponents' action profile* $x_{-i} = (x_1, x_2, \ldots, x_{i-1}, x_{i+1}, \ldots, x_n) \in X_{-i} = \prod_{j \in N \setminus \{i\}} X_j$. Hence, agent i adopts the following decision-making principle. A rational agent chooses actions from the set

(1) $BR_i(\theta, x_{-i}) = \operatorname*{Arg\,max}_{x_i \in X_i} f_i(\theta, x_i, x_{-i}), \quad i \in N$

(here BR designates the *best response*)[5].

---

[3]Game-theoretic models imply that agents' rationality (their adherence to the hypothesis of rational behavior) represents a common knowledge. We accept this throughout the book.

[4]The state of nature may be a vector whose components reflect individual characteristics (types) of the agents.

[5]This book has independent numbering of formulas in each section.

Let us discuss possible principles of agent's decision-making. Each principle generates a corresponding equilibrium concept, i.e., defines the meaning of stability in the predicted outcome of the game. In parallel, we focus on awareness required for implementation of a certain equilibrium.

Dominant strategy equilibria. Assume that (for a certain agent) the set (1) is independent of the opponents' action profile. Then it represents the set of his/her dominant strategies. Next, the set of dominant strategies of all agents is called a *dominant strategy equilibrium* (DSE). When each agent possesses a dominant strategy, the agents can make decisions independently (*viz.*, choose actions without any information and beliefs about the opponents' action profile). Unfortunately, many games admit no DSE.

To implement a DSE (if exists), it suffices that each agent knows his/her own goal function and the feasible sets X', Θ.

Guaranteeing equilibria. The same awareness of the agents is required for implementing a *guaranteeing (maximin) equilibrium*, which exists almost in any game:

$$(2) \quad x_i^g \in \text{Arg} \max_{x_i \in X_i} \min_{x_{-i} \in X_{-i}} \min_{\theta \in \Theta} f_i(\theta, x_i, x_{-i}), \quad i \in N.$$

Imagine that (at least) for one agent the set (1) depends on the opponents' action profile (*viz.*, there exists no DSE). This substantially complicates the matter. Let us describe the corresponding cases.

Nash equilibria. Define the multi-valued mapping of best responses

$$(3) \quad BR(\theta, x) = (BR_1(\theta, x_{-1}), BR_2(\theta, x_{-2}), \dots, BR_n(\theta, x_{-n})).$$

A *Nash equilibrium* under a state of nature θ (more specifically, a *parametric Nash equilibrium*) is a point $x^*(\theta) \in X'$ satisfying the following condition:

$$(4) \quad x^*(\theta) \in BR(\theta, x^*(\theta)).$$

The inclusion (4) can be rewritten as

$$\forall i \in N, \forall y_i \in X_i : f_i(\theta, x^*(\theta)) \geq f_i(\theta, y_i, x^*_{-i}(\theta)).$$

The set of Nash equilibria $E_N(\theta)$ can be defined as

$$(5) \quad E_N(\theta) = \{x \in X' | x_i \in BR_i(\theta, x_{-i}), i \in N\}.$$

Consider the case of two agents. Then the set $E_N(\theta)$ is equivalently defined as the set of point pairs $(x_1^*(\theta), x_2^*(\theta))$ meeting the conditions

$$(6) \quad x_1^*(\theta) \in BR_1(\theta, BR_2(\theta, BR_1(\theta, \dots BR_2(\theta, x_2^*(\theta)) \dots))),$$
$$(7) \quad x_2^*(\theta) \in BR_2(\theta, BR_1(\theta, BR_2(\theta, \dots BR_1(\theta, x_1^*(\theta)) \dots))).$$

What awareness should agents have for implementing a Nash equilibrium by simultaneous independent choice of their actions?

By definition, any unilateral deviation from a Nash equilibrium point appears unbeneficial for any agent (provided that the rest agents choose corresponding components of the Nash equilibrium vector). If agents choose actions repeatedly, a Nash point demonstrates stability in a certain sense. Moreover, similarly to a DSE, it is implementable if an agent knows merely his/her goal function and the feasible sets X', Θ. However, this requires additional assumptions concerning decision-making principles of agents depending on game history [63, 121].

In the sequel, we mostly analyze single stage games. Under one-time choice of actions by agents, the knowledge of their goal functions and the sets X', Θ becomes insufficient for implementing a Nash equilibrium. Thus, for further exposition we adopt the following *assumption*. The information on the game Γ, the set Θ and the rational behavior of agents is a common knowledge.

In other words, this assumption implies the following. Each agent appears rational and knows the set of game participants, the goal functions and feasible sets of all agents, as well as the value of the state of nature. In addition, the agent is aware of that the rest agents know this and they know that he/she is aware of this fact, and so on (generally, this line of reasoning could be infinite). In particular, such awareness can be formed by a public announcement (i.e., simultaneous revelation to all agents in the same place). This brings to the feasibility of attaining the infinite *rank of informational reflexion* by all agents. We emphasize that the above assumption declares nothing about the awareness of agents about a specific value of the state of nature.

If the state of nature is a common knowledge, this suffices for implementing a Nash equilibrium. For substantiation, consider a two-player game and model the reasoning of agent 1 (agent 1 acts totally the same, and we study his/her reasoning separately only if this not the case). Agent 1 rationcinates as follows (see (6)). "By virtue of (1), my action must be the best response to an action of agent 2 under a given state of nature. Hence, I should model his/her behavior. Since, the goal functions and feasible sets form a common knowledge, I know that he/she would act according to (1), i.e., search for the best response to my actions under a given state of nature (see (7)). For this, he/she should model my actions. Again (due to the above assumption of common knowledge), he/she would follow the same line of reasoning as I, and so on (this process is infinite, see (6))." In game theory, researchers often draw a physical analogy between such reasoning and reflections in mirrors – see [105].

Therefore, for implementing a Nash equilibrium all parameters of a game and the state of nature must be a common knowledge (a possible weakening of this condition is outlined in [9]). This book focuses on reflexive games, where the state of nature is not a common knowledge and generally each agent possesses individual beliefs about the state of nature, the beliefs of other agents, and so on.

Subjective equilibria. The discussed types of equilibria represent special cases of *subjective equilibria*. Notably, a subjective equilibrium is an agents' action vector, where each component describes the best response of a corresponding agent to an opponents' action profile (expected by the agent according to his/her subjective viewpoint). Let us consider possible cases.

Suppose that agent i reckons on implementation of an opponents' action profile $\widehat{x}_{-i}^{B}$ and a state of nature $\widehat{\theta}_i$. Note the superscript "B" means "*beliefs*"; the term "*conjecture*" is also applicable. Consequently, agent i chooses

(8) $\quad x_i^B \in BR_i(\widehat{\theta}_i, \widehat{x}_{-i}^{B}), \quad i \in N.$

The vector x^B is a *point-type subjective equilibrium*.

Such definition of an equilibrium enables avoiding *substantiation* of agents' beliefs about actions of the opponents. In other words, it may happen that $\exists i \in N: \hat{x}^B_{-i} \neq y^B_{-i}$. A substantiated subjective equilibrium $(\hat{x}^B_{-i} = y^B_{-i}, i \in N)$ represents a Nash equilibrium. In particular, it suffices that all parameters of the game form a common knowledge and each agent models rational behavior of the opponents to construct $\hat{x}^B_{-i}$. Assume that the best response of each agent does not depend on the beliefs about the opponents' action profile. Such subjective equilibrium is a dominant strategy equilibrium.

Now, consider a somewhat general case. Agent i expects the opponents to choose their actions from the set $X^B_{-i} \subseteq X_{-i}$ and counts on implementation of a state of nature from the set $\hat{\Theta}_i \subseteq \Theta, i \in N$. The best response lies in the *guaranteeing subjective equilibrium*:

$$(9) \quad x_i(X^B_{-i}, \hat{\Theta}_i) \in \text{Arg} \max_{x_i \in X_i} \min_{x_{-i} \in X^B_{-i}} \min_{\theta \in \hat{\Theta}_i} f_i(\theta, x_i, x_{-i}), \quad i \in N.$$

If $X^B_{-i} = X_{-i}, \hat{\Theta}_i = \Theta, i \in N$, then $x_i(X^B_{-i}) = x^g_i, i \in N$. Notably, the guaranteeing subjective equilibrium coincides with "classical" guaranteeing equilibrium.

Let us continue the generalization process. As the best response of agent i, one may take a probability distribution $p_i(x_i)$, where $p_i(\cdot) \in \Delta(X_i)$; the latter is the set of different distributions over X_i, maximizing the expected gain of the agent subject to his/her beliefs about (a) the probability distribution $\mu_i(x_{-i}) \in \Delta(X_{-i})$ of the actions chosen by other agents and (b) the probability distribution $q_i(\theta) \in \Delta(\Theta)$ of the state of nature. In fact, we obtain the *Bayesian principle of decision-making*:

$$(10) \quad p_i(\mu_i(\cdot), q_i(\cdot), \cdot) = \arg \max_{p_i \in \Delta(X_i)} \int_{X', \Theta} f_i(\theta, x_i, x_{-i}) p_i(x_i) q_i(\theta) \mu_i(x_{-i}) d\theta dx, \quad i \in N.$$

Therefore, to implement a subjective equilibrium, one needs the minimal awareness of the agents – each agent must know his/her goal function $f_i(\cdot)$ and the feasible sets Θ, X'. However, such awareness may lead to the *incompatibility* of agents' beliefs regarding the state of nature and behavior of the opponents. To ensure compatibility (to make beliefs substantiated), we should involve additional assumptions on mutual awareness of the agents. Probably, the strongest assumption presumes common knowledge (it transforms a point-type subjective equilibrium into a Nash equilibrium and the set of Bayesian principles of decision-making into a Bayes-Nash equilibrium).

Bayes-Nash equilibria. Consider a game with incomplete information (see [78]); then a Bayes game is described by the following parameters:

a set of agents N;

a set of agents' feasible *types* K, where the type of agent i is $k_i \in K_i, i \in N$, and the type vector makes up $k = (k_1, k_2, \ldots, k_n) \in K' = \prod_{i \in N} K_i$;

a set of feasible action vectors $X' = \prod_{i \in N} X_i$ of the agents;

a set of utility functions $u_i: K' \times X' \to \Re^1$;

agents' beliefs $\mu_i(\cdot | k_i) \in \Delta(K_{-i}), i \in N$.

A *Bayes-Nash equilibrium* in a game with incomplete information is defined as a set of agents' strategies in the form $\sigma_i: K_i \to X_i$, $i \in N$, maximizing the expected utilities

$$(11) \quad U_i(k_i, \sigma_i(\cdot), \sigma_{-i}(\cdot)) = \int\limits_{\substack{k_{-i} \in \prod\limits_{j \neq i} K_j}} u_i(k, \sigma_i(k_i), \sigma_{-i}(k_{-i})) \mu_i(k_{-i}|k_i) dk_{-i}, \quad i \in N.$$

Generally, Bayesian games proceed from that the beliefs $\{\mu_i(\cdot|\cdot)\}_{i \in N}$ are a common knowledge. It suffices that the beliefs are *compatible*, i.e., are deducible by each agent using the Bayes formula for the distribution $\mu(k) \in \Delta(K')$ (the latter represents a common knowledge).

Consider Bayesian games, where $\{\mu_i(\cdot|\cdot)\}_{i \in N}$ is a common knowledge. In [18, 142] one can find the notion of *rationalizable strategies* (see also [4, 144]) $D_i \subseteq \Delta(X_i)$, $i \in N$, such that $D_i \subseteq BR_i(D_{-i})$, $i \in N$. In two-player games, the set of rationalizable strategies coincides with the set of strategies yielded by iterative elimination of strictly dominated strategies[6] [121]. The structure of a subjective equilibrium can be complicated by banning certain combinations of agents' actions, etc.

Therefore, implementing a DSE, a guaranteeing equilibrium and a subjective equilibrium (if any) requires that each agent has information (at least) on his/her goal function and all feasible sets. Moreover, implementing a Nash equilibrium also requires that the values of all essential parameters form a common knowledge.

Once again, we emphasize the following aspect. Implementability of a Nash equilibrium implies the ability of agents (and a *principal* or operations researcher, if they possess appropriate information) to evaluate a priori and independently the Nash equilibrium and to choose immediately the Nash equilibrium actions in the single stage game. Here a separate question relates to the specific choice of an equilibrium by agents and principal (if there exist several Nash equilibria [79]). Actually, common knowledge guarantees that each agent (and a principal) can model decision principles of other agents, including proper consideration of his/her decision principles by other agents.

Hence, the following conclusion seems natural. **The solution concept of a game has close connection to agents' awareness.** In a certain sense, the solution concepts of a DSE and Nash equilibrium represent limiting cases. The first one requires minimal awareness, while the second concept calls for infinite rank of informational reflexion of all agents. Below we describe "intermediate" cases of agents' awareness – hierarchies of beliefs – and construct corresponding solutions of the game. Now, let us present famous models of common knowledge and hierarchy of beliefs.

---

[6]A strongly dominated strategy is an agent's strategy such that there exists another strategy of the agent, ensuring a strictly greater payoff to this agent under any opponents' action profile. Iterative elimination of strictly dominated strategies lies in their serial (generally, infinite) exclusion from the set of agent's strategies. This procedure leads to evaluating the "weakest" solution of the game (the set of undominated strategies).

1.3   GENERAL APPROACHES TO THE DESCRIPTION OF INFORMATIONAL AND STRATEGIC REFLEXION

In the previous section, we have studied equilibrium concepts without reflexion (probably, the only exceptions are Nash equilibria and Bayes-Nash equilibria that admit the existence of a common knowledge). Indeed, each agent does not endeavor "being in the shoes of opponents."

Reflexion takes place when an agent possesses a certain hierarchy of beliefs, applying it in his/her decision-making. For a given agent, an hierarchy of beliefs includes his/her beliefs about the beliefs of other agents, beliefs about their beliefs about his/her beliefs and mutual beliefs of other agents, etc. Analysis of beliefs about uncertain factors corresponds to informational reflexion, whereas analysis of beliefs about decision-making principles relates to strategic reflexion. In terms of subjective equilibria, strategic reflexion corresponds to agent's assumptions how an opponent would evaluate a specific equilibrium (e.g., a subjective guaranteeing equilibrium). On the other part, informational reflexion is connected with specific assumptions regarding an external environment to-be-adopted by an opponent.

Consider conventional[7] approaches to the description of hierarchies of beliefs and a common knowledge.

According to [9, 11, 80], there are two different approaches to the description of awareness structures, *viz.*, *syntactic* and *semantic* ones. Recall that syntactics means syntax of sign systems, i.e., the structure of sign combinations and rules of their formation, "translation" and interpretation irrespective of their values and functions of sign systems. Semantics studies sign systems as tools of meaning expression; here the basic subject lies in interpretations of signs and sign combinations. Foundations of these approaches were laid in mathematical logic [84, 94].

Within the framework of syntactic approach, an hierarchy of beliefs is described explicitly. Suppose that beliefs are defined by a probability distribution. Then hierarchies of beliefs (at a certain level) correspond to distributions on the product of the set of states of nature and distributions reflecting beliefs of preceding levels [114]. An alternative is using "logic formulas" – rules of transforming elements of an initial set based on logic operations and operators such as "player i believes the probability of event ... is not smaller than α" [80, 176]. A knowledge is modeled by propositions (formulas) constructed according to certain syntactic rules.

According to semantic approach, beliefs of agents are defined by probability distributions on the set of states of nature. Hierarchies of beliefs get generated only by virtue of these distributions. In the elementary (deterministic) case, a knowledge represents the set Θ of feasible values of an uncertain parameter and different partitions $\{P_i\}_{i \in N}$ of this set. An element of the partition P_i containing $\theta \in \Theta$ forms the knowledge of agent i, namely, the set of values of the uncertain parameter, being indistinguishable for this agent under a known fact θ [8, 11].

---

[7]Hierarchies of beliefs and a common knowledge have become the subject of research not long ago. The pioneering works include the cited book by D. Lewis (1969) and the paper by R. Aumann (1976). Analysis of existing publications (see References) indicates of the growing interest in this problem domain.

The correspondence (or "equivalence") between syntactic and semantic approaches was established in [9, 159] and other works.

We also cite experimental research on hierarchies of beliefs [30, 122, 163]; see the surveys in [132, 174] and references in Section 3.4.

The above overview points at two existing "extremes." The first one lies in common knowledge. Here J. Harsanyi's merits [78] are (a) reducing all information on an agent (determining the latter's behavior) to a single characteristic – agent's type – and (b) constructing a Bayes-Nash equilibrium by hypothesizing that the probability distribution of types is a common knowledge. The second "extreme" relates to infinite hierarchy of compatible or incompatible beliefs. (For an example, see the structure discussed in [114]. On the one hand, it describes all possible Bayesian games and all possible hierarchies of beliefs. On the other hand, it appears very general and, consequently, very cumbersome, thus interfering with constructive statement and solution of specific problems).

Most research on awareness seeks to answer the following question. When does an hierarchy of agents' beliefs describe a common knowledge and/or reflect adequately their awareness? [21, 56]. The dependence of game solutions on a finite hierarchy of compatible or incompatible beliefs of agents (the whole range between the above "extremes") has been almost not studied. First, the exception is the paper [152], where Bayes-Nash equilibria for three-level hierarchies of incompatible probabilistic beliefs of two agents were constructed by assuming that at the lower level beliefs coincide with beliefs at the preceding level. In addition, see the Π_m-type assumptions and corresponding equilibria in [136]. The second exception consists in Chapter 2 of this book. We describe arbitrary (finite or infinite, compatible or incompatible) hierarchies of point-type beliefs, construct and analyze an informational equilibrium, *viz.*, an equilibrium of a reflexive game. The feasibility and reasonability of extending the results to the case of interval or probabilistic uncertainties of agents is outlined in the Conclusion.

Therefore, relevant issues include studying strategic reflexion (Chapter 3), constructing solutions to a reflexive game and analyzing its dependence on the hierarchy of agents' beliefs (Chapter 2).

Informational and strategic reflexion. Traditionally, game-theoretic models and/or models of collective decision-making utilize one of two assumptions regarding mutual awareness of agents [132]. The first one implies that all essential information and decision principles adopted by agents are known to all agents, all agents know this fact and so on (such reasoning could be infinite). Actually, this is the concept of a *common knowledge*, which serves, e.g., in evaluating a Nash equilibrium. The second assumption claims that each agent (according to his/her awareness) follows a certain procedure of individual decision-making and has "almost no idea" of the knowledge and behavior of the rest agents. The first approach appears canonical in *game theory*, while the second approach has become popular in models of *collective behavior*. Yet, a variety of intermediate situations exists between these "extreme cases." Imagine that informational reflexion takes no place – a common knowledge on essential external parameters is observed. Let an agent have performed an act of *strategic reflexion*, i.e., an attempt to predict the behavior of other agents (not their awareness but decision principles). This agent chooses his/her actions using the forecast (we believe he/she possesses reflexion rank 1). Another agent (with reflexion rank 2) possibly knows about the existence of agents having reflexion rank 1. Consequently, such agent endeavors to predict their behavior, as well. Again, this line of reasoning could be infinite. A series of questions arises immediately. How does the behavior of a collective of agents depend on their distribution by reflexion rank (the number of agents with a specific rank in a collective)? Suppose that the shares of reflexing agents can be controlled. What are the optimal values of these shares? Here optimality is "measured" in terms of some criterion defined on the set of agents' actions.

Classic game-theoretic models proceed from the following. In a normal form game, agents choose Nash equilibrium actions. However, investigations in the field of *experimental economics* indicate this not always the case (e.g., see [171] and the overview [177]). The divergence between actual behavior and theoretical expectations has several explanations:

- limited cognitive capabilities of agents – see Section 3.3 and [90] (decentralized evaluation of a Nash equilibrium represents a cumbersome computational problem [126]). Furthermore, sometimes Nash equilibria provide no adequate description to the real behavior of agents in experimental single stage games (agents have not enough time for "correcting" their wrong beliefs about essential parameters of a game [18]). For instance, D. Bernheim's concept of rationalizable strategies requires unlimited rationality from agents (their high cognitive capabilities);
- agent's full confidence in that all the opponents would evaluate a Nash equilibrium;
- incomplete awareness;
- the presence of several equilibria.

Therefore, there exist at least two foundations ("theoretical" and "experimental" ones) for considering models of collective behavior of agents with different reflexion ranks.

Collective behavior. In contrast to game theory, the *theory of collective behavior* analyzes the behavior dynamics of rational agents under rather weak assumptions regarding their awareness. For instance, far from always agents need a common knowledge about the set of agents, sets of feasible actions and goal functions of opponents. Alternatively, agents may not predict the behavior of their opponents (as in game theory). Moreover, making decisions, agents may "know nothing about the existence of" specific agents or possess aggregated information about them.

The most widespread model of collective behavior dynamics is the *model of indicator behavior* (see references in [132]). The essence of the model consists in the following. Suppose that at instant t each agent observes the actions of all agents $\{x_i^{t-1}\}_{i \in N}$ that have been chosen at the preceding instant $t-1$, $t=1,2,\ldots$ The initial action vector $x^0 = (x_1^0, \ldots, x_n^0)$ is assumed known.

Each agent can evaluate his/her *current goal* – an action maximizing his/her goal function provided that at a current instant all agents choose the same actions as at the previous instant:

$$(1) \quad w_i(x_{-i}^{t-1}) = \arg \max_{y \in \Re^1} F_i(y, x_{-i}^{t-1}), \quad t=1,2,\ldots, \quad i \in N.$$

According to the hypothesis of indicator behavior, at each instant an agent makes a "step" from his/her previous action to the current goal:

$$(2) \quad x_i^t = x_i^{t-1} + \gamma_i^t \, [w_i(x_{-i}^{t-1}) - x_i^{t-1}], \quad i \in N, \ t=1,2,\ldots,$$

where $\gamma_i^t \in [0; 1]$ designate "the values of steps." For convenience, such collective behavior can be called "optimization behavior" (thus, we emphasize its difference from play behavior). Evidently, if $\gamma_i^t \equiv 0$, no dynamics can be observed. Under $\gamma_i^t \equiv 1$, at each step an agent chooses the best response (see (1.2.1)). Yet, the corresponding dynamics may appear unstable.

The approaches adopted by the theory of collective behavior and game theory agree in the following sense. The both study the behavior of rational agents (compare (1.2.1) with (2)), while game equilibria generally represent equilibria for dynamic procedures of collective behavior. For instance, the Nash equilibrium (1.2.2) specifies an equilibrium for the dynamics (2) of collective behavior.

To make the picture complete, note one more aspect, as well. The theory of collective behavior proposes another approach (going beyond the scope of this book), namely, *evolutionary game theory* [175]. This science studies the behavior of large homogeneous groups (populations) of individuals in typical repeated conflicts; each strategy is applied by a set of players, whereas a corresponding goal function characterizes the success of specific strategies (instead of specific participants of such interaction).

Thus, game theory often employs maximal assumptions regarding agents' awareness (e.g., the hypothesis of existing common knowledge), while the theory of collective behavior involves the minimal assumptions. The intermediate position belongs to reflexive models. And so, let us discuss the role of (informational and strategic) reflexion in decision-making by agents.

Reflexion in game theory and models of collective behavior: the structure of problem domain. Game theory and the theory of collective behavior analyze interaction models for rational agents. Approaches and results of these theories can be considered at three interconnected epistemological levels (that correspond to different functions of modeling [128]) – see Fig. 1.1:

- phenomenological level, where a model aims at describing and/or explaining the behavior of a system (a collective of agents);
- predictive level (the aim is forecasting the system behavior);
- normative level (the aim is ensuring a required system behavior).

In game theory, a common scheme consists in (1) describing the "model of a game" (phenomenological level), (2) choosing an equilibrium concept defining the stable outcome of a game (predictive level) and (3) stating a certain control problem – find values of controlled "game parameters" implementing a required equilibrium (normative level). An interested reader would find the corresponding illustration in Fig. 1.1.

Taking into account *informational reflexion* leads to the necessity of constructing and analyzing awareness structures. This enables defining an informational equilibrium, as well as posing and solving informational control problems – see Fig. 1.1.

Taking into account strategic reflexion generates a similar chain marked by heavy lines in Fig. 1.1: "models of strategic reflexion" – "reflexive structure" – "reflexive equilibrium" – "reflexive control."

A comparison of approaches to modeling of informational and strategic reflexion is given by Table 1.1.

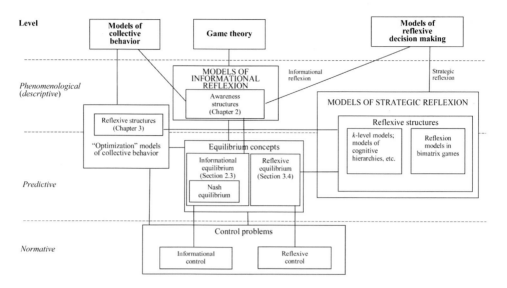

Figure 1.1 Descriptive and normative models of informational and strategic reflexion.

Table 1.1 Modeling of informational and strategic reflexion: a comparison of approaches.

Parameter	Informational reflexion (Chapter 2)	Strategic reflexion (Chapter 3)
Model of a "game"	Awareness structure	Reflexive structure
Equilibrium	Informational equilibrium	Reflexive equilibrium
Control	Informational control	Reflexive control

Thus, we have discussed the general approaches to the description of informational and strategic reflexion. Let us continue with systematic treatment of corresponding results. Chapter 2 focuses on informational reflexion and informational control, and Chapter 3 deals with strategic reflexion and reflexive control (see Table 1.1).

Chapter 2

Informational reflexion and control

The current chapter serves for defining an informational equilibrium and analyzing its properties. We start with describing informational reflexion in two-player games (Section 2.1). Then we provide the general model (Section 2.2), i.e., characterize the awareness structure adopted by reflexive game participants in their decision-making, as well as introduce the notion of awareness structure complexity. Section 2.3 presents the notion of an awareness structure as the solution concept for a reflexive game. Next, Section 2.4 describes graph of a reflexive game as an analysis tool for the properties of an informational equilibrium. In Section 2.5 we determine regular awareness structures and state sufficient conditions of existence for an informational equilibrium. Section 2.6 focuses on studying the influence of reflexion ranks on agents' gains, as well as on scrutinizing the relationship between an awareness structure and an informational equilibrium. In Sections 2.7–2.9 we deal with stability of an informational equilibrium. Sections 2.11–2.15 are intended for modeling of informational actions, as well as formulate informational control problems and examine their properties (consistency, etc.). Finally, Section 2.16 includes some results of analyzing reflexion effects in planning mechanisms.

2.1 INFORMATIONAL REFLEXION IN TWO-PLAYER GAMES

This section contains qualitative discussion of an hierarchy of beliefs and informational reflexion of two agents. Actually, Section 2.1 gives necessary background for the general model considered in Section 2.2.

We have emphasized that the assumption regarding the state of nature as a common knowledge appears "extreme." In other words, it requires infinite reflexion from agents and can be assigned to a classical Nash equilibrium. However, agents may have different awareness. Let us study possible cases.

Adopt the following system of symbols (see [136, 138]): θ_i means the information (*beliefs*) of agent i about the state of nature, θ_{ij} designates the information of agent i about the information of agent j about the state of nature (beliefs about beliefs or *second-order beliefs*) , $i \neq j$, θ_{iji} is the information of agent i about the information of agent j about the information of agent i about the state of nature[1] (third-order beliefs),

---

[1] Here we apply the left-to-right system of indices, being "opposite" to V. Lefebvre's system (right-to-left).

and so on, $i, j = 1, 2$. Suppose that, making decisions, each agent considers true only "his/her" information about the state of nature[2] (see the confidence principle in [136]).

Therefore, the *awareness* of agent i takes the form $I_i = (\theta_i, \theta_{ij}, \theta_{ijk}, \ldots)$, i.e., it comprises all information available at the moment of decision-making (the hierarchy of his/her beliefs, where levels are defined by the length of index sequence in awareness components). The set of I_1 and I_2 is called the awareness structure of a two-player reflexive game. The model of the awareness structure of a reflexive game involving an arbitrary number of agents is presented in Section 2.2. The *length* of the maximal *index sequence* characterizes (exceeds by unity) *reflexion rank* of an agent.

In terms of V. Lefevbre's *reflexive polynomials* [98], the unit length of index sequence corresponds to situation when (1) agent i "sees" merely the set T (the set of feasible states of nature) and (2) agent i possesses information on a concrete value of the state of nature (the agent has his/her own belief about the set $T + Ti$, but reflexion rank equals 0).

The maximal length of index sequence coinciding with 2 corresponds to the unit reflexion rank. Here the agent has information on the beliefs of other agents (probably, including his/her own beliefs – *self-reflexion Tii*) about the set $T + Ti + Tji$, and so on.[3]

Generally, interpreting θ as the set T results in that a finite awareness structure $I_i = (\theta_i, \theta_{ij}, \ldots, \theta_{i_1 i_2 \ldots i_k})$, $k < \infty$, of agent i corresponds to the reflexive polynomial $Ti + Tji + \cdots + Ti_k \ldots i = (T + Tj + \cdots + Ti_k \ldots j)i$. In other words, both awareness structures and reflexive polynomials describe agents' awareness; but awareness structures enable constructive consideration of the mutual awareness of different agents (see Section 2.2).

Accept the axiom of self-awareness: successive indices in formulas of agents' awareness do not coincide. Notably, we eliminate models of awareness with information of the following kind: "What I know (think) about what my opponent knows (thinks) about what he/she knows (thinks) about ..." and so on. For instance, we exclude combinations $\theta_{11} \theta_{211}$, θ_{1221}, etc.

The above classification system allows introducing the notation RG_{kl}, $k, l = 0, 1, 2, \ldots$, for reflexive two-player games. Here the first index by unity exceeds the

---

[2]How agent i corrects his/her beliefs θ_i about possible states of nature (e.g., based on θ_{iji}) deserves separate investigations.

[3]Reflexion of basic level can be interpreted as follows. Imagine there is a subject perceiving the surrounding world. One can distinguish several levels of such perception (levels of reflexion). At zero level (the entity level, nonreflexive level), a subject possesses certain beliefs about the surrounding world (arising by its reflection). However, a subject does not realize their possible incompleteness or incorrectness). Speaking figuratively, the surrounding world coincides with subject's beliefs of it. The next (actually, first) level corresponds to subject's comprehension of a possible divergence between the surrounding world and his/her beliefs of it. A subject is permitted to look at himself/herself from an outsider's viewpoint. Such comprehension may modify the beliefs about the surrounding world and the ways of its reflection. We call the first level of perception (admitting reflexion) by scientific level; indeed, the differences between the subjective and objective descriptions of reality appear exactly at this level. The second level of reflexion will be called philosophical level (it gets characterized by the appearance of beliefs about diverse ways of reflection and feasibility of their choice). Further growth of reflexion levels is possible. However, it would be difficult to provide practical interpretations to the third, fourth or higher levels.

reflexion rank (and corresponding awareness) of agent 1, whereas the second index by unity exceeds the reflexion rank (and corresponding awareness) of agent 2.

Evidently, within the framework of this model, awareness and reflexion rank are interconnected as follows. **For an agent, reflexion rank is by unity smaller than the maximal number of indices reflecting his/her awareness.** For instance, an agent with the awareness $I_i = (\theta_i, \theta_{ij}, \ldots, \theta_{i_1 i_2 \ldots i_k})$, where $i, j, i_1, i_2, \ldots, i_k \in N$ has reflexion rank $k - 1$.

The introduced assumptions impose certain constraints on the awareness structure of two agents. Notably, if $\theta_{i_1 i_2 \ldots i_k} \neq \varnothing$ reflects information of agent i, then $i_1 = i$ (the first index always coincides with the number of an agent possessing such information). In the case $k > 2$, indices alternate. Hence, under even k (odd reflexion ranks) one obtains $i_k = 3 - i$; actually, the first and last indices differ. On the other hand, for odd k (even reflexion ranks) we have $i_k = i$ (identical first and last indices).

Therefore, defining a reflexive game requires specifying the awareness of agents (e.g., $RG_{kl}(I_1, I_2)$). Of course, one should also determine goal functions and feasible sets.

An intricacy in modeling of reflexive games consists in the following. The presented description and classification system have been constructed according to the viewpoint of an operations researcher. In a concrete situation, agents may know nothing about the game played.

Let us adopt the model of agents' decision-making as follows. Each agent strives for choosing the best action (based on all available information). Due to insufficient awareness (the absence of a common knowledge), the actual action vector of agents possibly deviates from vectors evaluated by agents independently. That is, instead of a Nash equilibrium we observe an *informational equilibrium*[4]. The latter represents a subjective equilibrium of a reflexive game (in the common interpretation of an equilibrium).

To eliminate the existing uncertainty in models of reflexive games, agents employ two approaches. First, expect the worst-case values of uncertain parameters (the principle of maximal guaranteed result). Such approach implements a *subjective maximin (guaranteeing) equilibrium*. The latter is reflexive (as accounts decision principles of opponents), see Chapter 3. Second, assign certain awareness to other agents (e.g., the awareness which characterizes them). This approach implements a *subjective informational equilibrium*.

Here we underline the relevance of the term "subjective." *In reflexive games each agent evaluates "his/her" equilibrium, while the game outcome (the action vector of agents) is generally not an equilibrium[5] in the classic sense. The key idea concerns that each agent evaluates an "equilibrium" independently of the others. This aspect appreciably simplifies the description and analysis of their behavioral models.*

The last statement is of crucial importance. It enables studying decision principles of agents depending on information available to them at the moment of decision-making. Notably, instead of the reflexive game $RG_{kl}(I_1, I_2)$, we consider independently two reflexive models of agents' decision-making with the hierarchies of beliefs I_1 and

---

[4]A rigorous definition of an informational equilibrium is given in Section 2.3.
[5]A classic Nash equilibrium appears "objectively" rational, whereas an informational equilibrium is subjectively rational (under available awareness).

I_2 and reflexion ranks $k - 1$ and $l - 1$, respectively. Again, in this model the reflexion rank of an agent is fully described by his/her awareness. Furthermore, all knowledge of an agent (regarding the state of nature, beliefs and decision principles of the opponent, etc.) get included in his/her awareness – the hierarchy of beliefs.

We postpone a detailed treatment of these aspects to Section 2.2. (until systematic description of informational reflexion).

2.2 AWARENESS STRUCTURE OF GAMES

Consider the set of agents: $N = \{1, 2, \ldots, n\}$.

Denote by $\theta \in \Theta$ the uncertain parameter (we believe that the set Θ is a common knowledge for all agents). The *awareness structure* I_i of agent i includes the following elements. First, the belief of agent i about the parameter θ; denote it by θ_i, $\theta_i \in \Theta$. Second, the beliefs of agent i about the beliefs of the other agents about the parameter θ; denote them by θ_{ij}, $\theta_{ij} \in \Theta$, $j \in N$. Third, the beliefs of agent i about the beliefs of agent j about the belief of agent k; denote them by θ_{ijk}, $\theta_{ijk} \in \Theta$, j, $k \in N$. And so on (evidently, this reasoning is generally infinite). In the sequel, we employ the term "*awareness structure*," which is a synonym of "*informational structure*" and "*hierarchy of beliefs*."

Therefore, the awareness structure I_i of agent i is specified by the set of values $\theta_{ij_1 \ldots j_l}$, where l runs over the set of nonnegative integer numbers, $j_1, \ldots, j_l \in N$, while $\theta_{ij_1 \ldots i_l} \in \Theta$.

The *awareness structure I of the whole game* is defined in a similar manner; in particular, the set of the values $\theta_{i_1 \ldots i_l}$ is employed, with l running over the set of non-negative integer numbers, $i_1, \ldots, i_l \in N$, and $\theta_{i_1 \ldots i_l} \in \Theta$. We emphasize that the agents are not aware of the whole structure I; each of them knows only a substructure I_i.

Thus, an awareness structure is an infinite n-tree; the corresponding nodes of the tree describe specific awareness of real agents from the set N, and also phantom agents (complex reflexions of real agents in the mind of their opponents).

A *reflexive game* Γ_I is a game defined by the following tuple:

(1) $\Gamma_I = \{N, (X_i)_{i \in N}, f_i(\cdot)_{i \in N}, I\}$,

where N stands for a set of real agents, X_i means a set of feasible actions of agent i, $f_i(\cdot): \Theta \times X' \to \Re^1$ is his/her goal function ($i \in N$); Θ indicates a set of feasible values of the uncertain parameter and I designates the awareness structure.

Therefore, *a reflexive game generalizes the notion of a normal-form game (determined by the tuple $\{N, (X_i)_{i \in N}, f_i(\cdot)_{i \in N}\}$) to the case when agents' awareness is reflected by an hierarchy of their beliefs* (i.e., the awareness structure I). Within the framework of the accepted definition, a "classical" normal-form game is a special case of a reflexive game (a game under a common knowledge among the agents). Consider the "extreme" case when the state of nature appears a common knowledge; for a reflexive game, the solution concept (proposed in this book based on an informational equilibrium, see Section 2.3) turns out equivalent to the Nash equilibrium concept.

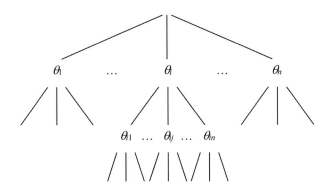

Figure 2.1 Tree of an awareness structure.

The set of relations among the elements characterizing the agents' awareness may be illustrated by a tree (see Fig. 2.1). Note that the awareness structure of agent i is represented by the corresponding subtree starting from the node θ_i.

Let us make the following important remark. Here and in the sequel, we mostly confine ourselves to the consideration of "point-type" awareness structures (the components consist of the elements belonging to the set Θ). More general models (e.g., an interval uncertainty, a probabilistic uncertainty or a fuzzy uncertainty) are considered in Sections 2.13–2.15. For the description of awareness models, see Chapter 1. Possible directions of further research are outlined in the Conclusion.

To proceed and formulate a series of definitions and properties, we introduce the following notation:

Σ_+ stands for a set of finite sequences of indexes belonging to N;

Σ is the sum of Σ_+ and the empty sequence;

$|\sigma|$ indicates the number of indexes in the sequence $\sigma \in \Sigma$ (for the empty sequence, it equals zero); this parameter is known as the length of an index sequence.

Imagine θ_i represents the belief of agent i about the uncertain parameter, while θ_{ii} means the belief of agent i about his/her own belief. It seems then natural that $\theta_{ii} = \theta_i$. In other words, agent i is well-informed on his/her own beliefs. Moreover, he/she assumes that the rest agents possess the same property. Formally, this means that the *axiom of self-awareness* is accepted.

The axiom of self-awareness:

$$\forall i \in N, \quad \forall \tau, \sigma \in \Sigma: \theta_{\tau i i \sigma} = \theta_{\tau i \sigma}.$$

In the sequel, we suppose validity of this axiom. In particular, being aware of θ_τ for all $\tau \in \Sigma_+$ such that $|\tau| = \gamma$, an agent may explicitly evaluate θ_τ for all $\tau \in \Sigma_+$ with $|\tau| < \gamma$.

In addition to the awareness structures I_i ($i \in N$), a researcher may also analyze the awareness structures I_{ij} (i.e., the awareness of agent j according to the belief of agent i), I_{ijk}, and so on. Let us identify the awareness structure with the agent being characterized by it. In this case, one may claim that *n real* agents (*i-agents*, where $i \in N$) having the awareness structures I_i also play with **phantom agents** (*τ-agents*, where $\tau \in \Sigma_+$, $|\tau| \geq 2$) having the awareness structures $I_\tau = \{\theta_{\tau\sigma}\}$, $\sigma \in \Sigma$. It should be emphasized that phantom agents exist merely in the minds of real agents; still, they have an impact on their actions; these aspects will be discussed below.

Introduce now *identical awareness structures*, a basic notion for Chapter 2.

Awareness structures I_λ and I_μ (λ, $\mu \in \Sigma_+$) are said to be *identical* if the following conditions are met:

1) $\theta_{\lambda\sigma} = \theta_{\mu\sigma}$ for any $\sigma \in \Sigma$;
2) the last indexes in the sequences λ and μ coincide.

Identity of these structures will be designated by $I_\lambda = I_\mu$.

Condition 1 in the definition above is transparent; yet, Condition 2 should be discussed in a greater detail. The matter is that in the sequel we consider actions of a τ-agent depending on his/her awareness structure I_τ and his/her goal function f_i (note the latter is determined by the last index in the sequence τ). Therefore, it appears convenient for us to believe that identity of the awareness structures means identity of the goal functions.

Assertion 2.2.1. $I_\lambda = I_\mu \Leftrightarrow \forall \sigma \in \Sigma \; I_{\lambda\sigma} = I_{\mu\sigma}$.

Proof. $I_\lambda = I_\mu \Rightarrow \forall \sigma, \kappa \in \Sigma \quad \theta_{\lambda\sigma\kappa} = \theta_{\mu\sigma\kappa} \Rightarrow \forall \sigma \in \Sigma \; I_{\lambda\sigma} = I_{\mu\sigma}$. The inverse implication is obvious: select σ as the empty sequence. $\bullet$[6]

Practical interpretation of Assertion 2.2.1 lies in the following. Identity of two awareness structures means identity of all their substructures.

The next result represents an alternative statement for the axiom of self-awareness.

Assertion 2.2.2. $\forall i \in N \quad \forall \tau, \sigma \in \Sigma \; I_{\tau i i \sigma} = I_{\tau i \sigma}$.

Proof. $\forall i \in N \; \forall \tau, \sigma \in \Sigma \; \theta_{\tau i i \sigma} = \theta_{\tau i \sigma} \Leftrightarrow \forall i \in N \quad \forall \tau, \sigma, \kappa \in \Sigma \; \theta_{\tau i i \sigma \kappa} = \theta_{\tau i \sigma \kappa} \Leftrightarrow \forall i \in N \; \forall \tau, \sigma \in \Sigma \; I_{\tau i i \sigma} = I_{\tau i \sigma}$. $\bullet$

Clearly, the definition of identical awareness structures (as well as other definitions in the current Section) can be reformulated in *τ-subjective* context, i.e., according to the beliefs of a τ-agent ($\tau \in \Sigma_+$). Awareness structures I_λ and I_μ (λ, $\mu \in \Sigma_+$) are called *τ-subjectively identical*, if $I_{\tau\lambda} = I_{\tau\mu}$.

In what follows, we state definitions and assertions directly in the τ-subjective sense for $\tau \in \Sigma$. Indeed, if τ is the empty sequence of indexes, "τ-subjectively" means "objectively."

A λ-agent is said to be τ-subjectively *adequately aware* of the beliefs of a μ-agent (or, in a compact form, of a μ-agent) if

$$I_{\tau\lambda\mu} = I_{\tau\mu}(\lambda, \mu \in \Sigma_+, \tau \in \Sigma).$$

---

[6]The symbol "$\bullet$" designates the end of an example or proof.

We will use the expression $I_\lambda >_\tau I_\mu$ to indicate the τ-subjective adequate awareness of the λ-agent about the μ-agent.

Assertion 2.2.3. Each real agent τ-subjectively considers himself/herself as being adequately aware of any agent, i.e.,

$$\forall i \in N \quad \forall \tau \in \Sigma \quad \forall \sigma \in \Sigma_+ \quad I_i >_{\tau i} I_\sigma.$$

Proof. By virtue of Assertion 2.2.2, one obtains the identity $I_{\tau i i \sigma} = I_{\tau i \sigma}$. According to the definition of τ-subjectively identical awareness structures, this implies that $I_i >_{\tau i} I_\sigma$. •

Substantially, Assertion 2.2.3 reflects the following fact. The awareness structure under consideration implicates the confidence of each agent in his/her adequate awareness of all elements belonging to the structure.

A λ-agent and a μ-agent are said to be τ-subjectively *mutually aware* of each other under the conditions

$$I_{\tau\lambda\mu} = I_{\tau\mu}, \quad I_{\tau\mu\lambda} = I_{\tau\lambda} \ (\lambda, \mu \in \Sigma_+, \tau \in \Sigma).$$

The τ-subjective mutual awareness of a λ-agent and a μ-agent will be denoted by $I_\lambda ><_\tau I_\mu$.

A λ-agent and a μ-agent are said to be τ-subjectively *identically aware of a σ-agent* if $I_{\tau\lambda\sigma} = I_{\tau\mu\sigma} \ (\sigma, \lambda, \mu \in \Sigma_+, \tau \in \Sigma)$.

The τ-subjective identical awareness of a λ-agent and a μ-agent of a σ-agent will be denoted by

$$I_\lambda >_\sigma <_\tau I_\mu.$$

A λ-agent and a μ-agent are said τ-*subjectively identically aware* if $\forall i \in N \ I_{\tau\lambda i} = I_{\tau\mu i} \ (\lambda, \mu \in \Sigma_+, \tau \in \Sigma)$.

The τ-subjective identical awareness of a λ-agent and a μ-agent will be denoted by $I_\lambda \sim_\tau I_\mu$.

Moreover, the relation of identical awareness of a certain agent and the relation of identical awareness represent equivalence relations (they enjoy the properties of reflexivity, symmetry and transitivity on a set of agents).

Below we show that identical awareness is equivalent to identical awareness of any agent.

Assertion 2.2.4. $I_\lambda \sim_\tau I_\mu \Leftrightarrow \forall \sigma \in \Sigma_+ I_\lambda >_\sigma <_\tau I_\mu$.

Proof. $I_\lambda \sim_\tau I_\mu \Leftrightarrow \forall i \in N \ I_{\tau\lambda i} = I_{\tau\mu i} \Leftrightarrow \{$due to Assertion 2.2.1$\} \Leftrightarrow \forall i \in N \forall \kappa \in \Sigma$ $I_{\tau\lambda i\kappa} = I_{\tau\mu i\kappa} \Leftrightarrow \{$by setting $\sigma = i\kappa\} \Leftrightarrow \forall \sigma \in \Sigma_+ \quad I_{\tau\lambda\sigma} = I_{\tau\mu\sigma} \Leftrightarrow \forall \sigma \in \Sigma_+ \ I_\lambda >_\sigma <_\tau I_\mu$. •

These definitions demonstrate that (in terms of adequate, mutual and identical awareness) a situation can be modeled via the identity of corresponding awareness structures. The next result concerns connections among the above notions.

Assertion 2.2.5. For any $\tau \in \Sigma$, the three following conditions are equivalent:

1. any two real agents are τ-subjectively mutually aware of each other;
2. all real agents are τ-subjectively identically aware;
3. for any $i \in N$, the value of $I_{\sigma i}$ τ-subjectively depends only on i.

In other words, for any $\tau \in \Sigma$ we have

$$(\forall i, j \in N \; I_i ><_\tau I_j) \Leftrightarrow (I_1 \sim_\tau \ldots \sim_\tau I_n) \Leftrightarrow (\forall i \in N \; \forall \sigma \in \Sigma I_{\tau \sigma i} = I_{\tau i}).$$

Proof.   Let us demonstrate the implications $1 \Rightarrow 2$, $2 \Rightarrow 3$, $3 \Rightarrow 1$.

$1 \Rightarrow 2$. For any $i, j, m \in N$, we have $I_i >_\tau I_m, I_j >_\tau I_m$, which means the identities $I_{\tau im} = I_{\tau m}$, $I_{\tau jm} = I_{\tau m}$. Hence, $I_{\tau im} = I_{\tau jm}$, and condition 2 is valid (see Assertion 2.2.4).

$2 \Rightarrow 3$. Condition 3 is trivial for the empty sequence σ. Thus, choose an arbitrary nonempty sequence $\sigma \in \Sigma_+$. Then $\sigma = i_1 \ldots i_l$ ($i_k \in N$, $k = 1, \ldots, l$), and for any $i \in N$ we obtain the following expressions:

$I_{\tau i} = \{$by virtue of Assertion 2.2.2$\} = I_{\tau i i} = \{$so long as $I_i \sim_\tau I_{i_l}\} = I_{\tau i_l i} = \{$by virtue of Assertion 2.2.2$\} = I_{\tau i_l i_l i} = \{$so long as $I_{i_l} \sim_\tau I_{i_{l-1}}$ and by virtue of Assertion 2.2.4$\} = I_{\tau i_{l-1} i_l i} = \cdots = I_{\tau i_1 \ldots i_l i} = I_{\tau \sigma i}$.

$3 \Rightarrow 1$. For any $i, j \in N$, we have $I_{\tau i j} = I_{\tau j}$, $I_{\tau j i} = I_{\tau i}$, and, consequently, $I_i ><_\tau I_j$. •

The notion of identical awareness structures allows for introducing another relevant property – the complexity of the structure. We underline that (along with the structure I) there exists a denumerable set composed of the awareness structures I_τ, $\tau \in \Sigma_+$ such that among them one may separate out certain classes of pairwise non-identical structures by the identity relation. The number of the mentioned classes is naturally referred to as the *complexity of the awareness structure.*

We will say that the awareness structure I has *finite complexity* $v = v(I)$, if there is a finite set of pairwise nonidentical structures $\{I_{\tau_1}, I_{\tau_2}, \ldots, I_{\tau_v}\}$, $\tau_l \in \Sigma_+$, $l \in \{1, \ldots, v\}$ such that any structure I_σ, $\sigma \in \Sigma_+$, has an identical structure from this set. Otherwise, the structure I possesses the infinite complexity: $v(I) = \infty$.

A finite-complexity awareness structure is called *finite* (however, the corresponding awareness tree is infinite). If this is not the case, an awareness structure is said to be *infinite*.

Obviously, the minimal possible complexity of an awareness structure equals the number of real agents participating in the game (one can check that awareness structures for real agents are pairwise nonidentical by definition).

Any (finite or denumerable) set of pairwise nonidentical structures I_τ, $\tau \in \Sigma_+$ such that any structure I_σ, $\sigma \in \Sigma_+$ is identical to one of them, is referred to as a *basis* of the awareness structure I.

Suppose that the awareness structure I has finite complexity; then it is possible to estimate the maximal length of the index sequence γ such that, given all structures I_τ, $\tau \in \Sigma_+$, $|\tau| = \gamma$, one can find the rest structures. In a certain sense, this length characterizes *reflexion rank* necessary to describe the awareness structure.

We will say that the awareness structure I, $v(I) < \infty$, has *finite depth* $\gamma = \gamma(I)$ when the following conditions hold true:

1) for any structure I_σ, $\sigma \in \Sigma_+$, there exists an identical structure I_τ, $\tau \in \Sigma_+$, $|\tau| \le \gamma$;
2) for any positive integer ξ ($\xi < \gamma$), there exists a structure I_σ, $\sigma \in \Sigma_+$, being identical to none of the structures I_τ, $\tau \in \Sigma_+$, $|\tau| = \xi$.

If $v(I) = \infty$, the depth is also considered infinite: $\gamma(I) = \infty$.

The description of an awareness structure being avaialble, one can study the process of joint decision-making by real and phantom agents. This naturally brings to the notion of an informational equilibrium.

2.3 INFORMATIONAL EQUILIBRIUM

Assume that the awareness structure I of a game is given; this means that the awareness structures are also defined for all (real and phantom) agents. Within the framework of the hypothesis of rational behavior, the choice of an action x_τ performed by a τ-agent is described by his/her awareness structure I_τ. Hence, the mentioned structure being available, one may model agent's reasoning and evaluate his/her action. On the other hand, while choosing his/her action, the agent models actions of the rest agents (i.e., performs reflexion). Therefore, estimating the game outcome, we should account for the actions of real and phantom agents.

A set of actions x_τ^*, $\tau \in \Sigma_+$, is called an **informational equilibrium**, if the following conditions are met:

1. the awareness structure I possesses finite complexity v;
2. $\forall \lambda, \mu \in \Sigma: I_{\lambda i} = I_{\mu i} \Rightarrow x_{\lambda i}^* = x_{\mu i}^*$;
3. $\forall i \in N, \forall \sigma \in \Sigma$:

$$(1) \quad x_{\sigma i}^* \in \operatorname*{Arg\,max}_{x_i \in X_i} f_i(\theta_{\sigma i}, x_{\sigma i 1}^*, \ldots, x_{\sigma i, i-1}^*, x_i, x_{\sigma i, i+1}^* \ldots, x_{\sigma i, n}^*).$$

Here Condition 1 claims that a reflexive game involves a finite number of real and phantom agents.

Condition 2 expresses the requirement that the agents with an identical awareness choose identical actions.

Finally, Condition 3 reflects rational behavior of agents – each agent strives for maximizing the individual goal function via a proper choice of his/her action. For this, an agent substitutes actions of the opponents into his/her goal function; the actions are rational in the view of the considered agent (according to the available beliefs of the rest agents).

The presence of Condition 3 in the above definition is beyond doubt. Consider two examples illustrating the relevance of Conditions 1–2.

Examples 2.3.1–2.3.2. There exist two agents with the goal functions

$$f_1(\theta, x_1, x_2) = (\theta - x_2)x_1 - \frac{x_1^2}{2}, \quad f_2(\theta, x_1, x_2) = (\theta - x_1)x_2 - \frac{x_2^2}{2},$$

where $x_i \in \Re^1$, $i = 1, 2$. Study different awareness structures of the agents.

Example 2.3.1. Let the awareness structure be defined by

$$\theta_1 = 1, \quad \theta_{12} = 3, \quad \theta_{121} = 5, \quad \theta_{1212} = 7, \ldots;$$

$$\theta_2 = 2, \quad \theta_{21} = 4, \quad \theta_{212} = 6, \quad \theta_{2121} = 8, \ldots.$$

(due to the axiom of self-awareness, elements with identical serial indexes can be eliminated from analysis).

This structure has infinite complexity. The system of equations (1) takes the form

$$
\begin{array}{ll}
x_1 = 1 - x_{12}, & x_2 = 2 - x_{21}, \\
x_{12} = 3 - x_{121}, & x_{21} = 4 - x_{212}, \\
x_{121} = 5 - x_{1212}, & x_{212} = 6 - x_{2121}, \\
x_{1212} = 7 - x_{12121}, & x_{2121} = 8 - x_{21212}, \\
\text{and so on;} & \text{and so on.}
\end{array}
$$

Clearly, this system includes a denumerable number of equations and admits infinitely many solutions. Indeed, choose arbitrarily the values of x_1 and x_2; consequently, express the rest variables through them. •

Example 2.3.2. Let the awareness structure be given by $\theta_\sigma = 1$ for any $\sigma \in \Sigma_+$. If Condition 2 is violated, the system (1) possesses a denumerable number of equations:

$$
\begin{array}{ll}
x_1 = 1 - x_{12}, & x_2 = 1 - x_{21}, \\
x_{12} = 1 - x_{121}, & x_{21} = 1 - x_{212}, \\
x_{121} = 1 - x_{1212}, & x_{212} = 1 - x_{2121}, \\
x_{1212} = 1 - x_{12121}, & x_{2121} = 1 - x_{21212}, \\
\text{and so on;} & \text{and so on.}
\end{array}
$$

By analogy to Example 2.3.1, we obtain infinitely many solutions. Choose arbitrarily the values of x_1, x_2 and express the rest variables through them. •

It would seem that Condition 2 makes necessary to solve an infinite (denumerable) number of equations (yielding infinitely many values x_τ^*) for evaluating an informational equilibrium. However, the actual number of equations and solutions turns out finite.

Assertion 2.2.6. Suppose there exists an informational equilibrium x_τ^*, $\tau \in \Sigma_+$. Then it consists of (at most) ν pairwise different actions, and the system (1) includes at most ν pairwise different equations.

Proof. Let x_τ^*, $\tau \in \Sigma_+$, be an informational equilibrium. Then finiteness of the awareness structure and Condition 2 immediately imply that the number of pairwise different values x_τ^* constitutes (at most) ν.

Now, consider two arbitrary identical awareness structures: $I_\lambda = I_\mu$. In this case, we have $\theta_\lambda = \theta_\mu$ and $x_\lambda^* = x_\mu^*$. Next, for any $i \in N$ the equality $I_{\lambda i} = I_{\mu i}$ holds; hence, $x_{\lambda i}^* = x_{\mu i}^*$. This means that two equations coincide in the system (1) with the actions x_λ^*

and x_μ^* in the left-hand side. Since there are ν pairwise different awareness structures, the number of pairwise different conditions (1) does not exceed ν. •

Therefore, to find an informational equilibrium x_τ^*, $\tau \in \Sigma_+$, one must state ν conditions of the form (1) for each of ν pairwise different values x_τ^* corresponding to the pairwise different awareness structures I_τ.

All agents being identically aware of the situation, complexity of the awareness structure is minimal and equals the number of agents. In this case, the system (1) defines a Nash equilibrium, while an informational equilibrium becomes a Nash equilibrium.

Let us summarize the results. When real agents have an identical awareness (i.e., reflexive reality is a common knowledge), an informational equilibrium turns into a Nash equilibrium (no phantom agents "arise"). However, even in the general case an informational equilibrium and a Nash equilibrium are closely connected.

Consider an awareness structure I of finite complexity ν having the basis $\{I_{\tau_1}, \ldots, I_{\tau_\nu}\}$. An informational equilibrium involves real and phantom agents from a set $\Xi = \{\tau_1, \ldots, \tau_\nu\}$, each choosing a corresponding action in the vector $\{x_{\tau_1}, \ldots, x_{\tau_\nu}\}$, $x_{\tau_l} \in X_{\omega(\tau_l)}$, $l \in \{1, \ldots, \nu\}$. Note $\omega(\sigma)$ designates the last index in a sequence σ, where $\sigma \in \Sigma_+$.

For each agent from the set Ξ, express the goal function as

$$(2) \quad \varphi_{\tau_l}(x_{\tau_1}, \ldots, x_{\tau_\nu}) = f_{\omega(\tau_l)}(\theta_{\tau_l}, x_{\sigma_1}, \ldots, x_{\sigma_n}),$$

where $I_{\tau_l i} = I_{\sigma_i}$, $\sigma_i \in \Xi$ for all $i \in N$, $l \in \{1, \ldots, \nu\}$. Since $I_{\sigma_{\omega(\tau_l)}} = I_{\tau_l \omega(\tau_l)} = I_{\tau_l}$, formula (2) can be rewritten as

$$(3) \quad \varphi_{\tau_l}(x_{\tau_1}, \ldots, x_{\tau_{l-1}}, x_{\tau_l}, x_{\tau_{l+1}}, \ldots, x_{\tau_\nu})$$
$$= f_{\omega(\tau_l)}(\theta_{\tau_l}, x_{\sigma_1}, \ldots, x_{\sigma_{\omega(\tau_l)-1}}, x_{\tau_l}, x_{\sigma_{\omega(\tau_l)-1}}, \ldots, x_{\sigma_n}).$$

In practice, the expressions (2) and (3) have the following interpretation. The goal function maximized by a τ_l-agent ($\tau_l \in \Xi$) in a reflexive game subjectively depends on his/her beliefs of the parameter θ, his/her action and actions of $(n - 1)$ agents from the set Ξ. In other words, the function φ_{τ_l} essentially depends on the variables $\{x_{\tau_1}, \ldots, x_{\tau_\nu}\}$ (and on the quantity θ_{τ_l} as a parameter); furthermore, this dependence coincides with the function f_i, where $i = \omega(\tau_l)$. Evidently, the function φ_{τ_l} "inherits" the properties of the function $f_{\omega(\tau_l)}$.

Anticipating things, we provide an example. Suppose that the graph of a reflexive game (see the next section) takes the form in Fig. 2.3. The goal functions of real agents are $f_i(\theta, x_1, x_2, x_3)$, $x_i \in X_i$, $i \in \{1, 2, 3\}$. Then an informational equilibrium engages five agents from the set $\Xi = \{1, 2, 3, 31, 32\}$, having the goal functions

$$\varphi_1(x_1, x_2, x_3, x_{31}, x_{32}) = f_1(\theta_1, x_1, x_2, x_3);$$
$$\varphi_2(x_1, x_2, x_3, x_{31}, x_{32}) = f_2(\theta_2, x_1, x_2, x_3);$$
$$\varphi_3(x_1, x_2, x_3, x_{31}, x_{32}) = f_3(\theta_3, x_{31}, x_{32}, x_3);$$
$$\varphi_{31}(x_1, x_2, x_3, x_{31}, x_{32}) = f_1(\theta_{31}, x_{31}, x_{32}, x_3);$$
$$\varphi_{32}(x_1, x_2, x_3, x_{31}, x_{32}) = f_2(\theta_{32}, x_{31}, x_{32}, x_3).$$

Using (3), the system of equations (2) for evaluating an informational equilibrium $(x^*_{\tau_1}, \ldots, x^*_{\tau_\nu})$ is representable as

$$x^*_{\tau_l} = \arg\max_{x_{\tau_l} \in X_{\omega(\tau_l)}} \varphi_{\tau_l}(x^*_{\tau_1}, \ldots, x^*_{\tau_{l-1}}, x_{\tau_l}, x^*_{\tau_{l+1}}, \ldots, x^*_{\tau_\nu}),$$

where l runs all integers from 1 to ν. Obviously, we arrive at the system of equations for evaluating a Nash equilibrium in a game with an identical awareness of τ_l-agents, $l \in \{1, \ldots, \nu\}$. This circumstance enables applying to an informational equilibrium the sufficient conditions of existence of a Nash equilibrium (with proper modifications).

For instance, the following result is well-known. Consider a continuous game with convex action sets X_i, where the goal function of each agent f_i is continuous in all arguments and strictly concave in x_i. Such game admits a pure strategies' Nash equilibrium.

This fact can be reformulated, leading to a sufficient condition of existence of an informational equilibrium in a reflexive game.

Assertion 2.2.7. Consider a reflexive game with an awareness structure of a finite complexity. Assume that actions sets X_i represent convex subsets of linear metric domains. Moreover, suppose that under any $\theta \in \Theta$ the goal functions of agents $f_i(\theta, x_1, \ldots, x_n)$ appear continuous in all variables and strictly concave in x_i. Then this game admits an informational equilibrium.

Proof. Continuity in all arguments and strict convexity in x_i of the function f_i implies that (a) the functions φ_{τ_l} ($\tau_l \in \Xi$, $\omega(\tau_l) = i$) defined by (3) are continuous in all arguments and (b) each function φ_{τ_l} is strictly concave in x_{τ_l}. Thus, Assertion 2.2.7 directly follows from the above fact. •

Evidently, these properties are enjoyed by the goal functions in Examples 2.4.1–2.4.3. Hence, reflexive games with such goal functions admit an informational equilibrium for any awareness structures of a finite complexity.

Informational equilibria (see (1)) form rather cumbersome structures. Establishing a connection between an awareness structure and an informational equilibrium may cause an embarrassment. The graph of a reflexive game provides a convenient language for describing the mutual awareness of agents and a powerful analysis tool for the properties of an informational equilibrium. We discuss it below.

2.4   GRAPH OF A REFLEXIVE GAME

If an awareness structure has a finite complexity, one may draw the **graph of a reflexive game**; it illustrates the interconnection among actions of real and phantom agents.

The nodes of such directed graph are represented by the actions x_τ, $\tau \in \Sigma_+$, that correspond to the pairwise nonidentical awareness structures I_τ (alternatively, by the components of the awareness structure θ_τ or simply by the number τ of a certain real or phantom agent, $\tau \in \Sigma_+$).

Nodes are connected via arcs according to the following rule. Each node $x_{\sigma i}$ has arcs coming from $(n-1)$ nodes that correspond to the structures $I_{\sigma ij}, j \in N\backslash\{i\}$. Two nodes being connected via two opposite arcs, the edge is represented by two arrows.

Let us emphasize an important aspect. The graph of a reflexive game meets the system of equations (2.3.1) (i.e., it satisfies the definition of an informational equilibrium). Nevertheless, the solution to this system possibly does not exist.

Thus, the graph G_I of the reflexive game Γ_I (see the notion of a reflexive game above) having a finite complexity of the awareness structure is defined in the following way:

– nodes of the graph G_I correspond to real and phantom agents participating in the reflexive game (in other words, to the pairwise nonidentical awareness structures);
– arcs of the graph G_I describe the mutual awareness of agents; if there exists a path from a certain (real or phantom) agent to another one, the latter is adequately aware of the former.

Suppose that nodes of the graph G_I describe beliefs of the corresponding agent about the state of nature. Then the reflexive game Γ_I with the finite awareness structure I may be specified by the tuple $\Gamma_I = \{N, (X_i)_{i \in N}, f_i(\cdot)_{i \in N}, G_I\}$, where N is a set of real agents, X_i stands for a set of feasible actions of agent i, $f_i(\cdot): \Theta \times X' \to \Re^1$ gives his/her goal function, $i \in N$, and G_I designates the graph of the reflexive game.

In many cases, one benefits by describing a reflexive game exactly in terms of the graph G_I (thus, not involving the tree of awareness structure).

We provide several examples of informational equilibrium evaluation using the graph of a reflexive game.

Examples 2.4.1–2.4.3. Consider three agents with goal functions of the form

$$f_i(\theta, x_1, x_2, x_3) = (\theta - x_1 - x_2 - x_3)x_i - \frac{x_i^2}{2},$$

where $x_i \geq 0$, $i \in N = \{1, 2, 3\}$, and $\theta \in \Theta = \{1, 2\}$.

The interpretation is the following: x_i represents the production output of agent i, θ describes consumer demand for the products. Then the first term in the goal function expresses the sales proceeds (price is multiplied by output by analogy to the Cournot oligopoly model); the second term serves to describe manufacturing costs. See models of the Cournot oligopoly in [108, 121, 147].

To simplify the exposition, call *a pessimist* any agent believing that demand is low ($\theta = 1$); on the other hand, *an optimist* believes demand is high ($\theta = 2$). Therefore, the cases of Examples 2.4.1–2.4.3 differ only by the awareness structures.

Example 2.4.1. Let agents 1–2 be optimists, while agent 3 represents a pessimist. Suppose that all agents possess an identical awareness. Then, according to Assertion 2.2.5, for any $\sigma \in \Sigma$ we obtain the identities $I_{\sigma 1} = I_1$, $I_{\sigma 2} = I_2$, $I_{\sigma 3} = I_3$.

According to Condition 2 in the notion of an informational equilibrium, similar properties hold for equilibrium actions x_σ^*.

Obviously, any awareness structure appears identical to one of those structures forming the basis $\{I_1, I_2, I_3\}$. Hence, complexity of the given awareness structure equals 3, while the depth makes 1. The corresponding graph of the reflexive game is illustrated by Fig. 2.2 below.

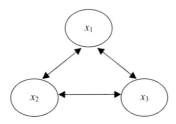

Figure 2.2 The graph of the reflexive game in Example 2.4.1.

Evaluation of the informational equilibrium requires solving the system of equations (see (2.3.1)) as follows.

$$
\begin{cases}
x_1^* = \dfrac{2 - x_2^* - x_3^*}{3}, \\[2mm]
x_2^* = \dfrac{2 - x_1^* - x_3^*}{3}, \\[2mm]
x_3^* = \dfrac{1 - x_1^* - x_2^*}{3},
\end{cases}
\Leftrightarrow
\begin{cases}
x_1^* = \dfrac{1}{2}, \\[2mm]
x_2^* = \dfrac{1}{2}, \\[2mm]
x_3^* = 0.
\end{cases}
$$

Thus, the actions of the agents in an informational equilibrium are defined by $x_1^* = x_2^* = 1/2, x_3^* = 0.$ •

Example 2.4.2. Assume that agents 1–2 are optimists, while agent 3 (being a pessimist) believes all agents are also pessimists and have an identical awareness. Agents 1-2 possess an identical awareness being adequately aware of agent 3.

We have $I_1 \sim I_2, I_1 > I_3, I_2 > I_3, I_1 \sim_3 I_2 \sim_3 I_3$. Fig. 2.3 shows the graph of the reflexive game.

These conditions can be stated as the following identities for any $\sigma \in \Sigma$ (see appropriate definitions and Assertions 2.2.1, 2.2.2 and 2.2.5):

$$I_{12\sigma} = I_{2\sigma}, \quad I_{13\sigma} = I_{3\sigma}, \quad I_{21\sigma} = I_{1\sigma}, \quad I_{23\sigma} = I_{3\sigma}, \quad I_{3\sigma1} = I_{31}, \quad I_{3\sigma2} = I_{32}, \quad I_{3\sigma3} = I_3.$$

Similar formulas hold true for equilibrium actions x_σ^*. The left-hand sides of the identities show that under $|\sigma| > 2$ any structure I_σ is identical to a certain structure I_τ, $|\tau| < |\sigma|$. Therefore, the depth of the structure I does not exceed 2 (this structure has a finite complexity). Their right-hand sides indicate that the basis includes the structures $\{I_1, I_2, I_3, I_{31}, I_{32}\}$ (clearly, such structures are pairwise different).

The complexity and depth of the stated awareness structure constitute 5 and 3, respectively.

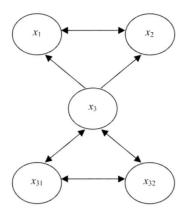

Figure 2.3 The graph of the reflexive game in Example 2.4.2.

The following system of equations should be solved to find an informational equilibrium (see (2.3.1)):

$$
\begin{cases}
x_1^* = \dfrac{2 - x_2^* - x_3^*}{3}, \\[2mm]
x_2^* = \dfrac{2 - x_1^* - x_3^*}{3}, \\[2mm]
x_3^* = \dfrac{1 - x_{31}^* - x_{32}^*}{3}, \\[2mm]
x_{31}^* = \dfrac{1 - x_{32}^* - x_3^*}{3}, \\[2mm]
x_{32}^* = \dfrac{1 - x_{31}^* - x_3^*}{3},
\end{cases}
\Leftrightarrow
\begin{cases}
x_1^* = \dfrac{9}{20}, \\[2mm]
x_2^* = \dfrac{9}{20}, \\[2mm]
x_3^* = \dfrac{1}{5}, \\[2mm]
x_{31}^* = \dfrac{1}{5}, \\[2mm]
x_{32}^* = \dfrac{1}{5}.
\end{cases}
$$

Consequently, the actions of real agents in the informational equilibrium are $x_1^* = x_2^* = 9/20$, $x_3^* = 1/5$. •

Example 2.4.3. Suppose that agents 1-3 represent optimists. Let agents 1 and 2, agents 2 and 3 be mutually aware. According to agent 1, agent 3 considers all agents as identically aware pessimists. According to agent 3, agent 1 believes all agents are identically aware pessimists.

Consequently, $I_1 > < I_2$, $I_2 > < I_3$, $I_1 \sim_{13} I_2 \sim_{13} I_3$, $I_1 \sim_{31} I_2 \sim_{31} I_3$.

Again, these conditions can be rewritten as a series of identities for any $\sigma \in \Sigma$ (by utilizing necessary definitions and Assertions 2.2.1, 2.2.2 and 2.2.5). As the result, we have

$$
I_{12\sigma} = I_{2\sigma}, \quad I_{13\sigma 1} = I_{131}, \quad I_{13\sigma 2} = I_{132}, \quad I_{13\sigma 3} = I_{13}, \quad I_{21\sigma} = I_{1\sigma},
$$
$$
I_{23\sigma} = I_{3\sigma}, \quad I_{31\sigma 1} = I_{31}, \quad I_{31\sigma 2} = I_{312}, \quad I_{31\sigma 3} = I_{313}, \quad I_{32\sigma} = I_{2\sigma}.
$$

Similar expressions are valid for equilibrium actions x_σ^*.

The left-hand sides of the identities show that under $|\sigma| > 3$ any structure I_σ is identical to a certain structure I_τ, $|\tau| < |\sigma|$. Hence, the depth of the structure I reaches (at most) 3. And so, this structure possesses a finite complexity. Their right-hand sides demonstrate that the basis consists only of the following awareness structures: I_1, I_2, I_3, I_{31}, I_{13}, I_{131}, I_{132}, I_{312}, I_{313}.

Next, for any $\sigma \in \Sigma$ we have $\theta_{131\sigma} = \theta_{31\sigma} = \theta_{313\sigma} = \theta_{13\sigma} = \theta_{123\sigma} = \theta_{213\sigma} = 1$. This result leads to the identities $I_{131} = I_{31}$, $I_{313} = I_{13}$, $I_{123} = I_{213}$.

Therefore, the basis includes the following pairwise different structures: $\{I_1, I_2, I_3, I_{31}, I_{13}, I_{132}\}$. The complexity of such awareness structure makes up 6, while its depth is 3. See Fig. 2.4 for the graph of the corresponding reflexive game.

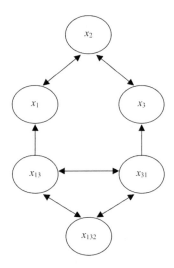

Figure 2.4 The graph of the reflexive game in Example 2.4.3.

Evaluating an informational equilibrium requires solving the system of equations (see (2.3.1)):

$$
\begin{cases}
x_1^* = \dfrac{2 - x_2^* - x_{13}^*}{3}, \\[2mm]
x_2^* = \dfrac{2 - x_1^* - x_3^*}{3}, \\[2mm]
x_3^* = \dfrac{2 - x_{31}^* - x_2^*}{3}, \\[2mm]
x_{31}^* = \dfrac{1 - x_{132}^* - x_{13}^*}{3}, \\[2mm]
x_{13}^* = \dfrac{1 - x_{31}^* - x_{132}^*}{3}, \\[2mm]
x_{132}^* = \dfrac{1 - x_{31}^* - x_{13}^*}{3},
\end{cases}
\Leftrightarrow
\begin{cases}
x_1^* = \dfrac{17}{35}, \\[2mm]
x_2^* = \dfrac{12}{35}, \\[2mm]
x_3^* = \dfrac{17}{35}, \\[2mm]
x_{31}^* = \dfrac{1}{5}, \\[2mm]
x_{13}^* = \dfrac{1}{5}, \\[2mm]
x_{132}^* = \dfrac{1}{5}.
\end{cases}
$$

Thus, the actions of real agents in the informational equilibrium are $x_1^* = x_3^* = 17/35$, $x_2^* = 12/35$. •

We have completed the description of the graph of a reflexive game. Now, revert to analyzing the properties of informational equilibria.

2.5   REGULAR AWARENESS STRUCTURES

In Section 2.2 we have introduced the notion of an awareness structure (an infinite tree reflecting an hierarchy of agents in a reflexive game). In Section 2.3 we have demonstrated that an informational equilibrium (as a solution to a reflexive game) exists in the case of a finite awareness structure. By definition, the finiteness of an awareness structure does not mean the finiteness of its tree. Instead, it signifies the existence a finite basis, where consideration of phantom agents (possessing the same awareness as other real or phantom agents) yields no new information. Thus, such consideration seems pointless.

Suppose there a priori is a finite tree reflecting several first levels of agents' beliefs (e.g., constructed from some interpretations). Generally, one cannot unambiguously determine the infinite awareness structure this tree belongs to. In other words, there may exist a set of awareness structures such that any finite number of their upper levels coincide.

Thus, defining an informational equilibrium by a finite tree of agents' beliefs requires additional assumptions. For instance, postulate that each phantom agent (corresponding to the lower level of a finite tree of beliefs) chooses his/her action as follows. This agent believes that an agent at the previous hierarchical level is adequately aware of him/her (see Assumption Π_m in [136] and subjective Bayes equilibria in [152]).

The present section deals with regular awareness structures with the following property. Under a given finite tree of beliefs and regularity of a corresponding awareness structure, an informational equilibrium is defined explicitly. Moreover, for regular awareness structures one succeeds in (a) establishing constructive existence conditions for an informational equilibrium, (b) analyzing the relationship between an informational equilibrium and an awareness structure (Section 2.6), (c) posing and solving reflexive control problems (Section 2.11).

The notion of an awareness structure appears rather general and embraces cases with nontrivial practical interpretations (see the discussion above). Therefore, consider the class of *regular awareness structures*, being sufficiently large to cover many real situations and allowing easy description. To specify such structures, let us introduce an auxiliary notion of a *regular finite tree* (RFT). Do this recursively.

Imagine that a game includes n agents. In the elementary case, all agents possess identical awareness; subsequently, the awareness structure has complexity n and depth 1. Describe this situation as a tree composed of a root note, n edges and n dangling nodes. Fig. 2.5 demonstrates such a tree for three agents (for better clarity, $\theta_1, \theta_2, \theta_{12}$ are replaced with 1, 2, 12).

This RFT corresponds to the reflexive game graph in Fig. 2.6.

Next, the RFT "grows" as follows. Each dangling node τi, $\tau \in \Sigma$, connects $(n-1)$ edges exactly; this generates $(n-1)$ dangling nodes τij, $j = 1, \ldots, i-1, i+1, \ldots, n$.

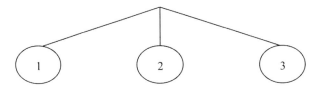

Figure 2.5 A regular finite tree.

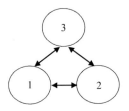

Figure 2.6 The reflexive game graph for the RFT shown in Fig. 2.5.

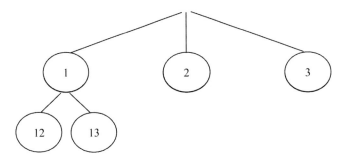

Figure 2.7 An example of an RFT of depth 2.

The resulting RFT has the following interpretation. If a dangling node τi, $\tau \in \Sigma$, exists, then τi-agent has identical awareness with τ-agent (in the case of empty sequence τ, τi-agent turns out real and his/her subjective beliefs coincide with objective ones).

As examples of regular awareness structures, we provide all possible structures of depth 2 (up to relabeling of agents).

First, take the RFT illustrated by Fig. 2.7.

Under $\theta_{12} = \theta_2$ and $\theta_{13} = \theta_3$, we again arrive at the graph in Fig. 2.7. If (at least) one equality is violated, one has the reflexive game graph presented by Fig. 2.8.

The next case of RFTs can be observed in Fig. 2.9.

Here we obtain two possible graphs of reflexive game (which are not reducible to the previous cases) – see Fig. 2.10 and Fig. 2.11.

The last case of an RFT is illustrated by Fig. 2.12.

This situation corresponds to three reflexive game graphs (which are not reducible to the previous ones) – see Fig. 2.13, Fig. 2.14 and Fig. 2.15. Obviously, the graphs in Fig. 2.14 and Fig. 2.15 appear disconnected.

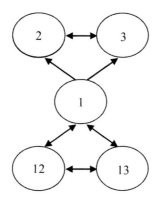

Figure 2.8 The reflexive game graph for the RFT shown in Fig. 2.7.

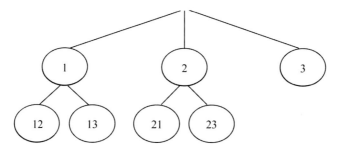

Figure 2.9 An example of an RFT of depth 2.

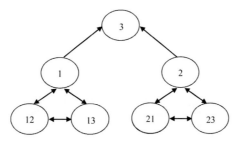

Figure 2.10 Reflexive game graph I for the RFT shown in Fig. 2.9.

Practical interpretations for the seven possible awareness structures of depth 2 discussed above cause no troubles. Consider three symmetrical structures (see Fig. 2.6, Fig. 2.13 and Fig. 2.15).

Fig. 2.6 agrees with identical awareness of agents. Their reflexive realities coincide. In fact, agents play the same game whose rules form a common knowledge.

Fig. 2.13 answers the "opposite situation." Agents have distorted and pairwise incompatible beliefs of each other. Each agent thinks that all agents possess an identical awareness (but all agents are mistaken). Actually, each agent plays his/her own game.

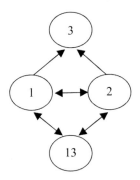

Figure 2.11 Reflexive game graph 2 for the RFT shown in Fig. 2.9.

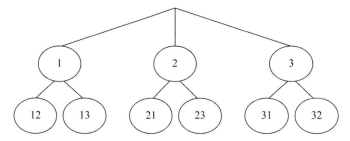

Figure 2.12 An example of an RFT of depth 2.

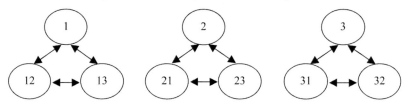

Figure 2.13 Reflexive game graph 1 for the RFT shown in Fig. 2.12.

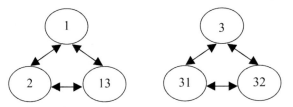

Figure 2.14 Reflexive game graph 2 for the RFT shown in Fig. 2.12.

Fig. 2.15 conforms to the following case. Each agent has a better awareness than the rest ones (according to his/her beliefs). For instance, agents have conducted negotiations, reporting their beliefs about an uncertain parameter to each other. However, all agents concealed their true beliefs (still, assuming truth-telling and trust by the

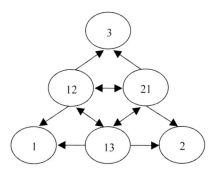

Figure 2.15 Reflexive game graph 3 for the RFT shown in Fig. 2.12.

opponents). Another interpretation of Fig. 2.15 is possible, as well. Agents have concluded an agreement, but each agent is going to transgress it (supposing that the rest would follow the agreement and expect the same behavior from the opponents).

The properties of regular awareness structures discussed in this section will be intensively used in analysis of reflexive control problems (see Section 2.11).

Finally, consider the issue regarding the existence of an informational equilibrium for regular awareness structures.

By RFT constructing procedure, equilibrium actions of agents (if any) can be evaluated by the "bottom-top" approach, i.e., by passing from dangling nodes to the RFT root. Imagine that for a certain $\tau \in \Sigma_+$ dangling nodes are $(n-1)$ nodes $\tau ij, j \in N \backslash \{i\}$. By definition of an RFT, exactly n agents from the set $\{\tau ij\}, j \in N$, have an identical awareness. Recall that, due to the axiom of self-awareness (see Section 2.2), identify between τi- and τii-agents. Thus, their equilibrium actions satisfy $x^*_{\tau ijk} = x^*_{\tau ik}, j, k \in N$,

$$(1) \quad x^*_{\tau ij} \in \operatorname*{Arg\,max}_{x_{\tau ij} \in X_i} f_j(\theta_{\tau ij}, x^*_{\tau i1}, \ldots, x^*_{\tau i,j-1}, x_{\tau ij}, x^*_{\tau i,j+1} \ldots, x^*_{\tau in}), \quad j \in N.$$

Interestingly, the rest agents "lie beyond the view" of n agents $\{\tau ij\}, j \in N$.

The system (1) represents a "standard" Nash equilibrium formula for a game with a common knowledge. If it admits a solution, one can evaluate the action of τi-agent.

Next, consider node $\tau, \tau \in \Sigma_+$, and nodes $\tau m, m \in N \backslash \{\omega(\tau)\}$, which contain node τi. Recall that $\omega(\tau)$ designates the last index in the sequence τ. All τm-agents can be divided into two sets, *viz.*, agents possessing an identical awareness with τ–agent and the rest agents (including τi-agent). For better clarity in such division, introduce the notation

$$\overline{N_\tau} = \{k \in N \mid I_{\tau k} \sim I_\tau\}.$$

We have observed that an equilibrium action of τi-agent (similarly, equilibrium actions of all τm-agents, $m \notin \overline{N_\tau}$) is defined regardless of the actions by other τm-agents, $m \in N$. Consequently, all τk-agents, $k \in \overline{N_\tau}$, may substitute the actions

$x^*_{\tau m}$, $m \notin \overline{N_\tau}$, into their goal functions. Therefore, evaluating the equilibrium actions $x^*_{\tau k}$, $k \in \overline{N_\tau}$ requires solving the system of equations

$$(2) \quad x^*_{\tau k} = \arg \max_{x_{\tau k} \in X_k} f_k(\theta_{\tau k}, x^*_{\tau 1}, \ldots, x^*_{\tau, k-1}, x_{\tau k}, x^*_{\tau, k+1} \ldots, x^*_{\tau n}), \quad k \in \overline{N_\tau}.$$

The system (2) determines a Nash equilibrium in the game of τk-agents, $k \in \overline{N_\tau}$. Its solution (if exists) enables finding the equilibrium action x^*_τ.

Moving from dangling nodes to the root node, one evaluates all equilibrium actions successively. For this, all systems of the form (1) and (2) must admit solutions. Thus, we state the following sufficient existence condition for an informational equilibrium in regular reflexive structures (the set of real agents N, their goal functions $\{f_i\}$ and sets of feasible actions $\{X_i\}$, as well as the set of feasible values Θ of an uncertain parameter are supposed fixed).

Assertion 2.5.1. For any nonempty set $\overline{N} \subseteq N$, assume the following. For any $\theta_k \in \Theta$, $k \in \overline{N}$, and any $x^*_m \in X_m$, $m \notin \overline{N}$, there exists a Nash equilibrium in the game with a common knowledge among k-agents. Notably, there are x^*_k, $k \in \overline{N}$, satisfying the formula

$$x^*_k \in \text{Argmax}_{x_k \in X_k} f_k(\theta_k, x^*_1, \ldots, x^*_{k-1}, x_k, x^*_{k+1} \ldots, x^*_n), \quad k \in \overline{N}.$$

Then an informational equilibrium exists for any finite awareness structure.

Let us summarize the outcomes. We have the language of agents' awareness description (awareness structures – see Section 2.2), the definition of a reflexive game solution (informational equilibria – see Section 2.3), as well as the properties of reflexive game graphs (Section 2.4) and regular awareness structures (the present Section). This enables analyzing the influence of agents' awareness (their reflexion ranks) on informational equilibria and their gains. By-turn, in Section 2.11 we proceed with reflexive control problems.

2.6 REFLEXION RANK AND INFORMATIONAL EQUILIBRIUM

Section 1.2 has defined a parametric Nash equilibrium, where the vector of equilibrium actions depends on the value of the state of nature (being a common knowledge). In Section 2.3, we have introduced the notion of an informational equilibrium as a subjective equilibrium depending on the awareness structure $I = (I_1, I_2, \ldots, I_n)$, where I_i specifies the awareness structure of agent i ($i \in N$).

Denote by $x^*_i(I_i)$ the set of subjectively equilibrium actions[7] of agent i, who has the awareness structure I_i, $i \in N$. Let $x^*(I_i)$ be the corresponding set of vectors of subjectively equilibrium actions. Next, designate by Ψ_i the set of all possible awareness

---

[7]Recall that a subjectively equilibrium action of an agent is a component of an informational equilibrium, which corresponds to his/her awareness.

structures[8] of agent i, and by $\psi_i^{k_i}$ the set of all possible finite awareness structures of agent i, $i \in N$ (with maximal depth k_i). According to the above definition, suppose that **an agent with a finite awareness structure of depth k possesses informational reflexion rank $k - 1$.** If awareness structures of all agents appear finite, the depth of the awareness structure I is also finite, constituting

$$\gamma(I) = 1 + \max_{i \in N}\{k_i\}.$$

Consider the set

(1) $\quad X_i^*(\Psi) = \bigcup_{I_i \in \Psi} x_i^*(I_i),$

where $\Psi \subseteq \Psi_i$ ($i \in N$). This set includes actions of agent i, that can be subjectively equilibrium (provided that his/her awareness structures belong to the set Ψ). In addition, construct the set of subjective equilibria under various awareness structures from the set Ψ:

(2) $\quad X^*(\Psi) = \bigcup_{I \in \Psi} x^*(I).$

For a fixed set Θ of feasible states of nature, we have the inclusions $\Psi_i^{k_i} \subseteq \Psi_i^{k_i+1}$, $k_i \in \aleph$, $i \in N$. Hence, the set of feasible subjective equilibria by no means gets narrow for higher depth of the awareness structure.

Therefore, we know the relationships (1)–(2) between the sets of potential equilibria and the set of possible awareness structures. Under infinite awareness structures of agent i ($I_i \in \Psi_i$), the set of possible subjective equilibria becomes $X^*(\Psi_i) = \bigcup_{I_i \in \Psi_i} x^*(I_i)$, $i \in N$. The following questions arise immediately. Is there a set of finite awareness structures yielding the same set of possible subjective equilibria for a given agent? What is the depth of such structures? We state a corresponding problem.

Assume that the goal functions and feasible sets of all agents[9], as well as the set of feasible states of nature are fixed and form a common knowledge. The *problem of maximal rational rank of informational reflexion* can be posed as

(3) $\quad k_i^* = \min\{k_i \in \aleph | X^*(\Psi_i) = X^*(\Psi_i^{k_i})\}, \quad i \in N.$

The problem (3) is formulated in the view of a researcher, who is analyzing the behavior of agents. This researcher seeks for the minimal reflexion rank such that any action of a given agent (representing a subjective equilibrium under a certain feasible awareness structure of this agent) is a subjective equilibrium in an awareness structure (whose depth exceeds the desired reflexion rank at most by unity). This aspect elucidates the term "objective." An alternative lies in acting as an agent to compare

---

[8]The values of the state of nature are elements of the awareness structure. Hence, the set of possible awareness structures depends on the set Θ (feasible valuesof the state of nature). Still, we do not show such dependence explicitly.

[9]By default, suppose that goal functions are continuous and feasible sets are compact.

his/her gains (guaranteed values of the goal functions) under different reflexion ranks. Again, we state a corresponding problem.

First, explain the notion of agent's gain. Denote by

$$(4) \quad v_i = \max_{x_i \in X_i} \min_{x_{-i} \in X_{-i}} \min_{\theta_i \in \Theta} f_i(\theta_i, x_i, x_{-i}), \quad i \in N,$$

the classic MGR[10] of agent i. Next, consider the subjective MGR of agent i with respect to the set of all subjective equilibria under all possible awareness structures from the set $\Psi \subseteq \Psi_i$:

$$(5) \quad v_i^s(\Psi) = \min_{y \in X(\Psi)} f_i(\theta_i, x), \quad i \in N.$$

It follows from (2), (4) and (5) that $v_i^s(\varphi) \geq v_i$, $\varphi \subseteq \Psi_i$, $i \in N$.

Let the goal functions and feasible sets of all agents, as well as the set of possible states of nature be fixed and form a common knowledge. *The maximal rational i-subjective rank of informational reflexion* lies in evaluation of

$$(6) \quad s_i^* = \min\{s_i \in \aleph \mid v_i^s(\Psi_i) = v_i^s(\Psi_i^{s_i})\}, \quad i \in N.$$

Formula (6) means it is necessary to find the minimal rank of agent's informational reflexion with the following property. For any higher rank of reflexion, there exists no informational equilibrium yielding a strictly smaller gain to the agent.

The result below is immediate from (3), (5) and (6), if one employs the monotonicity of subjective equilibria sets with respect to the depths of awareness structures.

Assertion 2.6.1. $s_i^* \leq k_i^*$, $i \in N$.

As a matter of fact, Assertion 2.6.1 claims the following. Suppose we understand agent's gain as the guaranteed value of his/her goal function (on the set of all possible subjective equilibria). Then the maximal rational subjective rank of informational reflexion of any agent does not exceed the maximal rational objective rank of his/her informational reflexion. In other words, assume there exists a certain rank of informational reflexion, "exhausting" the set of subjective equilibria. Then this rank represents the upper bound of the maximal depth of an awareness structure being rational in the view of the agent under consideration.

According to Assertion 2.6.1, one should study the problem (3). However, solving it in the general case appears awkward. Therefore, we analyze the special case of a reflexive two-player game with a regular awareness structure. By analogy, the results can be extended to the case of any finite number of agents.

Consider an RFT and corresponding graphs of the reflexive game. Recall that a subjective common knowledge arises at lower levels of the RFT. Notably, in a reflexive game with a regular awareness structure, agents at two lower levels may have identical or nonidentical beliefs about uncertain parameters. The former case is called a

---

[10]We believe that the set of awareness structures Ψ is such that $\theta_i \in \Theta$, $i \in N$.

symmetric common knowledge at lower levels, while the latter case is known as an *asymmetric common knowledge at lower levels*.

Denote by

$$(7) \quad E_N(\theta_1, \theta_2) = \{(x_1(\theta_1, \theta_2), x_2(\theta_1, \theta_2)) \in X' |$$

$$\forall y_1 \in X_1 \ f_1(\theta_1, x_1(\theta_1, \theta_2), x_2(\theta_1, \theta_2)) \geq f_1(\theta_1, y_1, x_2(\theta_1, \theta_2))$$

$$\forall y_2 \in X_2 \ f_2(\theta_2, x_1(\theta_1, \theta_2), x_2(\theta_1, \theta_2)) \geq f_2(\theta_2, x_1(\theta_1, \theta_2), y_2)\}$$

the classic Nash equilibrium in a two-player game, where information on the values of θ_1, $\theta_2 \in \Theta$ forms a common knowledge. Introduce the set of the best responses of agent i to opponent's choice of actions from the set X_{-i} (under the set Θ of possible states of nature) in the form

$$BR_i(\Theta, X_{-i}) = \bigcup_{x_{-i} \in X_{-i}, \theta \in \Theta} \text{Arg} \max_{x_i \in X_i} f_i(\theta, x_i, x_{-i}), \quad i = 1, 2.$$

In addition, consider the following quantities and sets:

$$(8) \quad E_N = \bigcup_{\theta_1, \theta_2 \in \Theta} E_N(\theta_1, \theta_2), \quad E_N^0 = \bigcup_{\theta \in \Theta} E_N(\theta, \theta),$$

$$(9) \quad X_i^0 = \bigcup_{\theta_1, \theta_2 \in \Theta} x_i(\theta_1, \theta_2) = Proj_i E_N, \quad i = 1, 2,$$

$$(10) \quad X_i^k = BR_i(\Theta, X_{-i}^{k-1}), \quad k = 1, 2, \ldots, \ i = 1, 2.$$

The mapping $BR_i(\Theta, X_{-i}): \Theta \times X_{-i} \to X_i$ is said the *reflexive mapping* of agent i, $i = 1, 2$.

Properties of these sets are characterized by Assertion 2.6.2. This result follows from the definitions (7)–(10).

Assertion 2.6.2. $E_N^0 \subseteq E_N, X_i^k \subseteq X_i^{k+1}, k = 0, 1, \ldots, i = 1, 2.$

Study the reflexive two-player game with a finite[11] regular awareness structure.

The reflexive mapping of agent i is called *stationary* if $X_i^k = X_i^{k+1}, k = 0, 1, \ldots,$ $i = 1, 2.$

Take agent i ($i = 1, 2$) and analyze his/her subjective equilibria under different ranks of informational reflexion. We will increase them successively (recall that informational reflexion rank is by unity smaller than the depth of a corresponding awareness structure).

Under a fixed depth of a regular awareness structure I_i, the graph of a reflexive game can be built by two methods. First, by assuming an asymmetric common knowledge at a lower level. Second, by assuming a symmetric common knowledge

---

[11]The requirement concerning the finiteness of an awareness structure seems natural. Human capabilities of data processing are limited (see the discussion above). Furthermore, the depth of an awareness structure can be bounded by a sufficiently large finite number.

at a lower level (for this, introduce an additional phantom agent with the same awareness as the agent corresponding to the lower level of the tree I_i). We deal with both cases.

1. Suppose that the depth of the awareness structure ψ_i^1 of agent i ($i = 1, 2$) makes up $k_i = 1$ (θ_i exists only). The corresponding graph of the reflexive game acquires the form $x_i \leftrightarrow x_{ij}$ (here and in the sequel, $j \neq i$). That is, real agent i plays the game with phantom agent ij, and they both possess the information θ_i according to the view of agent i (a symmetric common knowledge takes place at a lower level). Hence, $X^*(\psi_i^1) = E_N^0$.

2. Suppose that the depth of the awareness structure ψ_i^2 of agent i equals $k_i = 2$ (θ_i and θ_{ij} do exist). Two cases[12] are possible then.

In case 1, the appropriate graph of the reflexive game becomes $x_i \leftrightarrow x_{ij}$ (an asymmetric common knowledge at a lower level). Notably, real agent i (believing in the state of nature θ_i) thinks that he/she plays the game with phantom agent ij (possessing the information θ_{ij}). According to agent i, the set of feasible equilibria of he game constitutes $E_N \supseteq X^*(\psi_i^1)$.

In case 2 (a symmetric common knowledge at a lower level), the corresponding graph of the reflexive game takes the form $x_i \leftarrow x_{ij} \leftrightarrow x_{iji}$. Hence, real agent i (believing in the state of nature θ_i) thinks that he/she plays the game with phantom agent ij. By-turn, the latter plays the game with phantom agent iji (these agents possess the information θ_{ij} in the view of agent i). According to agent i, the set of feasible equilibria of the game played by agents ij and iji is E_N^0. Consequently, agent ij can (in the view of agent i) choose an action from the set $Proj_j E_N^0$. Thus,

(11) $X^*(\psi_i^2) = BR_i(\Theta, Proj_j E_N^0) \times Proj_j E_N^0 \subseteq X_i^1 \times X_j^0.$

Since $E_N \subseteq X_i^0 \times X_j^0$, then $X^*(\psi_i^1) \subseteq X^*(\psi_i^2)$. For agent i, increasing the reflexion rank from 0 to 1 in the problem (3) is rational.

3. Suppose that the depth of the awareness structure ψ_i^3 of agent i is $k_i = 3$ (we have θ_i, θ_{ij}, and θ_{iji}). Two cases exist here.

In case 1 (an asymmetric common knowledge at a lower level), the corresponding graph of the reflexive game is given by $x_i \leftarrow x_{ij} \leftrightarrow x_{iji}$. Hence, real agent i (believing in the state of nature θ_i) thinks that he/she plays the game with phantom agent ij (believing in the state of nature θ_{ij}). By-turn, the latter plays the game with phantom agent iji (possessing the information θ_{iji}). According to agent i, the set of feasible equilibria of the game played by agents ij and iji becomes E_N. This yields

$X^*(\psi_i^3) = BR_i(\Theta, X_j^0) \times X_j^0 = X_i^1 \times X_j^0.$

---

[12]For each reflexion rank, these cases correspond to a "symmetric" and "asymmetric" (subjective) common knowledge.

In case 2 (a symmetric common knowledge at a lower level), the corresponding graph of the reflexive game takes the form $x_i \leftarrow x_{ij} \leftarrow x_{iji} \leftrightarrow x_{ijij}$. Notably, real agent i (believing in the state of nature θ_i) thinks that he/she plays the game with phantom agent ij (believing in the state of nature θ_{ij}). By-turn, the latter plays the game with phantom agent iji. But this agent plays the game with phantom agent $ijij$ (they both possess the information θ_{ijij} in the mind of agent i). According to agent i, the set of feasible equilibria of the game played by agents ij and $ijij$ makes up $E_N^0 \subseteq X_i^0 \times X_j^0$. Hence, agent iji can choose an action from the set $Proj_i E_N^0$ (as agent i thinks). Then agent ij can choose an action from the set $BR_j(\Theta, Proj_i E_N^0)$ (again, in the view of agent i). Therefore,

$$X^*(\psi_i^3) = BR_i(\Theta, BR_j(\Theta, Proj_i E_N^0)) \times BR_j(\Theta, Proj_i E_N^0) \subseteq X_i^2 \times X_j^1.$$

As far as $X^*(\psi_i^2) \subseteq X^*(\psi_i^3)$, increasing the reflexion rank from 1 to 2 in the problem (3) seems rational for agent i.

By analogy, one would easily show the following. Under an asymmetric common knowledge at a lower level, the set of subjective equilibria of agent i is defined by

$$(12) \quad AX_i^*(\Psi_i^{k_i}) = X_i^{k_i - 2}, \quad k_i = 2, 3, \dots$$

Under a symmetric common knowledge at a lower level, the set of subjective equilibria of agent i has the representation

$$(13) \quad SX_i^*(\Psi_i^{k_i}) = BR_i(\Theta, \dots, BR_j(\Theta, Proj_i E_N^0), \dots), \quad k_i = 2, 3, \dots.$$

Since $SY_i^*(\Psi_i^{k_i}) \subseteq X_i^{k_i - 1} \subseteq X_i^{k_i}$, we have the result formulated in Assertion 2.6.3. In fact, it implies that any subjective equilibrium obtained within a symmetric common knowledge at a lower level can be made a subjective equilibrium within an asymmetric common knowledge at a lower level (by increasing the depth of an awareness structure by unity).

Assertion 2.6.3. $SX_i^*(\Psi_i^{k_i}) \subseteq AX_i^*(\Psi_i^{k_i + 1}), \quad k_i = 2, 3, \dots.$

Assertion 2.6.3 and the definition of a stationary reflexive mapping lead to the following. Reflexive two-player games with a regular awareness structure enjoy the property below (regardless of a symmetric or asymmetric common knowledge at a lower level).

Assertion 2.6.4. If reflexive mappings of agents are stationary, then the maximal rational subjective rank of informational reflexion equals 2 and

$$X_i^*(\Psi) = X_i^0, \quad i = 1, 2.$$

In the case of stationary reflexive mappings, Assertion 2.6.4 enables considering reflexive control as *informational regulation*, where a controlled object represents agents' beliefs about uncertain parameters [69, 136].

Generally, three situations are possible:

1) If reflexive mappings are stationary, the set of subjective equilibria forms $\prod_{i\in N} X_i^0 \subseteq$
   X', *viz.*, turns out a subset (probably, a proper subset – see Example 2.6.1) of the set X' comprising feasible actions of agents;
2) If reflexive mappings are nonstationary, the set of subjective equilibria probably coincides with the set X' of feasible actions of agents – see Example 2.6.2;
3) If reflexive mappings are nonstationary, the set of subjective equilibria can be strictly larger than $\prod_{i\in N} X_i^0$, but does not coincide with (is smaller than) the set X'
   of feasible actions of agents – see Example 2.6.3;

Example 2.6.1. Let $f_i(\theta, x_1, x_2) = x_i - x_i^2/2(\theta + \alpha x_j)$, where $\alpha \in (0;1), j \neq i, i = 1, 2,$ $\Theta = [0;1]$. Then $BR_i(\theta_i, x_j) = \theta_i + \alpha x_j, j \neq i, i, j = 1, 2$.

Evaluate a Nash equilibrium: $x_i^*(\theta_i, \theta_j) = (\theta_i + \alpha\theta_j/(1 - \alpha^2), j \neq i, i, j = 1, 2$. Define $X_i^0 = [0; 1/(1 - \alpha)], i = 1, 2$. Clearly, the reflexive mapping in this example is stationary, i.e., $X_i^k = X_i^0, k = 1, 2, \ldots, i = 1, 2$ (see Fig. 2.16). By varying θ_i and θ_j (performing informational regulation), a principal can implement any point from the set $[0; 1/(1 - \alpha)]^2$ as a subjective equilibrium. •

Under nonstationary reflexive mappings, an increase in the depth of an awareness structure does not narrow the set of subjective equilibria. Therefore, analysis of a game-theoretic model does not assist in assigning a priori bounds for the maximal rational rank of agents' reflexion (see the discussion of the role of informational constraints in Section 3.3). Below we provide an example.

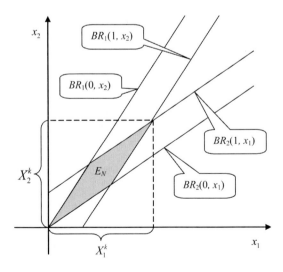

Figure 2.16 The set of subjective equilibria in Example 2.6.1.

Example 2.6.2. Let $f_1(\theta, x_1, x_2) = \theta(1 - x_2)x_1 - x_1^2/2$, $f_2(\theta, x_1, x_2) = \theta x_1 x_2 - x_2^2/2$, where $\Theta = [\frac{1}{2}; 1]$, $X_1 = X_2 = (0; 1)$. Then $BR_1(\theta, x_2) = \theta(1 - x_2)$, $BR_2(\theta, x_1) = \theta x_1$.

Obviously, the set E_N is a quadrilateral with nodes $(^2/_5, ^1/_5)$, $(^2/_3, ^1/_3)$, $(^1/_2, ^1/_2)$ and $(^1/_3, ^1/_3)$. Hence, $X_1^0 = [^1/_3; ^2/_3]$, $X_2^0 = [^1/_5; ^1/_2]$ (see Fig. 2.17).

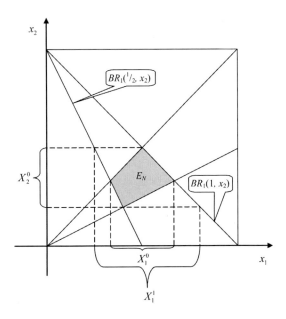

Figure 2.17 The set of subjective equilibria in Example 2.6.2.

Denote by $\alpha_{i,k}$ and $\beta_{i,k}$ $(i = 1, 2)$ the left and right limits of the segment X_i^k. Consequently, we have the expressions

$$\alpha_{2,k+1} = \frac{1}{2}\,\alpha_{1,k}, \quad \beta_{2,k+1} = \beta_{1,k}, \quad \alpha_{1,k+1} = \frac{1}{2}(1 - \beta_{2,k}), \quad \alpha_{1,k}, \quad \beta_{1,k+1} = 1 - \alpha_{2,k},$$

where $k = 0, 1, \ldots$ Simple manipulations lead to

$$\alpha_{i,k+4} = \frac{1}{4}\,\alpha_{i,k}, \quad \beta_{i,k+4} = \frac{3}{4} + \frac{1}{4}\,\beta_{i,k}, \quad i = 1, 2, \quad k = 0, 1, \ldots.$$

Therefore, $\alpha_{i,k} \to 0, \beta_{i,k} \to 1$ as $k \to \infty$, where $i = 1, 2$. And so, $\bigcup_{k \geq 0} X_i^k = (0; 1) = X_i$, $i = 1, 2$. This means that an appropriate increase in the depth of agent's reflexion guarantees any feasible action of the agent. •

Examples 2.6.1–2.6.2 show that the set of subjective equilibrium actions of i-agent (under all possible awareness structures of this agent) $\bigcup_{k \geq 0} X_i^k$ may coincide with X_i^0 (being rather narrow) or with X_i (being maximally wide). Make sure that "intermediate" situations exist, as well. Notably, the set of subjective equilibria can be strictly larger than the initial set E_N of Nash equilibria (but strictly narrower than the set of feasible actions).

Example 2.6.3. Suppose that the goal functions of agents and the set Θ are the same as in the previous example. Modify the sets of feasible actions merely: $X_1 = X_2 = (-c; 1 + c)$, $c > 0$. Then we still have $\bigcup_{k \geq 0} X_i^k = (0; 1)$, $i = 1, 2$, but $(0; 1) \subset X_i$. •

Consider another example illustrating the diversity of behavioral types of reflexing agents under growing ranks of their reflexion.

Example 2.6.4. Two agents choose actions from the unit segment. The agents possess the following goal functions: $f_1(\theta, x_1, x_2) = 4\theta x_1 x_2(1 - x_2) - x_1^2/2$, $f_2(\theta, x_1, x_2) = x_1 x_2 - x_2^2/2$. The state of nature takes values from the set $\Theta = (1/4; 1]$.

Assume that a common knowledge with a certain state of nature $\theta_0 \in \Theta$ takes place at a lower level of a finite RFT of depth m_0. Evaluate a Nash equilibrium in the game of phantom agents: $x_1(\theta_0) = x_2(\theta_0) = 1 - 1/(4\,\theta_0)$. Find the best responses of agents 1–2 to opponents' actions:

$$BR_1(\theta, x_2) = 4\theta x_2(1 - x_2), \quad BR_2(\theta, x_1) = x_1, \quad \theta \in \Theta.$$

Thus, the best responses of $\tau 1$-agents, $\tau \in \Sigma$, $|\tau| \leq m_0$, meet the logistic [165] mapping

$$(14) \quad x_1^m = 4\theta x_1^{m-1}(1 - x_1^{m-1}), \quad m = 1, 2, \ldots, [m_0/2],$$

with the initial point $x_1^0 = 1 - 1/(4\theta_0)$ (here $[\cdot]$ designates the integer part operator).

Analysis of (14) implies that, depending on the awareness θ_τ of τ-agent, real agent 1 has the following types of asymptotically (as $m_0 \to \infty$) stable strategies with weak dependence on the initial point: the choice of a unique action; periodic behavior; chaotic or periodic behavior (note that this awareness is identical for all agents at levels from 1 to (m_0-2), i.e., $\theta_\tau \equiv \theta$ for a certain value $\theta \in \Theta$ under $|\tau| \leq m_0 - 2$; in the case of different awareness of agents, one would observe even more sophisticated types of behavior). Practical interpretations and drawbacks of such uncertain collective behavior are obvious. •

The sets of subjective equilibria enjoy monotonicity with respect to reflexion rank (the depth of awareness structure) – see Assertion 2.6.2. Hence, promising directions of further research include:

1. searching for the minimal reflexion rank "exhausting" the set of subjective equilibria;
2. estimating the rate of change (alternatively, the rate of convergence) for the sequence $\{X_i^k\}$;
3. for a given action of an agent, defining the minimal rank of his/her informational reflexion, making this action a subjective equilibrium, etc.

For quadratic goal functions of agents (linear reflexive mappings), the stationarity conditions are established in [39].

Concluding this section, let us emphasize the following. Apparently, there is a definite analogy between reflexive games and informational extensions of games [67] (including *Howard's metagames* [86, 87]). Informational extensions of games presume the existence of agent's pre-ordering: choosing an action, an agent may know actions of the agents that have made their choice earlier (in this ordering). The described games (with a fixed sequence of moves) are called *hierarchical games* [67]. An important result has been derived for such games. Any action vector guaranteeing that agent's gains are not smaller than the corresponding maximin values in the original game can be implemented as an equilibrium in a certain informational extension of the original game. Generally speaking, reflexive games with RFTs do not have this property. Example 2.6.1 illustrates that in some cases the set of informational equilibria remains narrow. At the same time, the hierarchical property and the reflexive property of games by no means contradict each other. For instance, an hierarchical game can be reflexive. And so, synthesis of research results in the fields of metagames and reflexive games seems a challenging issue for future investigations.

This section has studied the relation between the set of informational equilibria and an awareness structure. This relation enables posing and solving reflexive control problems (control of an awareness structure and an informational equilibrium – see Section 2.11).

Below we describe specific types of informational equilibria [39]. Section 2.7 considers stability of an equilibrium [139], when each agent observes exactly the outcome expected at the moment of decision-making. Thus, the awareness structure of a game appears invariable.

In Section 2.8, we divide stable informational equilibria into true and false ones. Moreover, we state a sufficient condition ensuring the truth of all stable equilibria (Assertion 2.8.1).

Section 2.9 deals with the case when agents observe actions of each other (as a result of their game). A series of assertions elucidating the interconnection between equilibrium stability and some properties of the awareness structure of a game are provided.

Finally, Section 2.10 draws an analogy between reflexive games and Bayesian games (an alternative approach to modeling of decision process under incomplete awareness). We prove a theorem substantiating (in a certain mathematical sense) the necessity of bounding agent's reflexion rank.

2.7 STABLE INFORMATIONAL EQUILIBRIA

The "classical" concept of a Nash equilibrium is remarkable for its self-sustained nature. Notably, assume that a repeated game takes place and all agents (except agent i) choose the same equilibrium actions. Then agent i benefits nothing by deviating from his/her equilibrium action; evidently, this feature is directly related to the following. Beliefs of all agents about reality are adequate, i.e., the state of nature appears a common knowledge.

Generally speaking, the situation may change in the case of an informational equilibrium. Indeed, after a single play of the game some agents (or even all of them) may observe an unexpected outcome due to an inconsistent belief about the state of nature (or due to an inadequate awareness of opponents' beliefs). Anyway, the self-sustained nature of the equilibrium is violated; actions of agents may change as the game is repeated.

However, in some cases a self-sustained equilibrium takes place for differing (generally, incorrect) beliefs of the agents. As a matter of fact, such situation occurs when each agent (real or phantom) observes the game outcome he/she expects. To develop a formal framework, we need to refine the definition of a reflexive game.

Recall a reflexive game is defined by the tuple $\{N, (X_i)_{i \in N}, f_i(\cdot)_{i \in N}, \Theta, I\}$, where $N = \{1, 2, \ldots, n\}$ means a set of game participants (players, agents), X_i indicates a set of feasible actions of agent i, $f_i(\cdot): \Theta \times X' \to \Re^1$ represents his/her goal function, $i \in N$, and I is an awareness structure. Let us augment this structure with a set of functions $w_i(\cdot): \Theta \times X' \to W_i$, $i \in N$, each mapping the vector (θ, x) into an element w_i of a certain set W_i. The element w_i is exactly what agent i observes as the outcome of the game.

In the sequel, the function $w_i(\cdot)$ will be referred to as the *observation function* of agent i [39, 139]. Suppose that observation functions are a common knowledge of the agents.

Imagine that $w_i(\theta, x) = (\theta, x)$, i.e., $W_i = \Theta \times X'$; then agent i observes both the state of nature and the actions of all agents. On the contrary, the set W_i being composed of a single element, agent i observes nothing.

Suppose the reflexive game admits an informational equilibrium x_τ, $\tau \in \Sigma_+$ (recall τ is an arbitrary nonempty sequence of indexes belonging to N). Next, fix $i \in N$ and consider agent i. He/she expects to observe the following outcome of the game:

(1) $w_i(\theta_i, x_{i1}, \ldots, x_{i,i-1}, x_i, x_{i,i+1}, \ldots, x_{in})$.

Actually, he/she observes

(2) $w_i(\theta, x_1, \ldots, x_{i-1}, x_i, x_{i+1}, \ldots, x_n)$.

Therefore, the stability requirement for agent i implies the coincidence of the values (1) and (2) (again, we indicate these are the elements of a certain set W_i).

Assume that the values (1) and (2) are identical; in other words, agent i has no doubts regarding validity of his/her beliefs after the game. Meanwhile, is it enough for the agent to choose the same action x_i in the next game? Clearly, the answer is negative; the example below gives an illustration.

Example 2.7.1. Consider a reflexive bimatrix game, where $\Theta = \{1, 2\}$. The gains are specified by bimatrices, see Fig. 2.18 (agent 1 chooses the row, while agent 2 chooses the column: $X_1 = X_2 = \{1; 2\}$).

$$\theta = 1 \qquad\qquad \theta = 2$$

$$\begin{pmatrix} (1,1) & (0,0) \\ (0,1) & (2,0) \end{pmatrix} \begin{pmatrix} (0,1) & (1,2) \\ (1,1) & (2,2) \end{pmatrix}$$

Figure 2.18 The matrices of players' gains in Example 2.7.1.

The graph of the corresponding reflexive game is shown by Fig. 2.19.

Figure 2.19 The graph of the reflexive game in Example 2.7.1.

Moreover, set $\theta = \theta_1 = 1$, $\theta_2 = \theta_{21} = 2$ and suppose that each agent observes his/her gain (i.e., for any agent the observation function coincides with his/her gain function). Obviously, the informational equilibrium is provided by the combination $x_1 = x_2 = x_{21} = 2$; that is, agents 1 and 2 (as well as 21-agent and the rest phantom agents) choose the second actions. However, after the game the real state of nature ($\theta = 1$) becomes known to agent 2; note the latter gains 0 instead of the expected value of 2. Hence, next time agent 2 would choose the action $x_2 = 1$, motivating agent 1 to change his/her action (i.e., to choose $x_1 = 1$). •

Therefore, a stable equilibrium requires phantom ij-agent ($i, j \in N$) also to observe the "necessary" value. As the result of the game, the agent in question expects to observe

(3) $\quad w_j(\theta_{ij}, x_{ij1}, \ldots, x_{ij,j-1}, x_{ij}, x_{ij,j+1}, \ldots, x_{ijn})$.

Actually (in other words, i-subjectively, since ij-agent exists in the mind of i-agent) he/she observes the value

(4) $\quad w_j(\theta_i, x_{i1}, \ldots, x_{i,j-1}, x_{ij}, x_{i,j+1}, \ldots, x_{in})$.

Consequently, for ij-agent the stability requirement implies the coincidence of (3) and (4).

In general case (i.e., for τi-agent, $\tau i \in \Sigma_+$), we introduce the following definition of stability.

Definition. An informational equilibrium $x_{\tau i}$, $\tau i \in \Sigma_+$, is said to be *stable* under a given awareness structure I if for any $\tau i \in \Sigma_+$ the following equality holds:

(5) $\quad w_i(\theta_{\tau i}, x_{\tau i1}, \ldots, x_{\tau i,i-1}, x_{\tau i}, x_{\tau i,i+1}, \ldots, x_{\tau in}) = w_i(\theta_\tau, x_{\tau 1}, \ldots, x_{\tau,i-1}, x_{\tau i}, x_{\tau,i+1}, \ldots, x_{\tau n})$.

If an informational equilibrium is not stable in the mentioned sense, we will call it *unstable*. For instance, the informational equilibrium in Example 2.7.1 appears unstable.

Assertion 2.7.1. Consider an awareness structure I of complexity ν. Suppose there exists an informational equilibrium $x_{\tau i}$, $\tau i \in \Sigma_+$. Then the system (5) includes at most ν pairwise different conditions.

Proof. Consider two (arbitrary) identical awareness structures: $I_{\lambda i} = I_{\mu i}$. Since $x_{\tau i}$ is an equilibrium, we have $\theta_{\lambda i} = \theta_{\mu i}$, $x_{\lambda i} = x_{\mu i}$, $I_{\lambda ij} = I_{\mu ij}$, and $x_{\lambda ij} = x_{\mu ij}$ for any $j \in N$. Thus, the stability conditions (5) are identical for λi- and μi-agents. There are ν pairwise different awareness structures; hence, the number of pairwise different conditions (5) does not exceed v. •

2.8 TRUE AND FALSE EQUILIBRIA

We will divide stable informational equilibria into two classes, *viz*, true and false equilibria. Let us give an example.

Example 2.8.1. Consider a game of three agents with the goal functions

$$f_i(r_i, x_1, x_2, x_3) = x_i - \frac{x_i(x_1 + x_2 + x_3)}{r_i},$$

where $x_i \geq 0$, $i \in N = \{1, 2, 3\}$. The goal functions are a common knowledge (except the *types* of the agents – their individual parameters $r_i > 0$). The type vector $r = (r_1, r_2, r_3)$ may be treated as the state of nature. By supposition, each agent knows his/her own type.

The graph of the reflexive game is shown in Fig. 2.20; note that $r_2 = r_3 = r$, $r_{21} = r_{23} = r_{31} = r_{32} = c$. Each agent knows his/her type and observes the sum of the opponents' actions.

Obviously, this game has the informational equilibrium

(1) $x_2 = x_3 = (3r - 2c)/4$,
       $x_{21} = x_{23} = x_{31} = x_{32} = (2c - r)/4$,
       $x_1 = (2r_1 - 3r + 2c)/4$.

The stability conditions (see (5)) then yield

(2) $x_{21} + x_{23} = x_1 + x_3$, $x_{31} + x_{32} = x_1 + x_2$.

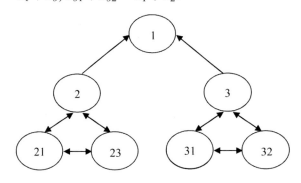

Figure 2.20 The graph of the reflexive game in Example 2.8.1.

We have formulated the conditions for 2- and 3-agents only, as far as their counterparts for 1-, 21-, 23-, 31-, and 32-agents are trivial.

Substitute (1) into (2) to obtain the following equality as a necessary and sufficient condition of stability:

(3) $2c = r_1 + r$.

Let the condition (3) be met. Then the equilibrium actions of real agents are determined by

(4) $x_2 = x_3 = (3r - r_1)/4, x_1 = (3r_1 - 2r)/4.$

Now, assume that agents' types are a common knowledge (see Fig. 2.21).

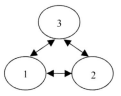

Figure 2.21 The common knowledge in Example 2.8.1.

One would easily ascertain that, in the common knowledge case, the unique equilibrium is specified by (4). •

Therefore, the condition (3) leads to a counter intuitive situation. Agents 2 and 3 have false beliefs (Fig. 2.20); nevertheless, their equilibrium actions (4) are exactly the same as under the identical awareness (Fig. 2.21). Call such a stable equilibrium a true equilibrium.

Suppose the set of actions $x_{\tau i}$, $\tau i \in \Sigma_+$, represents a stable informational equilibrium. It will be referred to as a *true* equilibrium if the set $(x_1, \ldots, x_n)$ is an equilibrium under the common knowledge about the state of nature θ (or about the agents' types $(r_1, \ldots, r_n)$).

In particular, the above definition implies that under a common knowledge any informational equilibrium is true. Let us show another case when any informational equilibrium is also true.

Assertion 2.8.1. Let the goal functions of the agents be defined by

$$f_i(r_i, x_1, \ldots, x_n) = \varphi_i(r_i, x_i, y_i(x_{-i})),$$

while the observation functions are $w_i(\theta, x) = y_i(x_{-i}), i \in N$; the corresponding interpretation is as follows. The gain of each agent depends on his/her type, action and observation function (which depends on actions of the opponents, but not on their types). Then any stable equilibrium is true.

Proof.   Assume that $x_{\tau i}$, $\tau i \in \Sigma_+$ is a stable informational equilibrium and the conditions of Assertion 2.8.1 hold. Then for any $i \in N$ we obtain

$$x_i \in \text{Arg} \max_{y_i \in X_i} f_i(r_i, y_i, x_{i,-i}) = \text{Arg} \max_{y_i \in X_i} \varphi_i(r_i, y_i, y_i(x_{i,-i})).$$

Due to the stability property, the equality

$$y_i(x_{i,-i}) = y_i(x_{-i})$$

takes place, resulting in

$$x_i \in \text{Arg} \max_{y_i \in X_i} \varphi_i(r_i, y_i, y_i(x_{-i})) = \text{Arg} \max_{y_i \in X_i} f_i(r_i, y_i, x_{-i}) .$$

Since $i \in N$ is arbitrary, the last formula means that the set $(x_1, \ldots, x_n)$ gives an equilibrium under complete awareness. •

A stable informational equilibrium which is not true in the above sense is said to be a *false* equilibrium.

In other words, a false equilibrium is a stable informational equilibrium, which is not an equilibrium in the case of an identical awareness of the agents (under a common knowledge).

Example 2.8.2. Consider a reflexive bimatrix game with $\Theta = \{1, 2\}$. The gains are specified by bimatrices, see Fig. 2.22 (agent 1 chooses the corresponding row, while agent 2 chooses the column: $X_1 = X_2 = \{1; 2\}$).

$$\theta = 1 \qquad\qquad \theta = 2$$

$$\begin{pmatrix} (2,2) & (4,1) \\ (1,4) & (3,3) \end{pmatrix} \quad \begin{pmatrix} (2,2) & (0,3) \\ (3,0) & (1,1) \end{pmatrix}$$

Figure 2.22 The matrices of players' gains in Example 2.8.2.

Next, let the actual situation be $\theta = 2$, while both agents consider $\theta = 1$ as a common knowledge. Each agent observes the pair (x_1, x_2), which provides the corresponding observation function.

In the informational equilibrium, both agents choose 1. Imagine the actual state of nature is a common knowledge; then each agent would choose the action equal to 2. Therefore, the agents' gains in the informational equilibrium are greater in comparison with the case when the actual state of nature is a common knowledge. •

2.9 THE CASE OF OBSERVABLE ACTIONS OF AGENTS

Section 2.2 has provided the notion of an informational equilibrium. It can be interpreted as a set of subjective equilibria. Real agent i $(i \in N)$ possesses the awareness structure I_i and defines a set of actions $(x_{i\sigma}^*(I_{i\sigma}))_{\sigma \in \Sigma}$, which is an equilibrium (according to his/her subjective view). In particular, this agent expects that real agent j $(j \in N)$ would choose the action $x_{ij}^*(I_{ij})$. Recall that phantom ij-agent represents the image of agent j in beliefs of agent i.

The present section deals with the following case. Observation function is the action vector of all agents:

$$w_i(\theta, x_1, \ldots, x_n) = (x_1, \ldots, x_n).$$

Then a *stable* informational equilibrium $x^* = (x_{\sigma i}^*)_{i \in N, \sigma \in \Sigma}$, meets the condition

(1) $\forall i \in N, \quad \forall \sigma \in \Sigma: x_{\sigma i}^* = x_i^*.$

Formula (1) implies that the action of a real agent coincides with the action expected from him/her by another (real or phantom) agent.

Introduce the following assumption regarding the goal functions $f_i(\cdot)$ and the sets Θ, X_i:

A1. For any $i \in N$, $\sigma \in \Sigma$, any beliefs $\theta_{\sigma i} \in \Theta$ and $\theta'_{\sigma i} \in \Theta$ such that $\theta_{\sigma i} \neq \theta'_{\sigma i}$, and any opponents' action profile

$$x^*_{\sigma i, -i} \in X_{-i} = \prod_{j \neq i} X_j$$

we have

$$(2) \quad BR_i(\theta_{\sigma i}, x^*_{\sigma i, -i}) \cap BR_i(\theta'_{\sigma i}, x^*_{\sigma i, -i}) = \varnothing,$$

where $BR_i(\theta_{\sigma i}, x^*_{\sigma i, -i}) = \operatorname*{Arg\,max}_{y_i \in X_i} f_i(\theta_{\sigma i}, x^*_{\sigma i 1}, \ldots, x^*_{\sigma i, i-1}, y_i, x^*_{\sigma i, i+1}, \ldots, x^*_{\sigma i n})$.

Assertion 2.9.1. Under Assumption A1, suppose there exists an informational equilibrium x^*. Then x^* is a stable informational equilibrium if the awareness structure of the game satisfies the condition

$$(3) \quad \forall i \in N, \quad \forall \sigma \in \Sigma: \theta_{\sigma i} = \theta_i.$$

Proof. Let (3) be true. Consequently, the awareness structure of the game possesses depth 1 and $\forall i \in N$, $\forall \sigma \in \Sigma$: $I_{\sigma i} = I_i$. This directly yields $x^*_{\sigma i} = x^*_i$ (see the second condition in the definition of an informational equilibrium). The necessity part is completed.

Sufficiency. Suppose that (1) takes place, but there are $i \in N$ and $\sigma \in \Sigma$ such that $\theta_{\sigma i} \neq \theta_i$.

Since x^*_i and $x^*_{\sigma i}$ are components of the informational equilibrium x^*, they match the inclusions

$$\begin{cases} x^*_i \in BR_i(\theta_i, x^*_{i, -i}), \\ x^*_{\sigma i} \in BR_i(\theta_{\sigma i}, x^*_{\sigma i, -i}). \end{cases}$$

By virtue of (1), the above system can be rewritten as

$$\begin{cases} x^*_i \in BR_i(\theta_i, x^*_{-i}), \\ x^*_i \in BR_i(\theta_{\sigma i}, x^*_{-i}), \end{cases}$$

which gives $BR_i(\theta_i, x^*_{-i}) \cap BR_i(\theta'_{\sigma i}, x^*_{-i}) \neq \varnothing$.

Thus, we have obtained a contradiction to (2). ●

Corollary. Under Assumption A1, stable informational equilibria may exist only in awareness structures meeting (3), i.e., in awareness structures of depth 1. In particular, false equilibria are impossible.

If one weakens the requirement (1), the result of Assertion 2.9.1 becomes invalid. For instance, consider "stable" an informational equilibrium x^* with the property

(4) $\forall i, j \in N \quad x_{ji}^* = x_i^*$

(the action of a real agent coincides with his/her action expected by another real agent). Within the framework of Assumption A1, there exist awareness structures that do not agree with (3), but the corresponding informational equilibria turn out "stable" in the sense of (4).

Assertion 2.9.1 seems relevant both in analysis and synthesis problems. Indeed, studying the properties of informational equilibria for a specific class of situations (see Assumption A1), one can separate out sets of awareness structures (using the condition (3)) with stable informational equilibria. In the context of informational control problems, Assertion 2.9.1 imposes definite constraints on the set of control actions leading to a stable equilibrium in the game of controlled subjects.

Now, suppose that agents' actions are still observable, i.e., stability is defined by the condition (1). However, each of n agents gets characterized by his/her type $r_i \geq 0$, $i \in N$. Generally speaking, other agents do not know the type of a given agent. Assume that the goal function of agent i has the form $f_i(r_i, x)$ (viz., it depends on the type of this agent only). Each agent possesses an hierarchy of beliefs I_i about agents' types. The hierarchy comprises infinitely many components, namely, r_{ij} – the belief of agent i about the type of agent j, r_{ijk} – the belief of agent i about the beliefs of agent j about the type of agent k, and so on $(i, j, k \in N)$.

A meaningful difference between interpretations in terms of the uncertain parameter θ and in terms of the agents' type vector $r = (r_1, r_2, \ldots, r_n) \in \mathfrak{R}_+^n$ consists in the following. In the former case, the assumption regarding observability of the state of nature θ seems more natural (i.e., this parameter represents a significant argument of agent's observation function). In the latter case (quite the contrary), the types of opponents are not observed – the observation function does not include them as arguments. According to Assertion 2.9.1, all stable equilibria are true. Thus, we fix the attention on stability analysis.

In this case, rewrite Assumption A1 and Assertion 2.9.1 as follows.

A1^r. For any $i \in N$, $\sigma \in \Sigma$, any beliefs $r_{\sigma i}$ and $r_{\sigma i}'$ such that $r_{\sigma i} \neq r_{\sigma i}'$, and any opponents' action profile $x_{\sigma i, -i}^* \in X_{-i}$ we have

$$BR_i(r_{\sigma i}, x_{\sigma i, -i}^*) \cap BR_i(r_{\sigma i}', x_{\sigma i, -i}^*) = \emptyset,$$

where $BR_i(r_{\sigma i}, x_{\sigma i, -i}^*) = \text{Arg} \max_{y_i \in X_i} f_i(r_{\sigma i}, x_{\sigma i 1}^*, \ldots, x_{\sigma i, i-1}^*, y_i, x_{\sigma i, i+1}^*, \ldots, x_{\sigma i n}^*)$.

Assertion 2.9.2. Under Assumption A1^r, suppose there exists an informational equilibrium x^*. Then x^* is a stable informational equilibrium if the awareness structure of the game satisfies the condition

$$\forall i \in N, \quad \forall \sigma \in \Sigma: r_{\sigma i} = r_i.$$

Proof of Assertion 2.9.2 repeats that of Assertion 2.9.1 *verbatim* (just replace θ with r and A1 with A1^r).

Define the following objects:

- the set Ψ of pairs (x, I) such that $x \in X'$, $I \in \mathfrak{I}$ and the action vector x of real agents represents an equilibrium[13] under the awareness structure I ($\mathfrak{I}$ indicates the set of different awareness structures).
- the set $\Psi_X(I) \subseteq X'$ of the action vectors of real agents, representing equilibria under the awareness structure I;
- the set $\Psi_I(x) \subseteq \mathfrak{I}$ of awareness structures, making the action vector x of real agents an equilibrium (solution to the inverse problem).

In addition, define the subsets of these sets, enjoying stability of an informational equilibrium:

- the set Ψ^s of pairs (x, I) such that $x \in X'$, $I \in \mathfrak{I}$ and the vector x is a stable equilibrium[14] under the awareness structure I;
- the set $\Psi_X^s(I) \subseteq X'$ of agents' action vectors being stable equilibria under the awareness structure I;
- the set $\Psi_I^s(x) \subseteq \mathfrak{I}$ of awareness structures ensuring that the action vector x of real agents is a stable equilibrium.

Denote by I_0 an awareness structure of depth 1, which corresponds to that the type vector r of agents is a common knowledge. Note that $\Psi_X^s(I_0) = \Psi_X(I_0)$, i.e., any equilibrium vector corresponding to a common knowledge represents a stable equilibrium.

In terms of the introduced sets, a *true equilibrium* is any pair $(x, I) \in \Psi^s$ such that $(x, I_0) \in \Psi$. In practice, this means that the action vector x remains a (stable) equilibrium, if the type vector becomes a common knowledge.

A *false equilibrium* is any pair $(x, I) \in \Psi^s$ such that $(x, I_0) \notin \Psi$. In other words, the action vector x looses its equilibrium property if the type vector turns out a common knowledge.

Suppose that $x^* = (x_1^*, \ldots, x_n^*)$ is a stable equilibrium vector. For each $i \in N$, define the following sets:

$$\rho_i = \{r_i \in \mathfrak{R}_+ \,|\, x_i^* \in BR_i(r_i, x_{-i}^*)\}.$$

They are independent of the awareness structure. Consequently, we adopt the above sets to formulate two results clarifying the connection between the awareness structure and stability of an equilibrium.

Assertion 2.9.3. Let x^* be a stable equilibrium vector of actions chosen by real agents. Assume that for any $i \in N$ the set ρ_i comprises one element exactly. Then the type vector represents a common knowledge (and the equilibrium is true).

---

[13]Notably, it forms the "real" part of an informational equilibrium, which is a set of actions of real and phantom agents.
[14]Notably, it forms the "real" part of a stable informational equilibrium.

Proof. Suppose that x^* is a stable equilibrium vector and for any $i \in N$ the set ρ_i includes only one element. Imagine there exist i and σ such that $r_{\sigma i} \neq r_i$. Stability of the equilibrium implies

$$\begin{cases} x_i^* \in BR_i(r_i, x_{-i}^*), \\ x_i^* \in BR_i(r_{\sigma i}, x_{-i}^*). \end{cases}$$

By definition of the set ρ_i, noncoinciding quantities $r_{\sigma i}$ and r_i belong to this set; however, it consists of one element. The resulting contradiction concludes the proof. •

Assertion 2.9.4. Assume that the action vector x^* of real agents is a stable equilibrium vector under a certain awareness structure. In this case, for any $i \in N$ and $\sigma \in \Sigma$ element of this structure satisfy $r_{\sigma i} \in \rho_i$.

Proof. Suppose that the equilibrium is stable. Then $\forall i \in N \ \forall \sigma \in \Sigma$ we have $x_i^* \in BR_i(r_{\sigma i}, x_{-i}^*)$, i.e., $r_{\sigma i} \in \rho_i$. •

Assertion 2.9.4 imposes rigid conditions on the awareness structure. If an equilibrium is stable, all types of real agents and beliefs about the types belong to the sets ρ_i.

2.10 REFLEXIVE GAMES AND BAYESIAN GAMES

In addition to reflexive games, a possible way of game-theoretic modeling under an incomplete awareness lies in *Bayesian games* proposed in the late 1960s by J. Harsanyi [78]. In Bayesian games, all individual information of an agent (not a common knowledge) being available at the moment of his/her decision-making is called agent's *type*. Moreover, each agent knows his/her type and possesses certain beliefs about the types of other agents (in the form of a probabilistic distribution). Formally, a Bayesian game is described by the following parameters:

- a set of agents N;
- a set of agents' feasible *types* R_i, where the type of agent i is $r_i \in R_i$, $i \in N$, and the type vector makes up $r = (r_1, r_2, \ldots, r_n) \in R' = \prod_{i \in N} R_i$;
- a set of feasible action vectors $X' = \prod_{i \in N} X_i$ of the agents;
- a set of goal functions f_i: $R' \times X' \to \mathfrak{R}^1$ (generally, the goal function of an agent depends on the types and actions of other agents);
- agents' beliefs $F_i(\cdot | r_i) \in \Delta(R_{-i})$, $i \in N$ (here R_{-i} designates the set of different combinations of types of all agents except agent i, $R_{-i} = \prod_{j \in N \setminus \{i\}} R_j$; the symbol $\Delta(R_{-i})$

stands for the set of different probabilistic distributions on R_{-i}).

A *Bayes-Nash equilibrium* in a Bayesian game is defined as a set of agents' strategies in the form x_i^*: $R_i \to X_i$, $i \in N$, maximizing the expected goal functions

$$(1) \quad \forall i \in N: x_i^*(r_i) \in \text{Arg} \max_{x_i \in X_i} \int_{r_{-i} \in \prod_{j \neq i} R_j} f_i(r, x_i, x_{-i}^*(r_{-i})) dF_i(r_{-i} | r_i),$$

where $x^*_{-i}(r_{-i})$ corresponds to the set of strategies of all agents except agent i. Interestingly, in a Bayesian game the strategy of an agent represents a "variable" action (as a function of his/her type – in contrast to fixed actions in common games).

J. Harsanyi's model may have different interpretations (see [78]). According to a possible interpretation, all agents know the a priori distribution of types $F(r) \in \Delta(R')$. Being aware of their own types, agents evaluate the conditional distribution $F_i(r_{-i}|r_i)$ by the Bayes formula. In this case, the agents' beliefs $\{F_i(\cdot|\cdot)\}_{i \in N}$ are said to be *compatible*. In particular, they form a common knowledge – each agent can find such beliefs and knows that the rest agents enjoy the same ability, etc.

Another interpretation concerns the following. Suppose there exists a certain set of potential participants of a game, having different types. Each "potential" agent chooses his/her strategy depending on his/her type. Then this agent chooses randomly n "actual" players for the game. Generally speaking, the beliefs of agents are not necessarily compatible (even representing a common knowledge). We also emphasize an important aspect. In [78] this interpretation is called *Selten's game*; R. Selten won the 1994 Nobel Memorial Prize in Economic Sciences (shared with J. Harsanyi and J. Nash).

Now, consider a situation when conditional distributions do not necessarily represent a common knowledge. Here a convenient approach consists in the following. Let agents' gains be dependent on their actions and on a certain parameter $\theta \in \Theta$ ("the state of nature"). The latter is not a common knowledge and can be viewed as the set of agents' types. Notably, the goal function of agent i takes the form $f_i(\theta, x_1, \ldots, x_n)$: $\Theta \times X' \to \Re^1$, $i \in N$. In Chapter 2 we have underlined that an agent chooses his/her strategy as the result of informational reflexion (an agent ponders over what each opponent knows or thinks about the parameter θ, the beliefs of other agents, etc.). Thus, we naturally arrive at the notion of an awareness structure of an agent, which reflects his/her awareness regarding an unknown parameter, beliefs of other agents, etc.

J.-F. Mertens and S. Zamir [114] constructed the universal beliefs space within the framework of probabilistic awareness; here beliefs of agents include the probabilistic distribution on the set of the states of nature, the probabilistic distribution on the set of the states of nature and also the distributions on the set of the states of nature characterizing the beliefs of other agents, and so on. The game formally comes to a "universal" Bayesian game, where the entire awareness structure of an agent is his/her type. However, the construction of [114] is so bulky that deriving a general solution to the "universal" Bayesian game seems impossible.

Therefore, in this section we get restricted to two-player games only. Agents' beliefs are defined by a "point-type" awareness structure (agents have certain beliefs about the value of an uncertain parameter, certain beliefs about the opponent, and so on). Under such simplifications, Bayes-Nash equilibrium evaluation comes to solving a system of two relations that define two functions each depending on a countable number of variables.

Let two agents play a game with the goal functions

(2) $\quad f_i(\theta, x_1, x_2), \quad \theta \in \Theta, x_i \in X_i, i = 1, 2.$

The functions f_i and the sets X_i, Θ represent a common knowledge. Agent 1 possesses the following beliefs: the uncertain parameter is $\theta_1 \in \Theta$; agent 2 believes that the uncertain parameter makes up $\theta_{12} \in \Theta$; agent 2 believes that agent 1 believes that the uncertain parameter equals $\theta_{121} \in \Theta$, and so on. Therefore, the point-type awareness structure of agent 1, denoted by I_1, is defined by an infinite sequence of elements of the set Θ. Similarly, let agent 2 have the point-type awareness structure I_2:

(3) $I_1 = (\theta_1, \theta_{12}, \theta_{121}, \ldots), \quad I_2 = (\theta_2, \theta_{21}, \theta_{212}, \ldots).$

Now, look at the reflexive game (2)–(3) from the "Bayesian" point of view [37]. Here, the awareness structure of an agent ($I_i, i = 1, 2$) is his/her type. To evaluate a Bayes-Nash equilibrium, one should find equilibrium actions of agents of all possible types (not only of the fixed types (3)).

The definition of an equilibrium (1) demonstrates what distributions $F_i(\cdot|\cdot)$ appear in this case. For instance, suppose that the type of agent 1 is $I_1 = (\theta_1, \theta_{12}, \theta_{121}, \ldots)$; then the distribution $F_1(\cdot|I_1)$ assigns probability 1 to the type $I_2 = (\theta_{12}, \theta_{121}, \theta_{1212}, \ldots)$ of the opponent and probability 0 to the remaining types. On the other hand, if the type of agent 2 becomes $I_2 = (\theta_2, \theta_{21}, \theta_{212}, \ldots)$, then the distribution $F_2(\cdot|I_2)$ assigns probability 1 to the type $I_1 = (\theta_{21}, \theta_{212}, \theta_{2121}, \ldots)$ of the opponent and probability 0 to the remaining types.

In the sequel, for simplicity we adopt the following notation:

$$BR_1(\theta, x_2) = \text{Arg} \max_{x_1 \in X_1} f_1(\theta, x_1, x_2), \quad BR_2^{-1}(\theta, x_2) = \{x_1 \in X_1 \,|\, x_2 \in BR_2(\theta, x_1)\},$$

$$BR_2(\theta, x_1) = \text{Arg} \max_{x_2 \in X_2} f_2(\theta, x_1, x_2), \quad BR_1^{-1}(\theta, x_1) = \{x_2 \in X_2 \,|\, x_1 \in BR_1(\theta, x_2)\}.$$

In addition, designate by $\varphi(\cdot)$ and $\psi(\cdot)$ the functions mapping a type to an equilibrium action:

$$x_1^*(I_1) = \varphi(I_1) = \varphi(\theta_1, \theta_{12}, \theta_{121}, \ldots), \quad x_2^*(I_2) = \psi(I_2) = \psi(\theta_2, \theta_{21}, \theta_{212}, \ldots).$$

With such system of symbols, the *point-type* Bayes–Nash equilibrium (1) can be rewritten as a pair of functions $(\varphi(\cdot), \psi(\cdot))$ meeting the conditions

(4) $\begin{cases} \varphi(\theta_1, \theta_{12}, \theta_{121}, \ldots) \in BR_1(\theta_1, \psi(\theta_{12}, \theta_{121}, \ldots)), \\ \psi(\theta_2, \theta_{21}, \theta_{212}, \ldots) \in BR_2(\theta_2, \varphi(\theta_{21}, \theta_{212}, \ldots)). \end{cases}$

We emphasize the following. Within the framework of a point-type awareness structure, agent i is sure that the uncertain parameter has value θ_i (regardless of the opponent's beliefs).

Thus, to evaluate an equilibrium, one should solve the system of functional equations (4). This gives the functions $\varphi(\cdot)$ and $\psi(\cdot)$, each depending on a denumerable number of variables.

Possible awareness structures may have a finite or infinite depth. Below we demonstrate that applying the concept of a *Bayes-Nash equilibrium* to agents with infinite-depth awareness structures leads to a paradoxical result (any feasible action of such agents is an equilibrium).

Define the notion of a finite depth of an awareness structure for a two-player game, where the awareness structure of each player represents an infinite sequence of elements from Θ.

Let the sequence $T = \{t_i\}_{i=1}^{\infty}$ of elements from Θ and a nonnegative integer k be given. The sequence $\omega_k(T) = \{t_i\}_{i=k+1}^{\infty}$ is called the k-*completion* of the sequence T.

We say that the sequence T has *infinite depth*, if for any n there exists $k > n$ such that the sequence $\omega_k(T)$ does not coincide with any sequence from the collection $\omega_0(T) = T, \omega_1(T), \ldots, \omega_n(T)$. Here the matter concerns standard elementwise coincidence. Otherwise, the sequence T has a *finite depth*.

In other words, a finite-depth sequence has a finite number of pairwise different completions, whereas an infinite-depth sequence has the infinite number of completions. For instance, the sequence $(1, 2, 3, 4, 5, \ldots)$ has infinite depth, and the sequence $(1, 2, 3, 2, 3, 2, 3, \ldots)$, finite.

Consider the game (2), where the goal functions f_1, f_2 and the sets X_1, X_2, Θ enjoy the property:

(5)   for any $x_1 \in X_1, x_2 \in X_2, \theta \in \Theta$, the sets
        $BR_1(\theta, x_2), BR_2(\theta, x_1), BR_2^{-1}(\theta, x_2)$ and $BR_1^{-1}(\theta, x_1)$ are nonempty.

The condition (5) implies the following. For any $\theta \in \Theta$ and any action $x_1 \in X_1$, agent 2 has (at least) one best response and, in turn, the action x_1 itself represents the best response to an action of agent 2; similarly, any action $x_2 \in X_2$.

Under the conditions (5) in the game (2), *any* action of an agent with an infinite-depth awareness structure is an equilibrium action (i.e., makes a component of some equilibrium (4)). This statement applies to both agents; for definiteness, formulate and prove it for agent 1.

Assertion 2.10.1. Consider the game (2) with the conditions (5). Suppose there exists (at least) one point-type Bayes–Nash equilibrium (4). Then, for any infinite-depth awareness structure I_1 and any $\chi \in X_1$, there exists an equilibrium $(x_i^*(\cdot), x_2^*(\cdot))$ such that $x_1^*(I_1) = \chi$.

The idea of proof relies on constructive definition of a corresponding equilibrium [37]. Fix an arbitrary equilibrium $(\varphi(\cdot), \psi(\cdot))$ and an arbitrary infinite-depth awareness structure I_1. By virtue of the conditions (4), the value of the function $\varphi(\cdot)$ on the awareness structure I_1 relates to the values of the functions $\varphi(\cdot)$ and $\psi(\cdot)$ on some other structures. We redefine the values of these functions on the above structures by preserving the conditions (4) (leaving unchanged the values on other structures). Moreover, let the "redefined" function $\tilde{\varphi}(\cdot)$ possess the value χ on the awareness structure I_1.

We precede the proof of Assertion 2.10.1 by four lemmas. Here we employ the following notation: if $p = (p_1, \ldots, p_n)$ and $T = \{t_i\}_{i=1}^{\infty}$ are, respectively, finite and infinite sequences of elements from Θ, then $pT = (p_1, \ldots, p_n, t_1, t_2, \ldots)$.

Lemma 2.10.1. Assume that the sequence T has infinite depth. In this case, for any finite sequence p and any k the sequence $p\omega_k(T)$ is also of infinite depth.

Proof. Since T is of infinite depth, it has the infinite number of pairwise different completions. Transition from T to $\omega_k(T)$ decreases their number at most by k (still, this number remains infinite). On the other hand, the number of pairwise completions is not reduced by passing from $\omega_k(T)$ to $p\omega_k(T)$. •

Lemma 2.10.2. Let the sequence T be representable as $T = ppp\ldots$, where p indicates some nonempty finite sequence. Then T has a finite depth.

Proof. Suppose that p takes the form $p = (p_1,\ldots,p_n)$. Hence, elements of the sequence T are interconnected by $t_{i+nk} = t_i$ for all integers $i \geq 1$ and $k \geq 0$. Choose an arbitrary j-completion, $j \geq n$. The number j is uniquely expressible as $j = i + nk$, where $i \in \{1,\ldots,n\}, k \geq 0$. One would readily show that $\omega_j(T) = \omega_i(T)$: the equality $t_{j+m} = t_{i+nk+m} = t_{i+m}$ holds true for any integer $m \geq 0$.

Recall that j is arbitrary. Thus, we have demonstrated that the sequence T admits at most n pairwise different completions, i.e., its depth is finite. •

Lemma 2.10.3. Let the sequence T satisfy the identity $T = pT$, where p is some nonempty finite sequence. Then T has a finite depth.

Proof. Set $p = (p_1,\ldots,p_n)$. We get $T = pT = ppT = pppT = ppppT = \ldots$. Thus, for any integer $k \geq 0$ the fragment $(t_{nk+1},\ldots,t_{nk+n})$ coincides with $(p_1,\ldots,p_n)$. Therefore, T is representable as $T = ppp\ldots$ and (by Lemma 2.10.2) has a finite depth. •

Lemma 2.10.4. let the sequence T satisfy the identity $pT = qT$, where p and q are some nonidentical nonempty finite sequences. Then T has a finite depth.

Proof. Set $p = (p_1,\ldots,p_n)$ and $q = (q_1,\ldots,q_k)$. Obviously, for $n = k$ the identity $pT = qT$ cannot be satisfied. And so, consider the case of $n \neq k$. For definiteness, imagine that $n > k$. Consequently, $p = (q_1,\ldots,q_k,p_{k+1},\ldots,p_n)$, and the condition $pT = qT$ yields $dT = T$, where $d = (p_{k+1},\ldots,p_n)$. Apply Lemma 2.10.3 to obtain that the sequence T is of a finite depth. •

Proof of assertion 2.10.1. Consider an arbitrary infinite-depth awareness structure of agent 1. For uniformity with Lemmas 2.10.1–2.10.4, denote it by $T = (t_1, t_2, \ldots) = \{t_i\}_{i=1}^{\infty}$ (instead of I_1). According to a premise of the Theorem, there exists (at least) one pair of the functions $(\varphi(\cdot), \psi(\cdot))$ meeting (4); fix any pair of such functions. Assume that the function $\tilde{\varphi}(\cdot)$ possesses the value χ on the sequence T: $\tilde{\varphi}(T) = \chi$ (here and in the sequel, the "redefined" functions are denoted by $\tilde{\varphi}(\cdot)$ and $\tilde{\psi}(\cdot)$). By substituting T as the argument of the function $\tilde{\varphi}(\cdot)$ into (4), we obtain the following result. On the strength of (4), the value $\tilde{\varphi}(T) = \chi$ is related to the values of the function $\tilde{\psi}(\cdot)$ on the sequence $\omega_1(T)$ and also on all sequences T' such that $\omega_1(T') = T$.

Choose the values of $\tilde{\psi}(\cdot)$ on these sequences to satisfy the conditions (4):

$$\tilde{\psi}(t_2, t_3, t_4, \ldots) \in BR_1^{-1}(t_1, \chi), \quad \tilde{\psi}(t_1^1, t_1, t_2, \ldots) \in BR_2(t_1^1, \chi).$$

where $t_1^1 \in \Theta$; it follows from (5) that this choice is feasible. If the set $BR_1^{-1}(t_1, \chi)$ or $BR_2(t_1^1, \chi)$ contains more than one elements, take any.

Next, by substituting into (4) the sequence $(t_2, t_3, \ldots)$ as the argument of the function $\tilde{\psi}(\cdot)$, we choose

$$\tilde{\varphi}(t_3, t_4, \ldots) \in BR_2^{-1}(t_2, \tilde{\psi}(t_2, t_3, \ldots)), \tilde{\varphi}(t_2^1, t_2, t_3, \ldots) \in BR_1(t_2^1, \tilde{\psi}(t_2, t_3, \ldots)).$$

Similarly, by substituting $(t_1^1, t_2, t_3, \ldots)$, we choose
$$\tilde{\varphi}(t_1^2, t_1^1, t_1, t_2, \ldots) \in BR_1(t_1^2, \tilde{\psi}(t_1^1, t_1, t_2, \ldots)) \text{ (here, } t_2^1, t_1^2 \in \Theta).$$

By performing this procedure for the obtained values, one can successively determine the values of $\tilde{\varphi}(\cdot)$ on all sequences of the form

(6) $(t_m^k, t_m^{k-1}, \ldots, t_m^2, t_m^1, t_m, t_{m+1}, \ldots), \qquad t_m^k, t_m^{k-1}, \ldots, t_m^2, t_m^1 \in \Theta,$

where $(m + k)$ is odd, and the values of the function $\tilde{\psi}(\cdot)$ on sequences of the form (6) with even $(m + k)$. We assume further that $t_m^1 \neq t_{m-1}$ is satisfied in (6) for $m > 1$. In this case, the representation (6) becomes unique.

The definition algorithm for the functions on the sequences (6) comprises two steps. At Step 1, set $\tilde{\varphi}(T) = \chi$ and determine the values of the corresponding functions on the sequences $\omega_m(T) = (t_m, t_{m+1}, \ldots)$, $m > 1$ (i.e., for $k = 0$), by applying alternatively the mappings BR_1^{-1} and BR_2^{-1}.

At Step 2, determine the values of the corresponding functions on the sequences (6) for $k \geq 1$ based on the value on the sequence $(t_m, t_{m+1}, \ldots)$ defined at Step 1 (by applying alternatively the mappings BR_1 and BR_2).

According to Lemma 2.10.1, all sequences of the form (6) have infinite depth. By Lemma 2.10.4, they are all pairwise different (if any two sequences of the form (6) coincided, this would contradict infiniteness of depth). Therefore, by determining the values of the functions $\tilde{\varphi}(\cdot)$ and $\tilde{\psi}(\cdot)$, we take no risk of assigning different values to the same argument.

We have defined the values of $\tilde{\varphi}(\cdot)$ and $\tilde{\psi}(\cdot)$ on the sequences (6) so that these functions still meet the conditions (4) (i.e., are a point-type Bayes–Nash equilibrium) and at the same time $\tilde{\phi}(T) = \chi$. The proof of Assertion 2.10.1 is completed. •

Thus, the present section has introduced the notion of a point-type Bayes-Nash equilibrium. Moreover, it has been demonstrated that any feasible action of an agent having an infinite-depth awareness structure represents an equilibrium under the additional conditions (5). (Analysis has been focused on two-player games. Nevertheless, one can hypothesize that the result obtained admits generalization to games with an arbitrary number of participants). Probably, this circumstance attests to the inexpediency of considering infinite-depth structures (both in terms of an informational equilibria and Bayes-Nash equilibria).

Generally speaking, the established assertion serves as an argument in favor of inevitable boundedness of informational reflexion rank of decision makers (for other arguments, see Sections 2.6 and 3.2).

2.11   INFORMATIONAL CONTROL

From the viewpoint of systems analysis, any system can be defined by specifying its staff, structure and functions. Therefore, one easily determines the model of an *organizational system* (OS), as well as the model of a social or economic (active) system, by specifying [135]:

-   *staff of the OS* (elements or participants of the OS);
-   *structure of the OS* (a set of informational, control, technological and other relations among OS participants);

- *sets of feasible strategies* (constraints and norms of activity imposed on OS participants); these sets reflect institutional, technological and other constraints and norms of their joint activity;
- *goal functions* (interests, purposes, etc.) of OS members;
- *information* (*awareness*), i.e., data regarding essential parameters being available to OS participants at the moment of decision-making (choosing the strategies);
- *the sequence of moves* (the order of data acquisition and strategy choice by OS participants).

The staff determines "who" is included into the system; the structure describes "who interacts with whom, who is subordinate to whom, etc." Finally, feasible sets define "who can do what," goal functions represent "who wants what," and the awareness states "who knows what."

Control is interpreted as an influence exerted on an organizational system to ensure its required behavior. Control may affect each of the parameters listed, called objects of control. These parameters of an OS are modified during the process of control and as the result of control. Hence, using the object of control as the first basis for classification of control mechanisms in OS, we obtain the following methods (types) of control:

- staff control;
- structure control;
- institutional control (control of constraints and norms of activity);
- motivational control (control of preferences and goals). For a detailed treatment of these methods, see [28, 113, 135];
- informational control (control of information available to OS participants at the moment of decision-making);
- "move" sequence control (control of the sequence of data acquisition and strategy choice by the OS participants).

Note that generally game-theoretic models consider move sequence control as a special case of structure control. Thus, we will not focus on it.

Let us briefly discuss specific features of different methods (types) of control[15].

Institutional control appears to be the most stringent–the principal seeks to achieve his/her goals by restricting the sets of feasible actions and the results of activity of his/her subordinates. Such restriction may be implemented via an explicit or implicit influence (legal acts, directives, orders) or mental and ethical norms, corporate culture, etc.

Motivational control seems "softer" than institutional control, and consists in purposeful modification of the preferences (utility functions) of subordinates. This modification is implemented via a certain system of penalties and/or incentives stimulating the choice of a specific action and/or attaining a definite result of activity. A wide class of motivational control problems includes planning problems and incentive problems [135]. For instance, in the case of incentive problems, motivational control

---

[15]In practice, it may be difficult to choose explicitly a type of control (some of them could and should be used simultaneously).

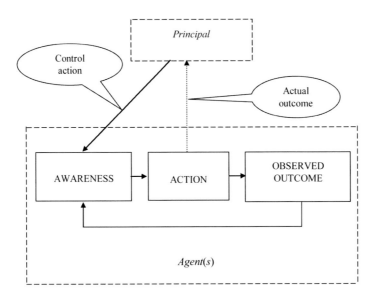

Figure 2.23 The model of informational control.

lies in direct stimulation of an agent for specific actions (an incentive enters the utility function additively).

As compared with institutional and motivational ones, *informational control* appears the "softest" (most indirect) type. Yet, its formal models have been least investigated.

This book is dedicated to one type of control actions (informational control), i.e., a purposeful impact on the awareness of agents. The suggested model of informational control is illustrated by Fig. 2.23 [39].

The model includes an agent (or several agents) and a principal. Each agent is characterized by the cycle "awareness of the agent → action of the agent → result observed by the agent → awareness of the agent." Generally speaking, these components vary for different agents. At the same time, the cycle could be viewed common for the whole controlled subsystem (i.e., for the complete set of agents). This feature is indicated by the word "*Agent(s)*" in Fig. 2.23.

The interaction between an agent (agents) and the principal is characterized by the following elements:

- an informational impact of the principal, which forms a certain awareness of an agent (agents). It seems possible to study the principal's influence on the outcome observed by an agent (agents), see the chain "principal → observed outcome" in Fig. 2.23. However, such investigation goes beyond the scope of this book (in a certain sense, it reduces the differences between informational control and motivational control);
- an actual outcome of the agent's action (or agents' actions), which has an impact on the preferences of the principal.

Let us discuss this model in detail.

The mathematical framework used to model the game-theoretic interaction of agents is provided by *reflexive games*, where agents choose actions based on their *awareness structures* – hierarchical beliefs about essential parameters of the situation ("state of nature"), beliefs about the beliefs of the opponents (other agents) and so on. Therefore, in terms of reflexive games, agent's awareness is characterized by his/her awareness structure (the awareness structure of the game is just the union of agents' awareness structures).

An agent chooses an action basing on his/her awareness structure. Under the given awareness structure, the actions of agents are, in fact, the components of an *informational equilibrium*, which is a solution of the reflexive game. The notion of an informational equilibrium generalizes the concept of a Nash equilibrium (probably, the most popular solution concept for noncooperative games).

In many cases, agent's awareness of the situation and of the beliefs of opponents may be inadequate. Hence, the result of a reflexive game observed by an agent either meets his/her expectations or not. This is defined by the following factors:

1)   how adequate is agent's awareness at the moment of action choice?
2)   how complete is the information observed by an agent about the outcome of the game?

For instance, the observed outcome could be the value of his/her goal function, the actions of opponents, the true value of an uncertain parameter (state of nature), etc. Generally, an agent observes the value of some function, which depends on the state of nature and the actions of opponents. This function is known as the *observation function*, and its impact on the awareness is illustrated by the chain "observed action → awareness." Imagine that all agents observe the result they reckon on (notably, for each agent the actual value of the observation function coincides with the expected one). In this case, the assumption regarding fixed (invariable) awareness structure seems natural, and an informational equilibrium appears *stable* (see Section 2.7).

Now, consider the interaction between agents and the principal. Implementing informational control, the principal (as usual) strives to maximize his/her utility. Assume the principal can form any awareness structure from a certain feasible set. The problem of informational control may be posed as follows. Find an awareness structure from the set of feasible structures, which maximizes the principal's utility in a corresponding informational equilibrium (perhaps, taking into account the principal's costs to form such an awareness structure). A rigorous statement of informational control problem is provided below.

Let us underline an important aspect. Within the proposed model, we adopt the assumption that the principal may form *any* awareness structure of agents. The issue how should the principal "convince" agents that specific states of nature and beliefs of the opponents take place is not discussed here. This issue requires particular analysis and knowledge accumulated by psychologists and sociologists (see Sections 2.12–2.14 and Chapter 4).

Let us give a formal statement to the control problem. Assume that the goal function of the principal, $\Phi(x, I)$, is defined on a set of real agents' actions and awareness structures. Next, suppose that the principal can form any awareness structure from a certain set $\mathfrak{I}'$. Under the awareness structure $I \in \mathfrak{I}'$, the action vector of real agents is an element of the set of equilibrium vectors $\Psi_X(I)$. We emphasize that the set $\Psi_X(I)$ may be empty; in the case of a missed equilibrium, the principal cannot predict the outcome of a game. To avoid this problem, introduce the set of feasible structures leading to the non-empty set of equilibria: $\mathfrak{I} = \{I \in \mathfrak{I}' \mid \Psi_X(I) \neq \emptyset\}$.

Imagine that, under the specified awareness structure $I \in \mathfrak{I}$, the set of equilibrium vectors $\Psi_X(I)$ includes (at least) two elements. As a rule, one of the following assumptions is then adopted [135]:

1) *the hypothesis of benevolence* (HB), which implies that agents always choose the equilibrium desired by the principal;
2) *the principle of maximal guaranteed result* (PMGR), i.e., the principal expects the worst-case equilibrium of the game.

Using either the HB or the PMGR, one has the *problem of informational control* in two settings as follows:

(1) $\max\limits_{x \in \Psi_X(I)} \Phi(x, I) \xrightarrow[I \in \mathfrak{I}]{} \max;$

(2) $\min\limits_{x \in \Psi_X(I)} \Phi(x, I) \xrightarrow[I \in \mathfrak{I}]{} \max.$

Naturally, if for any $I \in \mathfrak{I}$ the set $\Psi_X(I)$ consists of a single element, formulas (1) and (2) coincide.

In the sequel, the problem (1) (alternatively, (2)) will be called the *informational control problem in the form of the goal function*.

Now, provide an alternative formulation to the problem of informational control (being independent from the goal function of the principal). Assume that the principal wants agents choose an action vector $x \in X'$. The question arises, "For which vectors and by which awareness structure I would the principal achieve this?" In other words, the second possible formulation of the informational control problem is to find the following components. First, the *attainability set*, *viz.*, the one composed of the vectors $x \in X'$ such that for each of them the set of awareness structures $\Psi_I(x) \cap \mathfrak{I}$

(3)   is nonempty

or

(4)   consists of a single element.

Second, the corresponding feasible *awareness structures* $I \in \Psi_I(x) \cap \mathfrak{I}$, meeting the above property for each vector x. Note that the condition (3) "corresponds" to the HB, while the one of (4) "corresponds" to the PMGR.

The problem (3) (alternatively, (4)) will be referred to as *the problem of informational control in the form of the attainability set*.

Once again, we underline that the second formulation of the problem does not depend on the goal function of the principal. It merely reflects the possibility of bringing the system to a certain state by informational control.

In both formulations of the problem, the principal may be interested in stability of the resulting informational equilibrium or not. A stable informational equilibrium being required (i.e., when the system has to be rendered stable), one should substitute Ψ^s for Ψ in the formulas above. And the term "equilibrium" should be replaced by "stable equilibrium."

Let us summarize the aforesaid. The problem of informational control can be treated

1) in the form of the goal function or in the form of the attainability set;
2) involving the hypothesis of benevolence (HB) or the principle of maximal guaranteed result (PMGR);
3) with or without the stability requirement.

The choice of a specific setting depends on the actual situation modeled (there exist 8 different settings totally). Anyway, it appears necessary to establish a relation between the awareness structure and the action vector of agents (to analyze informational stability). The experience indicates that this stage is the most complicated and time-consuming for a researcher.

Chapter 4 discusses a series of particular models of informational control in different applications. For each model, we explain the most adequate setting of the problem (according to our viewpoint).

In the present section, readers can find the general analysis scheme for informational control problems. Of course, this scheme does not exhaust all possible cases. Instead, it elucidates the general logic of game-theoretic analysis (see Fig. 2.24) [39].

- **Step 1.** Description of the set of controlled subjects (agents), their feasible actions and goal functions. Interestingly, this step is necessary under game-theoretic approach to any control in socioeconomic systems.
- **Step 2.** Formalization of existing uncertainty – an uncertain parameter whose value is not a common knowledge among agents. Without a priori constraints on the set of feasible values of an uncertain parameter, the latter can be treated as an argument of agents' goal functions.
- **Step 3.** Definition of the set of awareness structures (see Section 2.1) that can be formed by a principal.

Steps 1–3 make up the **preliminary stage** of analysis. They yield the game-theoretic description of the game. In this book, we often hypothesize that the principal knows the actual value of the uncertain parameter.

- **Step 4.** Evaluation of an informational equilibrium, i.e., the relationship between the awareness structures and actions of agents (for the awareness structures defined at Step 3).
- **Step 5.** Stability analysis of the informational equilibrium. Truth/falsity verification (for stable equilibria only).

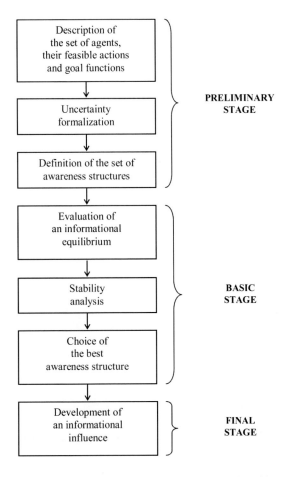

Figure 2.24 Analysis steps of informational control problems.

- **Step 6.** Choice of the most reasonable setting of informational control problem. Finding of an awareness structure solving this problem (choice of the "best" awareness structure). Solution gets simplified when reflexive partitions appear stationary (leading to simple organization of the desired awareness structure).

Steps 4–6 make up the **basic stage** of analysis. They yield an awareness structure to-be-formed by the principal among agents for attaining his/her goals.

- **Step 7.** Development of an informational influence on agents, which forms the awareness structure found at Step 6.

Step 7 makes up the **final stage** of analysis. To a large extent, this step goes beyond the game-theoretic approach. Rather, it represents the sphere of interest of psychologists and sociologists. This book describes just some types of informational impacts that modify certain components of the awareness structure of a game (see Sections 2.12–2.14 and Chapter 4).

2.12 MODELING OF INFORMATIONAL IMPACT

Within the framework of the model of decision-making adopted in this book (more specifically, in this chapter), agent's actions are defined by his/her awareness about the state of nature and the beliefs of opponents (other agents). Therefore, the issue regarding the informational impact of a principal on these beliefs seems relevant. In other words, we face the following question. How is the awareness structure of a game formed depending on certain informational impacts of a principal?

We acknowledge that providing a somewhat exhaustive answer would be impossible by treating only mathematical (in particular, game-theoretic) models. In the first place, the matter concerns the following. To a considerable degree, the process of information perception and adoption by humans gets conditioned by socio-psychological factors.

Readers can easily imagine obstacles connected with formalization of this process for an intelligent rational decision-maker. The latter represents the subject of game theory (e.g., see [121, p. 1]). All game solution concepts developed to-date (explicitly or implicitly) proceed from an a priori existing awareness structure[16]. What had been "before a game started"? How was a certain awareness structure formed? These questions go beyond the scope of discussion. Apparently, this is the boundary between a real person and an intelligent rational decision-maker.

Being aware of the limitations of mathematical modeling of human behavior (especially, that of the game-theoretic approach in informational control), let us consider possible types of informational impact.

Consider the following classification of messages [39], which a principal can send to agents for exerting an impact on their behavior.

1) input data – bare facts;
2) input data – logical inferences and analytical judgments based on a definite set of facts;
3) input data – appellative appraisals like "good/bad," "moral/immoral," "ethical/depraved," etc.
   New information employed by agents to make decisions can be divided into
4) "hard" information, which includes only actual data and facts;
5) "soft" information, which includes forecasts and appraisals.

One definitely draws an analogy between items 1 and 4, as well as between items 2 and 5. This issue will be analyzed in the sequel; now we concentrate on item 3.

This item, *in medias res*, relates to the ethical aspect of information (the ethical aspect of choice). Probably, the only attempt to provide a formal description to the

---

[16]Multi-step games are not an exception, either. The awareness of game participants may change during play (e.g., due to data exchange among them). Indeed, a multi-step game has a certain initial instant and, accordingly, an initial awareness of players.

ethical aspect was made by V. Lefebvre [96] (and the other investigators who generalized his theory of ethical choice [169]). His approach proceeds from the assumption
that a decision-maker performs *first-kind reflexion* [127], i.e., acts as an observer of
his/her own behavior, thoughts and feelings. The internal structure of an agent consists
of several interconnected levels (in particular, a special level "being responsible for"
the ethical aspect of choice). The final decision of an agent is determined by the mutual
influence of an external environment and the states of these internal levels.

In game theory (and in this textbook, as well), an agent is viewed as an individual
(i.e., an "indivisible" person); he/she performs *second-kind reflexion*, which relates to
decisions made by the opponents. Hence, we leave item 3 beyond our consideration
and address items 1, 4 and 2, 5.

An awareness structure of agent i (see Section 2.1) includes beliefs about:

–   the state of nature (θ_i);
–   beliefs of opponents ($\theta_{i\sigma}, \sigma \in \Sigma_+$).

Reporting either θ_i or $\theta_{i\sigma}$ is an informational impact. In other words, a principal
may report to an agent (or agents) information on the state of nature (i.e., the value
of an uncertain parameter) and on the beliefs of the opponents.

Thus, we have the following types of informational impact:

(i)   *informational regulation*;
(ii)   *reflexive control*.

Roughly speaking, they correspond to items 1, 4.

Concerning items 2 and 5, we should underline that they correspond to the
following type of informational impact:

(iii)   *active forecast*.

It consists in reporting information on the future values of specific parameters that
depend on the state of nature and actions of the agents [136].

In the forthcoming sections, we study types (i–iii) in a greater detail.

The elementary case of informational impact lies in the following. A principal
invites all agents and announces the value $\tilde{\theta}$ of the uncertain parameter θ publicly.
Here, a natural supposition is that the resulting awareness structure of a game has
depth 1 and consists of one element, i.e., for any $i \in N$ and $\sigma \in \Sigma$: $\theta_{i\sigma} = \tilde{\theta}$. In other
words, a common knowledge of agents is the value $\tilde{\theta}$ of the uncertain parameter.

Such **informational regulation** is called ***homogeneous*** [69, 136]. Denote by $\mathbf{A}_0$
the assumption that, being announced by a principal, the value $\tilde{\theta}$ becomes a common
knowledge. Actually, assumption $\mathbf{A}_0$ implies that agents trust the principal and this
represents a common knowledge.

Now, take a more sophisticated situation of informational regulation. The princi-
pal reports to agent i a value θ_i of the uncertain parameter θ. In this case, we accept
assumption $\mathbf{A}_1$: the resulting awareness structure of a game consists of n elements
and for any $i \in N$, $\sigma \in \Sigma$: $\theta_{i\sigma} = \theta_i$. In other words, agent i believes that the common
knowledge is the value θ_I of the uncertain parameter. Generally speaking, the resulting
awareness structure possesses depth 2.

Such **informational regulation** is called ***inhomogeneous*** [69, 136]. Clearly, homogeneous informational regulation represents a special case of inhomogeneous one (in this case, $\theta_1 = \theta_2 = \cdots = \theta_n = \tilde{\theta}$ and a corresponding awareness structure has depth 1).

Suppose that the principal reports to agent i the value of the uncertain parameter and the beliefs of other agents about this value. Therefore, agent i is informed of a set of numbers θ_{ij} ($j = 1, 2, \ldots, n$), i.e., the beliefs of each agent j about the uncertain parameter. Let us adopt **assumption A_2** which states the following. Agent i believes that each agent j ($j \neq i$) subjectively considers the value of the uncertain parameter θ_{ij} as a common knowledge. Consequently, the resulting awareness structure of the game possesses depth 3. Moreover, for any $i, j \in N$, $j \neq i$, and $\sigma \in \Sigma$ one obtains the equality $\theta_{ij\sigma} = \theta_{ij}$.

This situation gives an example of **informational control**.

Consider the following illustrative example. The book [36: p. 218] describes a psychological experiment conducted by the owner of a beef-importing company in the USA. "The company's customers – buyers for supermarkets and other retail food outlets – were called on the phone as usual by a salesperson and asked for a purchase in one of three ways. One set of customers heard a standard sales presentation before being asked for their orders. Another set of customers heard the standard sales presentation plus information that the supply of imported beef was likely to be scarce in the upcoming months. A third group received the standard sales presentation and the information about a scarce supply of beef, too; however, they also learned that the scarce supply news was not generally available information – it had come, they were told, from certain exclusive contacts that the company had.

... Compared to the customers who got only the standard sales appeal, those who were also told about the future scarcity of beef bought more than twice as much... The customers who heard of the impending scarcity via "exclusive" information ... purchased six times the amount that the customers who received only the standard sales pitch did. Apparently, the fact that the news about the scarcity information was itself scarce made it especially persuasive."

This example clearly illustrates the application of informational regulation ("the supply of imported beef was likely to be scarce in the upcoming months") and reflexive control ("the supply of imported beef was likely to be scarce ... the scarce supply news was not generally available information"). For its detailed discussion, see Section 4.7.

A series of model examples of informational regulation and reflexive control are studied in Chapter 6. A better classification and analysis of reflexive control seems a promising direction of further research.

Let us revert to an important issue. How can the principal form certain awareness structures? For this, select the model of the so-called simple messages. Within the framework of the model, investigators succeed in characterizing the set of implementable awareness structures.

Simple messages. A *simple message* consists in the following. Some agents get together, and the principal[17] reports to them the value of the uncertain parameter $\theta \in \Theta$. By supposition, all agents trust the principal. Moreover, each agent observes that his/her neighbor also knows the value of the uncertain parameter (the neighbors

---

[17]In particular, the principal making the message can be an agent.

of his/her neighbor are aware of this value, and so on). In other words, the fact that the message with the value of the uncertain parameter is trusted by all agents forms a common knowledge among these agents.

Formally, a simple message means the following. Take the set $G \subseteq N$ of agents and define

$$\tilde{G} = \{\sigma \in \Sigma \mid \sigma \text{ includes indexes from } G \text{ only}\}.$$

A *simple informational impact* (a *simple message*), denoted by $s \equiv G[\theta]$, where $G \subseteq N$ and $\theta \in \Theta$, is assigning a certain value θ to elements of the awareness structure θ_σ for each nonempty sequence of indexes $\sigma \in \tilde{G}$.

Agents may meet several times (and in different compositions). Thus, messages with the value of the uncertain parameter possibly vary. By supposition, agents trust new information (if they attended a previous "meeting"). Say that the *procedure of simple messages* lies in the following. The principal forms an awareness structure by choosing a finite sequence of simple informational impacts

$$\Pi \equiv \{s_1, s_2, \ldots, s_k\} \equiv \{s_i\} \equiv \{G_i[\theta_i]\},$$

where $G_i \subseteq N$, $\theta_i \in \Theta$. Such impacts modify the awareness structure sequentially (according to a specific law defined by a simple message).

However, is the principal able to form all feasible awareness structures? What awareness structures can he/she actually create? How can we describe them? What are their depth, complexity, etc.? Let us begin with one relevant result.

Assertion 2.12.1. For any sequence Π, there exists a corresponding sequence which forms the same awareness structure and G_q is not a subset of G_m in the case of $q < m$.

Proof. The whole essence of Assertion 2.12.1 concerns the following. Inviting a small group of agents "earlier" than the larger one (which includes the former) makes no sense. If $G_q \subseteq G_m$, then $\tilde{G}_q \subseteq \tilde{G}_m$. Hence, all elements of the awareness structure (that have been modified by the simple informational impact $G_q[\theta_q]$) are assigned a new value by the impact $G_m[\theta_m]$. In this case, the impact $G_q[\theta_q]$ can be not applied (still, the same awareness structure will be formed). Eliminate s_q from the sequence of impacts $\{s_i\}$ and proceed by analogy in all similar cases. The resulting sequence enjoys the properties listed in the statement of Assertion 2.12.1. •

Imagine that some agents from the set N did not attend the general meeting. Then (at least) one element $\theta_{12\ldots n}$ of the awareness structure remains uncertain. Therefore, in the sequel we presume that $G_1 = N$, i.e., the first group of agents makes the whole set N. In addition, suppose that the sequence Π agrees with Assertion 2.12.1 (this sequence is not a subset of G_m under $q < mG_q$). And so, it is reasonable to invite the largest group of agents at first meeting.

The definition of a simple informational impact implies the following. A given sequence of simple informational impacts Π uniquely forms a certain awareness structure I_Π.

Along with the objective sequence Π, consider subjective sequences Π_σ for $\sigma \in \Sigma_+$, taking place in minds of real and phantom agents. Suppose that a certain message s_i is missed according to the viewpoint of agent σ (the latter was absent at a corresponding "meeting"). In this case, assign the empty set symbol $\emptyset$ to element i of the sequence Π_σ; otherwise, write down the message s_i there and use the notation $\Pi_\sigma^i = s_i$. Under the introduced assumptions, we have $\Pi_\sigma^1 = s_1$.

Define the intersection of sequences of simple informational impacts: $\Pi_\sigma \cap \Pi_{\sigma'}$. The result represents a sequence of messages such that $(\Pi_\sigma \cap \Pi_{\sigma'})^i = s_i$ if $\Pi_\sigma^i = s_i$ and $\Pi_{\sigma'}^i = s_i$; otherwise, put the symbol $\emptyset$ instead of element i. In other words, this operation can be defined as the elementwise intersection according to the rules

$$s_i \cap s_i = s_i; \qquad s_i \cap \emptyset = \emptyset \cap s_i = \emptyset \cap \emptyset = \emptyset.$$

Form the sequence of messages Π_j (observed by real agent $j \in N$) as follows: $\Pi_j^i = s_i$, if $j \in G_i$, and $\Pi_j^i = \emptyset$, otherwise. Thus, we believe that an agent knows about meetings he/she actually participated in.

For instance, analyze Π_{12} (the sequence of messages "observed" by agent 2 according to the viewpoint of agent 1). This sequence includes meetings from Π_1 attended by agent 2, i.e., $\Pi_{12}^i = s_i$, if $\{1, 2\} \subseteq G_i$, and $\Pi_{12}^i = \emptyset$, otherwise. Clearly, $\Pi_{12} = \Pi_1 \cap \Pi_2$.

Let us generalize this definition to Π_σ with arbitrary $\sigma \in \Sigma_+$. Denote by $M_\sigma \subseteq N$ the agents being present in the sequence of indexes $\sigma \in \Sigma_+$. For each $\sigma \in \Sigma_+$, define $\Pi_\sigma = \bigcap_{j \in M_\sigma} \Pi_j$.

Recall that a given sequence of simple informational impacts Π uniquely forms a certain awareness structure I_Π. This assertion brings to the following conclusion. Imagine that two (real or phantom) agents, λ and μ, have observed the same sequence of messages, i.e., $\Pi_\lambda = \Pi_\mu$. Consequently, the same awareness structure is formed in their minds: $I_{\lambda i} = I_{\mu i}$ for any $i \in N$. As a matter of fact, we have provided the definition of an identical awareness of agents λ and μ (see Section 2.2). Below we reformulate it as an assertion.

Assertion 2.12.2. Let $\lambda, \mu \in \Sigma_+$. If $\Pi_\lambda = \Pi_\mu$, then agents λ and μ possess an identical awareness (i.e., $I_{\lambda i} = I_{\mu i}$ for any $i \in N$).

Generally, the converse is incorrect, since some messages do not modify an awareness. Consider the following example.

Example 2.12.1. Set $N = \{1, 2\}$, $s_1 = \{1, 2\}[5]$, $s_2 = \{1\}[5]$. The awareness structure is trivial – all elements equal 5, i.e., $\theta_\sigma = 5$ $\forall \sigma \in \Sigma_+$. And so, $I_{1i} = I_{2i}$ $\forall i \in N$, agents 1 and 2 have an identical awareness. At the same time, $\Pi = \Pi_1 = \{s_1, s_2\} \neq \Pi_2 = \{s_1, \emptyset\}$. •

If $M_\lambda = M_\mu$, by virtue of the definition we obtain $\Pi_\lambda = \Pi_\mu$. Hence, agents λ and μ possess an identical awareness. Example 2.12.1 illustrates that Assertion 2.12.2 generally takes no place ($M_\lambda \neq M_\mu \Rightarrow$ agents λ and μ have different awarenesses). However, we can state

Assertion 2.12.3. Let $\lambda, \mu \in \Sigma_+$. If $M_\lambda \neq M_\mu$, there exists a sequence of simple messages Π such that agents λ and μ have different awarenesses.

Proof.   The condition $M_\lambda \neq M_\mu$ means that $\exists\, i \in N: i \notin M_\lambda \cap M_\mu$. For definiteness, assume that $i \in M_\lambda$, but $i \notin M_\mu$. In this case, consider the following sequence of simple messages: $\Pi = \{s_1, s_2\}$, where $s_1 = N[\theta_1]$, $s_2 = M_\lambda[\theta_2]$ $(\theta_1 \neq \theta_2)$. Then $\theta_{\lambda i} = \theta_2$, whereas $\theta_{\mu i} = \theta_1$. This implies that $I_{\lambda i} \neq I_{\mu i}$, i.e., agents λ and μ have different awarenesses. •

Assertion 2.12.3 can be concretized. Notably, it is possible to show that (under certain conditions) messages "observed" by agent λ and not "observed" by agent μ make their awarenesses differ. The elementary option lies in requiring that (in all messages) all values of the uncertain parameter θ_i are pairwise different. We slightly weaken such requirement.

Assertion 2.12.4.   Let λ, $\mu \in \Sigma_+$. Moreover, suppose that $\exists\, s_i = G_i[\theta_i]$: $M_\lambda \subseteq G_i$, $M_\mu \not\subseteq G_i$ and $\neg\exists\, s_k = G_k[\theta_k]$: $M_\mu \cup G_i \subseteq G_k$ and $\theta_k = \theta_i$. Then agents λ and μ have different awarenesses.

Proof.   Denote $G_i \backslash M_\lambda = P$. Choose an arbitrary agent $\sigma \in \Sigma$ such that $M_\sigma = P$ (if $P = \varnothing$, then σ forms an empty sequence). Below we show that $\theta_{\lambda\lambda\sigma} \neq \theta_{\mu\lambda\sigma}$. First, $M_{\lambda\lambda\sigma} = M_\lambda \cup M_\lambda \cup M_\sigma = G_i$. Second, we study sequences of messages, where larger groups meet earlier than smaller ones. And so, the result of assigning $\theta_{\lambda\sigma} = \theta_i$ by message s_i will not be modified by subsequent messages (the group which includes G_i will not meet). The equality $\theta_{\mu\lambda\sigma} = \theta_i$ requires that $\exists\, s_k = G_k[\theta_k] : M_{\mu\lambda\sigma} \subseteq G_k$ and $\theta_k = \theta_i$. However, $M_{\mu\lambda\sigma} = M_\mu \cup M_\lambda \cup M_\sigma = M_\mu \cup G_i$, and such message does not exist by the premise of Assertion 2.12.4. •

We summarize the outcomes as follows. In the case of simple messages, the condition $M_\lambda = M_\mu$ (and only this condition!) guarantees that agents λ and μ possess an identical awareness. This suggests the idea of another definition to-be-discussed in the next subsection.

The graph of worlds. We say that agents $\lambda \in \Sigma_+$ and $\mu \in \Sigma_+$ belong to the same world (equivalently, enter the same world), if they have an identical awareness under any sequence of messages Π.

A subset of the set of real and phantom agents, $W \subseteq \Sigma_+$, is called a *world*, if for any λ, $\mu \in W$ we obtain $I_\lambda \sim I_\mu$ for any sequence of messages Π.

Assertions 2.12.2–2.12.3 lead to the following result.

Assertion 2.12.5.   Under the procedure of simple messages, agents $\lambda \in \Sigma_+$ and $\mu \in \Sigma_+$ belong to the same world if $M_\lambda = M_\mu$.

The procedure of simple messages being considered, we say that a given world corresponds to the set $Z \subseteq N$, if $Z = M_\lambda$ (here λ represents an agent from this world). It is convenient to designate a world by a corresponding set Z and say, e.g., that the world $\{1, 2\}$ contains agents 12, 21, 121, 212, 1212, 2121, ...

The number of different worlds is the number of nonempty subsets of the set of real agents N, i.e., $2^n - 1$. Thus, it appears a finite number. And so, an awareness structure generated by simple messages enjoys a finite complexity. Clearly, the basis of an awareness structure includes a finite number of elements. We should demonstrate that the number of pairwise nonidentical awareness structures is finite. For this, we formulate a trivial result.

Assertion 2.12.6. Suppose that phantom agents λ, $\mu \in \Sigma_+$ enter the same world and $\omega(\lambda) = \omega(\mu)$ (i.e., they have the same last index). In this case, $I_\lambda = I_\mu$.

Proof. If agents belong to the same world, they *a fortiori* possess an identical awareness. In particular, this means that $I_{\lambda\omega(\lambda)} = I_{\mu\omega(\mu)}$. On the other hand, the last two indexes in the sequences $\lambda\omega(\lambda)$ and $\mu\omega(\mu)$ coincide. Thus, they "collapse" by the axiom of self-awareness: $I_{\lambda\omega(\lambda)} = I_\lambda = I_\mu = I_{\mu\omega(\mu)}$. •

Now, we provide an upper bound for the number of pairwise nonidentical awareness structures in each specific world.

Assertion 2.12.7. Let $\lambda \in \Sigma_+$ and consider awareness structures corresponding to agents from the world M_λ. Among them, there exist pairwise nonidentical awareness structures, whose number does not exceed the cardinal number of the set M_λ (by turn, the latter is not higher than n).

Proof. Suppose that agent σ enters a given world. By definition, the expression for σ must involve all indexes from the set M_λ (and only them). In particular, the last index in the sequence σ can be from the set M_λ exclusively. Therefore, Assertion 2.12.6 directly leads to Assertion 2.12.7. •

Assertion 2.12.8. Simple messages form an awareness structure I with depth $\gamma(I) \leq n$ and a finite basis. The number of elements in the basis is bounded above by $\Sigma_{k=1}^n C_n^k k = 2^{n-1}n$. Furthermore, there exists a sequence of simple messages Π ensuring these upper estimates.

Proof. The set M_λ ($\lambda \in \Sigma_+$) represents a subset of the set N composed of n elements. Next, k gives the cardinal number of the set M_λ. Depending on $\lambda \in \Sigma_+$, $k \in \{1, 2, \ldots, n\}$. As is generally known, a set of the cardinal number n admits C_n^k different subsets having the cardinal number k. By applying Assertion 2.12.7, we derive the upper estimate for the number of elements in the basis of the awareness structure: $\Sigma_{k=1}^n C_n^k k$.

Show that $\gamma(I) \leq n$. Select an arbitrary agent $\sigma \in \Sigma_+$, $|\sigma| = m > n$. Then the sequence of indexes σ includes (at least) one pair of identical ones. In this pair, choose the first index and eliminate it from σ; this yields the sequence σ', $|\sigma'| = m - 1$. The second index remains in the sequence. This means that $M_\sigma = M_{\sigma'}$, i.e., agents σ and σ' belong to the same world. We have not modified the last index; therefore, $\omega(\sigma) = \omega(\sigma')$. Yet, in this case $I_\sigma = I_{\sigma'}$ by Assertion 2.12.6. Thus, we have demonstrated that $\forall \sigma \in \Sigma_+$: $|\sigma| > n \; \exists \; \sigma' \in \Sigma_+$: $|\sigma'| < |\sigma|$ and $I_\sigma = I_{\sigma'}$. Hence, $\gamma(I) \leq n$.

Consider a sequence of simple messages $\Pi = \{s_1, s_2, \ldots, s_{n+1}\}$, where $s_1 = N[\theta_0]$, $s_2 = N\backslash\{1\}[\theta_1], s_3 = N\backslash\{2\}[\theta_2], \ldots, s_{n+1} = N\backslash\{n\}[\theta_n]$, and all $\theta_i \in \Theta$ are pairwise different. This sequence possesses the following interpretation. Each agent attended one meeting with the rest agents as participants. Evidently, agents from different worlds observe different sequences of messages. Different value of θ_i serve for meeting the conditions of Assertion 2.12.4. According to the latter, in our example any two agents from different worlds have different awarenesses. Consequently, the above estimates take place for the awareness structure generated by the sequence of simple messages. •

The number of elements in the basis (as well as the number of worlds comprising agents with an identical awareness) is bounded. And so, awareness structures generated by the procedure of simple messages can be analyzed using the graph of worlds.

First, analyze how an identical awareness reflects the reflexive game graph. Nodes of the graph correspond to agents $\sigma \in \Sigma_+$, whose awareness structures I_σ enter the basis of the general awareness structure I. In the reflexive game graph, an incoming arc (an agent) of a node indicates the adequate awareness about this agent. Imagine that an arc comes from a certain agent σ to another agent in a group of agents with an identical awareness. In this case, we have arcs from agent σ to the rest agents belonging to the group. This circumstance suggests the idea of uniting all agents with an identical awareness in the graph. Recall that such agents enter the same world. Thus, we introduce the *graph of worlds*. Its nodes correspond to different worlds. Here arcs are constructed by the following rule. Suppose that a given world corresponds to the set $Z \subseteq N$ with cardinal number $m \leq n$ ($Z = M_\lambda$, where λ means an agent from this world). Then this world has $(n - m)$ incoming arcs from worlds corresponding to the sets $Z \cup \{j\}$, $j \in N \backslash Z$. As an example, we present the graph of worlds in the cases of $n = 2$ and $n = 3$ (see Fig. 2.25).

The graph of worlds enjoys the same property as the reflexive game graph. Notably, if a path exists from one world to another, agents belonging to the latter world are adequately aware of agents entering the former world. We emphasize that agents from one world have an identical awareness. Hence, the graph of worlds completely describes the reflexive awareness of agents about each other, being defined by the awareness structure generated by the procedure of simple messages.

Clearly, $\sigma \in \tilde{G} \Leftrightarrow M_\sigma \subseteq G$. Then the definition of a simple message implies the following. If we bind $\theta_\sigma = \theta$ for agent σ, the same takes place for the rest agents from his/her world (and for all agents belonging to worlds connected to his/her world via a path on the graph). And so, the value of an uncertain parameter $\theta \in \Theta$ in a node of the graph applies to the whole world. In other words, $\theta_\tau = \theta$ for any τ from this world.

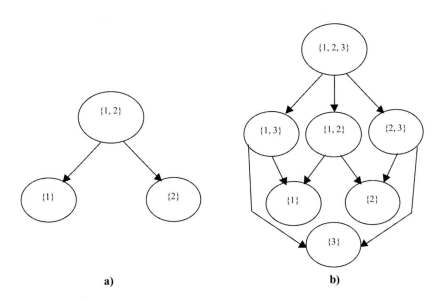

a)   b)

Figure 2.25 The graph of worlds: (a) two agents and (b) three agents.

Now, we direct the reader's attention to the following. The graph is drawn such that "larger" worlds lie above "smaller" ones. The reason concerns a possible interpretation. Assign different colors to different values of the uncertain parameter $\theta \in \Theta$. Accordingly, fill nodes with these colors. Suppose that a node has a certain color. Then $\theta_\sigma = \theta$ for each agent σ from this world one (θ is the value of the uncertain parameter corresponding to the color). By the impact $s \equiv G[\theta]$, the principal "applies a drop" of color θ to the world connected with the set $G \subseteq N$. Subsequently, the drop fills this world with color θ and (by "flowing down through arrows") colors all worlds having a path from the given world. Indeed, agents σ from these worlds meet the inclusion $M_\sigma \subseteq G$. Assume that a given node has been colored earlier. Then the impact fills it with a new color. After the last informational impact, we obtain the "colored" graph of worlds.

Let us elucidate the process of coloring for three agents.

Example 2.12.2. Set $N = \{1, 2, 3\}$, $s_1 = \{1, 2, 3\}$["no hatching"], $s_2 = \{1, 2\}$["diagonal hatching"], $s_3 = \{1, 3\}$["horizontal hatching"]. Fig. 2.26 shows the graphs of worlds after the messages s_2 and s_3.

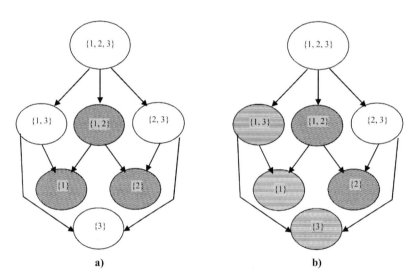

Figure 2.26 The graph of worlds: (a) after s_2 and (b) after s_3.

The "colored" graph of worlds enables recovering uniquely the awareness structure I, i.e., all elements θ_σ, $\sigma \in \Sigma_+$. For this, use the graph to find a world comprising a specific agent $\sigma \in \Sigma_+$, and check the color of the corresponding node.

The principal can create any coloring in the graph of worlds (if necessary). It suffices "to apply a drop of an appropriate color." In terms of the sequence of messages Π, this means the presence of the message $s_i = Z[\theta]$, where Z is the corresponding set of agents and θ designates the desired color. For instance, invite sequentially all groups of agents (nonempty subsets of N) according to the above rule (first, larger groups and then smaller ones). Then any coloring of worlds can be obtained (in particular, each world may have been filled with a unique color).

The "colored" graph reflects the mutual awareness of agents. This is an attractive approach to modeling due to its universalism (despite that it may have redundant description, e.g., just one message $s_1 = N[\theta]$).

Simple messages serve for forming rather simple awareness structures. Reformulate this statement as

Assertion 2.12.9. An awareness structure I can be constructed by the procedure of simple messages if $\forall \lambda, \mu \in \Sigma_+: M_\lambda = M_\mu \Rightarrow \theta_\lambda = \theta_\mu$.

Proof.   The necessity part has been discussed above. Indeed, under the procedure of simple messages, agents from the same world have identical beliefs about the value of an uncertain parameter. For the sufficiency part, construct a sequence of simple messages Π, which implements the awareness structure I_Π (consisting of elements θ'_σ) coinciding with I (consisting of elements θ_σ). Invite sequentially all nonempty subsets of real agents ($2^n - 1$ subsets totally), first larger groups and then smaller ones. In other words, consider the following sequence of messages Π: $s_1 = N[\theta_{1...n}]$, $s_2 = N \backslash \{1\}[\theta_{2...n}]$, $s_3 = N \backslash \{2\}[\theta_{13...n}], \ldots, s_{n+1} = N \backslash \{n\}[\theta_{1...n-1}]$, $s_{n+2} = N \backslash \{1, \ 2\} \ [\theta_{3...n}], \ldots, s_{\frac{n^2+n+2}{2}} = N \backslash \{n-1, n\}[\theta_{1...n-2}], \ldots, s_{2^n-n} = \{1\}[\theta_1], \ldots, s_{2^n-1} = \{n\}[\theta_n]$.

In the sense of color interpretation, we have the following result. Moving down the graph of possible worlds, we apply a drop of necessary color to each node-world. "Flowing down," this drop cannot change of color of nodes located above. Therefore, this sequence of messages assigns to each world the same color as in the awareness structure I (according to the premise, in the structure I agents entering the same world possess identical beliefs about the value of the uncertain parameter). In other words, $\forall \sigma \in \Sigma_+: \theta'_\sigma = \theta_\sigma$, i.e., $I = I_\Pi$. •

A couple of examples, where the principal succeeds in forming a desired awareness structure by the mechanism of simple messages (and does not), can be found in Section 4.4.

Thus, in the present section we have described the procedure of forming an awareness structure in reflexive games (by simple messages of the principal about the value of an uncertain parameter). Necessary and sufficient conditions to-be-imposed on an awareness structure in order to form it by the procedure of simple messages have been established. Moreover, we have developed and described a convenient tool of mutual awareness representation, *viz.*, the graph of worlds.

A natural area of further investigations relates to identifying more complicated types of messages. This would provide a better description to informational impacts used in practice.

2.13   SET-TYPE AWARENESS STRUCTURES

In this book, we *par excellence* study point-type awareness structures. They characterize the situation of correct (or, probably, incorrect) knowledge of an uncertain parameter by agents, the beliefs of opponents, the beliefs about beliefs, and so on. Sections 2.13–2.15 describe a more complicated model of awareness, where agents' beliefs have higher complexity [38, 41].

To begin, let us formulate a certain problem. It could be suggested to high school students at a competition in mathematics.[18]

Statement of the problem. Three friends participate in a game with the following rules. Friend 3 thinks of two (probably, identical) integer numbers between 1 and 9 inclusive. Then he reports the sum of the numbers to friend 1. Friend 2 is provided with information on the product of the numbers. Next, friend 3 (player 3 of the game) puts the question, "What numbers have been thought of?" Players 1 and 2 have to give the numbers or reply, "I have no idea." Note that they answer simultaneously without any sharing of information.

Imagine that both friends say, "I have no idea." Friend 3 puts the same question again. Going over possible options for a while, players 1 and 2 answer, "I have no idea." Once again, player 3 repeats the question and gets the same replies. The situation recurs seven times; at iteration 8, friend 1 guesses right the numbers.

The *problem* is to find what numbers have been thought of (in the sequel, we use italic type to indicate it).

This *problem* serves for elucidating the notions and structures introduced below.

Obviously, to solve the *problem*, it is necessary to describe the process of changes occurring in the awareness of players 1 and 2. In particular, how the first one goes from an incomplete awareness (in the beginning, he knows only the sum of the numbers) to the complete awareness. This requires describing the awareness including its reflexive component (the beliefs about opponent's beliefs) and the relation between the awareness of the players and their answers. In comparison with the previous sections of the book, a fundamental innovation appears in consideration of set-type awareness structures that model the dynamics (see Section 2.14). An alternative approach to modeling of incomplete awareness involves Bayesian games; for details, we refer to [10, 11].

Awareness structure. Let us describe the awareness structure in the case of an incomplete awareness. First, we give a formal description in terms of sets, their elements and mappings. Second, we illustrate the introduced notions by examples with regard to the *problem*.

Suppose that n subjects are involved in the interaction; call them *real agents*. Introduce the following notions and sets (the sets are considered to be finite):

Θ means the set of feasible states of nature;

A_i stands for the set of feasible samples of agent i, $i \in N = \{1, \ldots, n\}$; one of them is real, whereas the rest agents are *phantom*[19];

$A = A_1 \cup \ldots \cup A_n$ denotes the set of all agents;

$\Omega \subset \Theta \times A_1 \times \ldots \times A_n$ is the set of *possible worlds*.

In each possible world $\omega = (\omega_0, \omega_1, \ldots, \omega_n)$, there exists a definite state of nature $\omega_0 \in \Theta$ and definite samples $\omega_i \in A_i$ of each agent. We will say that the agent ω_i belongs to the world ω or enters into the world ω.

---

[18]The author does not know whether the problem was actually given to the students or not.
[19]As a rule, throughout the book we adopt the term "agent" for samples of agents.

In addition, we use the following notation:

η is up the *awareness function* of a certain agent; this function maps each agent $a \in A$ into the set of worlds $\eta(a) \subseteq \Omega$ that are considered possible according to his/her awareness;

$\omega^* \in \Omega$ represents the *real world*. One of possible worlds is real; i.e., it is characterized by the actually existing state of nature ω_0^* and agents ω_i^*.

Agents belonging to the real world are real themselves; the rest samples of the agents are phantom.

Assume that the following conditions hold true.

Condition 1 (agent's identity). $\forall i \in N, \forall a_i \in A_i, \forall \omega \in \eta(a_i): \omega_i = a_i$, i.e., each agent enters into all worlds that are considered possible by him/her.

Next, for each world ω, define the set of worlds and agents $I(\omega)$ *connected with the world* ω.

A world ω' is connected with a world ω^1 if there exist finite sequences of worlds $\omega^2, \ldots, \omega^m$ and agents $a_{i_1}, \ldots, a_{i_m}$ such that

$$a_{i_k} = \omega_{i_k}^k, \quad k = 1, \ldots, m, \quad \omega^{k+1} \in \eta(a_{i_k}), \quad k = 1, \ldots, m-1, \quad \omega' \in \eta(a_{i_m}).$$

An agent is connected with a world ω' if he/she enters into a world connected to the world ω'.

The notion of worlds/agents connected with a given world allows formulating the next condition.

Condition 2 (world's unity). For any world $\omega \in \Omega$ and any agent $a \in A$, we have $\omega \in I(\omega^*), a \in I(\omega^*)$; i.e., each world and each agent are connected with the real world.

A (*set-type*) *awareness structure* is a set of the form $(\Theta, A_1, \ldots, A_n, \omega, \omega^*, \eta(\cdot))$, where $\Omega \subset \Theta \times A_1 \times \cdots \times A_n, \omega^* \in \Omega, \eta: A_1 \times \cdots \times A_n \to 2^\Omega$, and the conditions of agent's identity and world's unity are satisfied. Here 2^Ω indicates the set of all subsets of Ω.

In this chapter (except Section 2.12), we have studied *point-type* awareness structures, where each agent considers possible a single world only; i.e., for each $a \in A$ the set $\eta(a)$ comprises a single element.

An awareness structure is said to be *correct* if for any agent there exists (at least) one world considered possible by the agent, i.e., $\forall a \in A: \eta(a) \neq \emptyset$.

An awareness structure is said to be *regular* if an agent considers possible exactly the world he/she enters into, i.e.,

$$\forall \omega \in \Omega, \quad \forall i \in N: \omega \in \eta(\omega_i).$$

In other words, correctness means the absense of an agent in the completely indefinite situation. Regularity implies the absence of an agent mistakenly thinking that he/she enters no world at all.

Clearly, each regular awareness structure is correct. Indeed, consider an arbitrary agent. The world's unity condition leads to the existence of a world being entered by the agent. According to the regularity condition, this world appears possible for the agent. Since the agent has been selected arbitrarily, the correctness of this structure is immediate.

The graph of an awareness structure. One can illustrate awareness structures by directed graphs having nodes of two types, namely, worlds (marked by rectangles) and agents (marked by circles). In this case, the real world is often darkened. The arrow from a certain agent to a world means that the agent enters into the world. By analogy, the arrow from a certain world to an agent indicates that the latter consider the former possible. A double arrow represents two single arrows, i.e., from an agent to a world and from the world to the agent. Evidently, (at least) one single arrow goes to each agent in a correct awareness structure. In regular awareness structures, all arrows turn out double.

Let us address the *problem* and study the following example of an awareness structure.

Example 2.13.1. Assume that the pair (6, 6) has been thought of. Then the awareness structure acquires the form shown in Fig. 2.27. Each circle has an index $i \in N = \{1, 2\}$. This means that an agent is the sample of agent i. Each rectangle contains a pair of numbers thought of. The node with the pair (6, 6) is darkened (as the real world). Imagine that another pair (from the ones in Fig. 2.27) has been thought of instead. The figure would remain almost the same (the only difference concerns the real world rectangle).

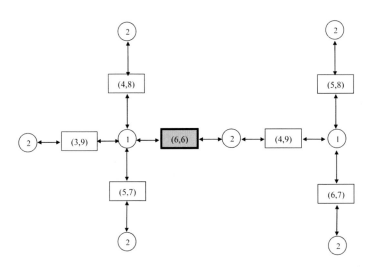

Figure 2.27 The pair (6, 6) has been thought of.

Real agent 2 knows the product of the numbers in question (36). Therefore, in addition to the actual pair (6, 6), he/she considers possible the pair (4, 9). By analogy, real agent 1 is aware of their sum (12) and considers possible the pairs (3, 9), (4, 8), (5, 7), and (6, 6).

In the world, where the pair (4, 9) has been thought of, (phantom) agent 1 considers possible the pairs (5, 8) and (6, 7).

Take the world, where one of the pairs (3, 9), (4, 8), (5, 7), (5, 8) and (6, 7) has been thought of. Here agent 2 knows such pair for sure (as being uniquely defined by the known product of the numbers). •

Example 2.13.2. Let the pair (4, 4) have been thought of. Then the awareness structure is represented by Fig. 2.28. Interested readers would easily verify this (by examining all pairs with the given sum and product).

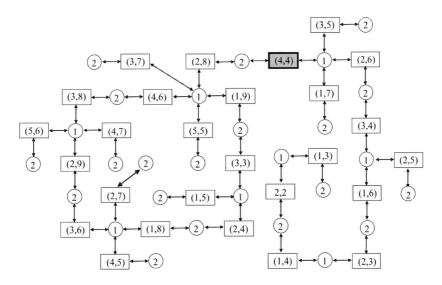

Figure 2.28 The pair (4, 4) has been thought of.

Different levels of awareness. In terms of awareness structures, one may formalize different levels of agents' awareness. In this book, we focus on three of them.

Let agents i and j belong to a world ω.

Identical awareness of the agents. The agents are said to possess an *identical awareness*, if the sets considered possible by them coincide: $\eta(\omega_i) = \eta(\omega_j)$.

Adequate awareness of agent 1 about agent 2. Imagine there exists a set of worlds considered possible by agent i; each of these worlds is entered by a certain sample of agent j. The samples in question may coincide (or not) with each other and with agent j. Agent i is said to possess the *adequate awareness* of agent j if such coincidence takes place, i.e.,

$$\forall \xi \in \eta(\omega_i) : \xi_j = \omega_j.$$

Major or minor awareness of agent 1 (against the awareness of agent 2). Clearly, in a given world, an agent considering this world as the only possible one appears the most aware (if such agent exists). Comparing agents by their awareness may fail in more complicated cases. However, it is natural to consider agent i as *having more awareness* than agent j if the following conditions take place: (a) $\omega \in \eta(\omega_i)$ (agent i considers possible the world he/she belongs to) and (b) $\eta(\omega_i) \subset \eta(\omega_j)$ (the set of possible worlds of agent j is larger and the uncertainty is higher).

Information equilibrium. Imagine that, in addition to an awareness structure (describing the awareness of agents), we have defined goal functions (describing the interests of agents) and their feasible actions. One may put the following question (traditional for game theory): "What actions would be chosen by the agents?"

Let $\theta \in \Theta$ be a state of nature and $x_i \in X_i$ be the action chosen by agent i. We consider a normal-form game, *viz.*, agents choose actions simultaneously and independently. To proceed, assume that $f_i(\theta, x_1, \ldots, x_n)$, $i \in N$ are the goal functions of agents and the corresponding awareness structure appears correct[20].

Then an *informational equilibrium* is the set of functions

$$\chi_i : A_i \rightarrow X_i, \quad i \in N,$$

such that

$$\chi_i(a_i) \in \text{Arg} \max_{x \in X_i} \min_{\omega \in \eta(a_i)} f_i(\omega_0, \chi_1(\omega_1), \ldots, \chi_{i-1}(\omega_{i-1}), x, \chi_{i+1}(\omega_{i+1}), \ldots, \chi_n(\omega_n)).$$

And so, each agent strives for maximizing his/her worst-case result within all worlds considered possible by him/her.

We emphasize that this notion of an informational equilibrium extends the corresponding notion in the case of point-type awareness structures (see Section 2.3).

To illustrate the notion of an informational equilibrium, we revert to the *problem* again. Each agent names the numbers being thought of or answers, "I have no idea" (to indicate the last option, we use the symbol $\{-\}$). Thus, the sets of feasible actions of both agents have the form

$$X_1 = X_2 = \Theta \cup \{-\},$$

where $\Theta = \{(a, b) \mid a \in \{1, \ldots, 9\}, b \in \{1, \ldots, 9\}\}$.

Let $i = 1, 2$; define the (identical) goal functions of the agents by

$$f_i(\theta, x_1, x_2) = \begin{cases} 1, & \text{if } (x_1 = x_2 = \theta) \text{ or } (x_1 = \theta, x_2 = \{-\}) \text{ or } (x_1 = \{-\}, x_2 = \theta); \\ 0, & \text{if } x_1 = x_2 = \{-\}; \\ -1, & \text{otherwise.} \end{cases}$$

In other words, the agents have the gain of 1, if (at least) one agent names the correct numbers and the opponent makes no mistake. When both answer, "I have no idea," each agent obtains zero gain. If (at least) one of agents names the wrong numbers, both receive the gain of -1.

Then an informational equilibrium takes the following form. A certain agent names the pair of numbers if and only if he considers a single world possible (i.e., the agent knows for sure the pair being thought of). If not, the agent replies, "I have no idea."

We proceed with transformation of awareness structures.

---

[20]If the structure is not correct, there exists an agent considering all worlds impossible. Modeling of such agent goes beyond the scope of this book.

2.14 TRANSFORMATION OF AWARENESS STRUCTURE

An awareness structure is, *sui generis*, a "snapshot" of the mutual awareness of agents. Obviously, with the course of time an awareness may change. We have earlier described the models of changes in awareness structures under the influence of messages (see Section 2.12) or observations of certain results of the game by agents. Although, the models supposed agents' radical refusal to use new awareness instead of the existing awareness. As a matter of fact, agents have been considered forgetful and not sure of their awareness.

In this section, we describe transformation of an awareness structure of a game when agents observe the outcome of the game. We assume that the information (being available to the agents and noncontradictory to new observations) is totally retained.

Recall that we study a normal-form game, i.e., agents choose their moves simultaneously and independently. Imagine that the game changes the awareness of an agent. Then each subsequent game (if any) is played by agents with new awareness structures irrespective of the previous or subsequent ones.

Let real agent i have the *observation function* $w_i = w_i(\theta, x_1, \ldots, x_n)$. Note this function is a common knowledge among the agents; for a detailed discussion of observation functions in the point-type case, see Sections 2.7 and 2.13. It could be interpreted as follows. Suppose that the environment θ exists in the world entered by agent[21] $a_i \in A_i$ and agents have chosen the actions $(x_1, \ldots, x_n)$. Then agent a_i observes the value $w_i \in W_i$, where W_i represents the set of feasible observations of the samples of agent i.

Transformation of an awareness structure has the following essence (if outlined in brief). For each (real or phantom) agent $a \in A$, the set of worlds $\eta(a)$ undergoes only those modifications that are considered possible by the agent. A modification lies in eliminating the worlds described by observation functions with values differing from the ones observed by the agent. It may happen that an agent receives different "signals" (i.e., different values of observation functions) from different worlds. In this case, the agent "disappears" and a few agents "arise" instead (each agent has his/her own awareness – see Fig. 2.29, where the values of observation functions are in rectangles).

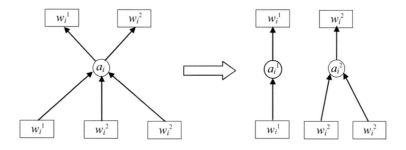

Figure 2.29 Transformation of an awareness structure changes the number of agents.

---

[21]Recall that just one sample of agent i belongs to each world, $i \in N$.

Now, let us specify the transformation rule for an awareness structure. Suppose there exists a unique information equilibrium χ, whose implementation leads to the following. The observation function of each agent possesses a certain value in each world w: $w_i = w_i(\omega_0, \chi_1(\omega_1), \ldots, \chi_n(\omega_n))$. Then the value of observation function depends on the world ω only, i.e., $w_i = w_i(w)$.

Assume there exists agent $a_i \in A_i, i \in N$. Below we describe the transformation procedure for his/her awareness. Denote by

$$H(a_i) = \{\omega \in \Omega \mid \omega_i = a_i\}$$

the set of worlds entered by agent a_i. Next, let $M = M(a_i)$ be the number of pairwise distinguishable values of the observation function w_i over the worlds from the set H; the values themselves are indicated by $w_i^1, w_i^2, \ldots, w_i^M$.

Then, as the result of such transformation, agent a_i is replaced by M agents (being generated and added to the set A_i); denote them by $a_i^1, a_i^2, \ldots, a_i^M$. We underline that, for each $k \in \{1, \ldots, M\}$, the relation between these agents and the worlds is defined by the following formulas:

$$H(a_i^k) := \left\{\omega \in H(a_i) \mid w_i(\omega) = w_i^k\right\};$$

$$\eta(a_i^k) := \left\{\omega \in \eta(a_i) \mid w_i(\omega) = w_i^k\right\}.$$

Agent a_i is eliminated from the set A_i.

Apply the described procedure to all agents $a \in A$. Consequently, all worlds and agents having no relation to the real world are eliminated from the sets Ω and A, respectively. This finalizes the modification of the awareness structure due to observing the results of interaction among the agents.

For regular awareness structures, the sets $H(a_i)$ and $\eta(a_i)$ coincide; therefore, the sets $H(a_i^k)$ and $\eta(a_i^k)$ are identical, as well. This implies that the transformation preserves the regularity of an awareness structure; i.e., regular structures are always transformed into regular structures (see Fig. 2.30).

Get back to the *problem* (see Section 2.13) and continue discussing Example 2.13.1 (see Fig. 2.27). The observation functions of both agents are as follows. An agent knows the reply of the opponent and his/her own gain. Recall that an agent names a specific pair of numbers only if he/she surely knows it (i.e., he/she considers a single world possible). Thus, after 1 question and answers, the awareness structure takes the form

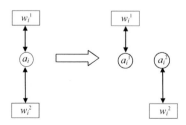

Figure 2.30 Transformation preserves the regularity of an awareness structure.

presented by Fig. 2.31. In particular, we have eliminated the worlds, where an agent has named a specific pair of numbers. Obviously, agent 1 knows for sure the numbers thought of.

Now, consider an extension of Example 2.13.2 (see Fig. 2.28). After 1 question and answers, the awareness structure acquires the form presented in Fig. 2.32.

Clearly, it takes 7 questions and replies for an agent to reach the complete awareness (see Fig. 2.33).

We have, *eo ipso*, found the solution to the *problem* – the pair (4, 4) has been thought of.

Strictly speaking, the comprehensive answer to the *problem* requires studing all feasible pairs of numbers thought of. However, a reader would easily verify that only for the pair (4, 4) the complete awareness is attained after 7 questions and answers.

In Sections 2.13–2.14, we have discussed the awareness structure of agents participating in a reflexive game. Moreover, we have demonstrated possible changes in this structure when agents observe the results of their actions. A promising direction of further research (including informational control) consists in modeling of changes in an awareness structure due to the messages from external subjects (for a given set of agents) and due to communication among agents.

In addition, it has been demonstrated that transformations preserve the regularity of an awareness structure. Analyzing other properties of an awareness structure or informational equilibria and conditions ensuring them under such transformations seems attractive.

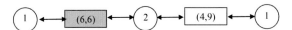

Figure 2.31 The awareness structure after one question and answers (the pair (6, 6) has been thought of).

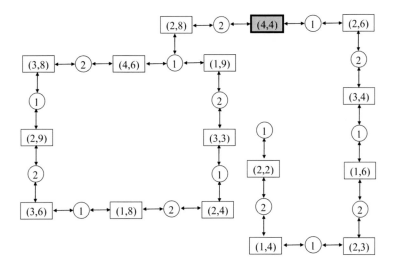

Figure 2.32 The awareness structure after 1 question and answers (the pair (4, 4) has been thought of).

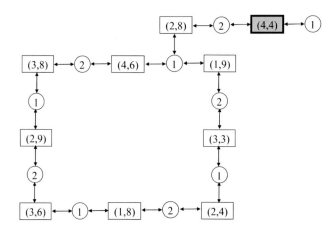

Figure 2.33 The awareness structure after 7 questions and answers (the pair (4, 4) has been thought of).

2.15 CONCORDANT INFORMATIONAL CONTROL

This section focuses on a model of concordant informational control, where agents are informed about the fact of being controlled by a principal. Still, agents trust messages of the principal. The existence conditions of such control are established and several properties are demonstrated [38].

We have earlier studied the problem of informational control under the following assumption. The principal may form an arbitrary awareness structure for agents (from a given class of structures). The simplest situation lies in the full trust of an agent to the principal. In other words, each agent believes all messages of the principal are true (see the surveys of decision procedures based on revelation of information in [26, 118, 160]). In this case, we will say that *unconcordant informational control* takes place. Indeed, control is implemented and an agent has no idea that the principal reports some information, pursuing individual interests.

In what follows, we suggest a model of *concordant informational control*; here agents are informed about the fact of control implementation by a principal. Still, they trust messages of the principal [38]. Evidently, implementing such control requires specific conditions. Establishing and analyzing the latter represent the scope of this section.

Unconcordant informational control. Using the following example, let us study an elementary scheme of unconcordant informational control.

Example 2.15.1. Consider an agent with the goal function (also known as the utility function) defined by

$$f(\theta, x) = \theta x - x^2/2.$$

In the formula above, θ stands for an uncertain parameter (a random variable possessing values from the set $\Theta = \{1, 3, 7\}$ with the identical probability of 1/3), and

$x \in [0; +\infty)$ means an action chosen independently by the agent. We provide a possible economic interpretation. The agent manufactures a certain product, and the product's market price θ is a priori unknown (thus, representing a random variable). The agent's costs to manufacture x units of the product constitute $x^2/2$. Subsequently, the goal function $f(\theta, x)$ specifies the agent's profits, and the agent strives for maximizing their expected value $E_\theta f(\theta, x)$.

The function $f(\theta, x)$ is linear in θ. Hence, the expected value of this function satisfies the following expression:

$$(1) \quad E_\theta f(\theta, x) = E\theta \cdot x - \frac{x^2}{2},$$

where $E\theta$ designates the mathematical expectation of the random variable θ. By maximizing the function (1), the agent evaluates his/her optimal action: $x^* = E\theta = \frac{1}{3} \cdot 1 + \frac{1}{3} \cdot 3 + \frac{1}{3} \cdot 7 = \frac{11}{3}$.

Now, assume that the principal participates in the situation, applying informational control by reporting a feasible range of the uncertain parameter. We believe that the principal knows the value θ, while the agent fully trusts his/her messages. For instance, suppose that the principal reports the set $\{1, 3\}$ (i.e., the value $\theta = 7$ is impossible and the probabilities of the values $\theta = 1$ and $\theta = 3$ equal 1/2 according to the Bayes formula). Then the agent computes his/her optimal action by $x^* = \frac{1}{2} \cdot 1 + \frac{1}{2} \cdot 3 = 2$.

Consideration of all feasible messages of the principal enables defining all actions chosen by the agent as the result of certain informational control – see Table 2.1.

Table 2.1   Messages of the principal and actions of the agent in Example 2.15.1.

Messages of the principal	Actions of the agent
{1}	1
{3}	3
{7}	7
{1, 3}	2
{1, 7}	4
{3, 7}	5
{1, 3, 7}	11/3

Therefore, by a proper message the principal ensures an arbitrary action of the agent from the set $\{1, 2, 3, 11/3, 4, 5, 7\}$. Obviously, the principal should choose a message such that the corresponding action of the agent is the most beneficial to the principal. •

Concordant informational control. In this subsection, consider an alternative situation which is less beneficial to the principal. The agent does not take on trust any messages of the principal. Such "suspicious" agent adheres to the following line of reasoning: "By reporting a specific message, the principal wants me choose a corresponding action. However, this action is beneficial to the principal. Would it be beneficial to me, as well?"

In this case, implementing informational control requires taking into account the interests of both the principal and agent. To formalize this requirement, we introduce the principal's goal function $F(\theta, x)$. Actually, it depends on the same arguments, i.e., on θ (the uncertain parameter representing a random variable with a known probability distribution) and on x (an action of the agent). Denote by s a message of the principal and suppose it belongs to a fixed set of feasible messages S.

Then the principal's strategy (control) lies in choosing a message depending on a known state of nature. In other words, the principal selects the function $s(\theta)$. On the other hand, the agent's strategy is to choose an action x depending on the principal's message. As a result, the agent selects the function $x(s)$.

To proceed, we formalize the sequence of interaction between the principal and agent (the set Θ and the probability distribution on Θ are considered as a common knowledge).

Step 1. The principal reports the function $s(\theta) : \Theta \to S$ to the agent.
Step 2. The principal finds out the actual value of θ.
Step 3. The principal reports the value $s \in S$.
Step 4. The agent chooses the action $x = x(s)$.

We emphasize the following aspect. The agent is concerned with messages of the principal merely, since the former may refine the set of feasible values for the uncertain parameter θ. Notably, having received the message s at Step 3, the agent is interested only in the set $\{\theta \in \Theta \,|\, s(\theta) = s\}$. Therefore and without loss of generality, we believe that at Step 1 the principal reports to the agent a certain partition of the set Θ. For the time being, the set Θ is supposed to be finite. Consequently, the partition takes the form $S = \{\Theta_1, \ldots, \Theta_m\}$, where $\Theta_1 \cup \ldots \cup \Theta_m = \Theta$, $\Theta_i \neq \emptyset$, $i \in M \in \{1, \ldots, m\}$, $\Theta_i \cap \Theta_j = \emptyset$ for $i \neq j$. The sets Θ_i are called *parts* of the partition S.

At Step 3, the principal informs the agent of a set $\Theta_i \in S, i \in \{1, \ldots, m\}$. Imagine that the agent receives the principal's message $\Theta_i \subset \Theta$ and trusts it. Then his/her optimal action maximizes the conditional expectation of the goal function (provided that the set of values θ is restricted from Θ to Θ_i):

$$(2) \quad X_i^* = \underset{x \in X}{\text{Argmax}} \; E_{\theta \in \Theta_i} f(\theta, x), \quad i = 1, \ldots, m.$$

Consider an arbitrary principal's message belonging to the partition S; let X^* be the set of all optimal actions of the agent:

$$X^* = X_1^* \cup \ldots \cup X_m^*.$$

Informational control is said to be *concordant* if for any value $\theta \in \Theta$ the principal benefits from reporting to the agent a part of the partition S, which contains θ (and the agent benefits from trusting the principal).

This statement could be formulated as

$$(3) \quad \forall i \in M \; \forall \theta \in \Theta_i \; \forall x^* \in X_i^* \; \forall x \in X^* : \text{ either } x \in X_i^*, \text{ or } F(\theta, x^*) \geq F(\theta, x).$$

The requirement (3) is called the *concordance condition*.

Under the condition (3) and the agent trusting the principal, the latter benefits from truth-telling. Moreover, if the condition (3) is valid and the principal makes true messages, the agent benefits from trusting the principal.

Whether the condition (3) is satisfied or not depends on the partition S (indeed, it uniquely defines the number of parts m, the parts Θ_i themselves, and the sets X_i^*, $i = 1, \ldots, m$). Hence, meeting the condition (3) means the *concordance* of the partition S and, at large, the *consistency* of informational control based on the partition S.

Remark. We have supposed that the set Θ is finite. In particular, such assumption implies the finiteness of the partition S. Obviously, the arguments would be the same for the infinite set Θ. In this case, the partition S may be (not necessarily) infinite: $S = \{\Theta_\alpha\}, \alpha \in A$, where A is a certain set of indices. However, one should require that each part Θ_α of the partition S is Borel measurable (to apply the mathematical expectation operator for evaluating the optimal action of the agent).

Example 2.15.1 (continuation). Assume that the agent is no more "credulous". Thus, the principal has to perform concordant informational control. Let the goal function of the principal be described by $F(\theta, x) = \gamma \theta x - x^2/2$, where $x \geq 0$ is an action of the agent, θ indicates an uncertain parameter (possessing values from the set $\Theta = \{1, 3, 7\}$ with the equal probabilities of $1/3$), and $\gamma > 0$ stands for a fixed parameter representing the level of similarity between the interests of both players (the principal and agent).

In practice, a possible interpretation of such situation lies in the following. The principal is aware of the market situation determined by the parameter θ. Being informed of the value θ, the principal reports it to the agent. The corresponding costs are shared equally by the principal and agent, while the profits are allocated in the proportion of $\gamma : 1$.

Let us analyze what conditions on the parameter γ make the partition $S = \{\{1\}, \{3\}, \{7\}\}$ concordant.

In the present case, we have $m = 3$, $\Theta_1 = \{1\}$, $\Theta_2 = \{3\}$, $\Theta_3 = \{7\}$, $X_1^* = \{1\}$, $X_2^* = \{3\}$, $X_3^* = \{7\}$. Moreover, the concordance condition (3) is rewritten as the following system of inequalities:

$$F(1,1) \geq F(1,3) \Leftrightarrow \gamma - \frac{1}{2} \geq 3\gamma - \frac{9}{2};$$

$$F(1,1) \geq F(1,7) \Leftrightarrow \gamma - \frac{1}{2} \geq 7\gamma - \frac{49}{2};$$

$$F(3,3) \geq F(3,1) \Leftrightarrow 9\gamma - \frac{9}{2} \geq 3\gamma - \frac{1}{2};$$

$$F(3,3) \geq F(3,7) \Leftrightarrow 9\gamma - \frac{9}{2} \geq 21\gamma - \frac{49}{2};$$

$$F(7,7) \geq F(7,1) \Leftrightarrow 49\gamma - \frac{49}{2} \geq 7\gamma - \frac{1}{2};$$

$$F(7,7) \geq F(7,3) \Leftrightarrow 49\gamma - \frac{49}{2} \geq 21\gamma - \frac{9}{2}.$$

The derived system is solved within the segment $\frac{5}{7} \leq \gamma \leq \frac{5}{3}$. Therefore, for any γ above, principal's reporting the true value of the uncertain parameter forms a concordant informational control. •

We will say that a set partition is *complete* if each part of the partition coincides with a specific element of the set. The example shows that for $\gamma = 1$ the goal functions of the principal and agent coincide and the complete partition $\{\{1\}, \{3\}, \{7\}\}$ is concordant. This fact takes place in the general case, as well.

Assertion 2.15.1. Suppose that the goal functions of the principal and agent coincide. Then the complete partition is concordant.

Proof.   Apparently, in the case of identical goal functions of the principal and agent, the concordance condition (3) always holds for the complete partition:

$$\forall \theta \in \Theta \ \ \forall x^* \in X_\theta^* \ \ \forall x \in X^* : \text{ either } x \in X_\theta^*,$$
$$\text{or } f(\theta, x^*) \geq f(\theta, x),$$

where $X_\theta^* = \underset{x \in X}{\text{Argmax}} \, f(\theta, x), \ X^* = \underset{\theta \in \Theta}{\bigcup} X_\theta^*. \ \bullet$

In some respect, the complete partition represents the limiting case. Another limiting case consists in the so-called *trivial partition* which includes a single part: $S = \{\Theta\}$. For the trivial partition, we have $X_1^* = X^*, X^* \backslash X_1^* = \emptyset$; consequently, the condition (3) is always true.

Assertion 2.15.2. The trivial partition is concordant.

Assertion 2.15.2 indicates of an obvious fact. Imagine that the principal reports the whole set Θ to the agent (thus, refusing of informational control). Then there is nothing else left for the agent to do but to "trust" the principal. In other words, the agent should choose his/her action based on an a priori known probabilistic distribution on the set Θ.

Optimality of concordant informational control. We have proven that concordant informational control always exists (at least, the trivial one – see Assertion 2.15.2). Suppose there are several concordant informational controls. A natural question arises concerning evaluation of *optimal concordant informational control*. Its rigorous formalization requires introducing a certain criterion of optimality. Let us assume that the principal maximizes the expected guaranteed result (i.e., the mathematical expectation of his/her goal function under the least favorable (worst-case) action of the agent). Denote by $\sigma(\theta, S)$ a part of the concordant partition S, which contains θ (the principal's message under true value of the uncertain parameter θ). Next, denote

$$X^*(\theta, S) = \text{Arg} \max_{x \in X} E_{\nu \in \sigma(\theta, S)} f(\nu, x).$$

Then the optimal concordant informational control S^* is the solution to the optimization problem

$$E_\theta F(\theta, x^*(\theta, S)) \xrightarrow[S]{} \max,$$

where

$$x^*(\theta, S) \in \underset{x \in X^*(\theta, S)}{\text{Arg} \min} \, F(\theta, x).$$

Generally, evaluating optimal concordant informational control is an intricate problem. A constructive approach could be suggested merely under specific conditions. A condition of that kind consists in a sufficiently small number of elements in the set Θ (going over all feasible partitions S seems then possible).

Example 2.15.1 (ending). For $\Theta = \{1, 3, 7\}$, there exist five different partitions. To evaluate optimal concordant informational control, one should verify the concordance condition and compute the expected value of the principal's goal function (for each partition). Previously, we have shown that the partition $S = \{\{1\}, \{3\}, \{7\}\}$ is concordant if $5/7 \leq \gamma \leq 5/3$. For informational control corresponding to this partition, find the mathematical expectation of the principal's goal function:

$$E_\theta F\left(\theta, x^*(\theta, S)\right) = \frac{1}{3}F(1, 1) + \frac{1}{3}F(3, 3) + \frac{1}{3}F(7, 7)$$

$$= \frac{1}{3}\left(\gamma - \frac{1}{2}\right) + \frac{1}{3}\left(9\gamma - \frac{9}{2}\right) + \frac{1}{3}\left(49\gamma - \frac{49}{2}\right) = \frac{59}{3}\gamma - \frac{59}{6}.$$

Similarly, one may study the remaining partitions (see Table 2.2); recall that we consider $\gamma > 0$. Table 2.2 shows that (due to the inequalities $9/14 < 5/7 < 13/2 < 5/3 < 3$ and $51 < 169/3 < 57 < 59$),

– for $0 < \gamma < 9/14$ and $\gamma > 5/3$, the partition $S^* = \{1, 3, 7\}$ is optimal;
– for $9/14 \leq \gamma < 5/7$, the partition $S^* = \{\{7\}, \{1, 3\}\}$ is optimal;
– for $5/7 \leq \gamma \leq 5/3$, the partition $S^* = \{\{1\}, \{3\}, \{7\}\}$ is optimal. •

Table 2.2 Concordant partitions and gains of the principal in Example 2.15.1.

Partition	The values of γ, ensuring concordant partition	The expected value of the principal's goal function (for concordant partitions)
$\{\{1\}, \{3\}, \{7\}\}$	$\dfrac{5}{7} \leq \gamma \leq \dfrac{5}{3}$	$59\left(\dfrac{\gamma}{3} - \dfrac{1}{6}\right)$
$\{\{1\}, \{3, 7\}\}$	$1 \leq \gamma \leq 3$	$51\left(\dfrac{\gamma}{3} - \dfrac{1}{6}\right)$
$\{\{3\}, \{1, 7\}\}$	$\varnothing$	$-$
$\{\{7\}, \{1, 3\}\}$	$\dfrac{9}{14} \leq \gamma \leq \dfrac{3}{2}$	$57\left(\dfrac{\gamma}{3} - \dfrac{1}{6}\right)$
$\{1, 3, 7\}$	$\gamma > 0$	$\dfrac{169}{3}\left(\dfrac{\gamma}{3} - \dfrac{1}{6}\right)$

Finally, we observe the following fact.

Assertion 2.15.3. Suppose that the goal functions of the principal and agent coincide. Then the complete partition is optimal (it corresponds to optimal concordant informational control).

Proof.   According to Assertion 2.15.1, the complete partition is concordant. Next, the expected value of the agent's gain attains maximum under his/her correct knowledge about the actual value of the uncertain parameter θ. On the other hand, the goal functions of the principal and agent are identical. Hence, the expected value of the principal's goal function takes the maximal value, as well. •

Concordant informational control (the case of several agents). In the previous subsections, we have analyzed the situation with two participants, i.e., a principal (a control subject) and an agent (a controlled subject). However, in many cases there exist several controlled subjects. In the sense of informational control, here a fundamental difference lies in the following. The principal's message could be addressed to all agents or to a certain agent (alternatively, to a certain subset of agents). Moreover, the principal's message may contain the value of an uncertain parameter or the awareness of other agents. Generally, one faces with a rather complex awareness structure.

A comprehensive treatment of informational control for several agents goes beyond the scope of this book. We confine ourselves to the case when identical information is simultaneously reported to all agents.

Thus, suppose there are n agents with the goal functions $f_i(\theta, x_1, \ldots, x_n), i=1, \ldots, n$. The probabilistic distribution of a random variable θ (i.e., an uncertain parameter possessing the values from a set Θ) represents a common knowledge. Each agent chooses his/her action x_i from the set X^i.

The Nash equilibrium concept prevails in solution of noncooperative games. A Nash equilibrium is an agents' action vector such that none of them benefits by unilaterally changing his/her action. In the absence of informational control, an equilibrium action vector of the agents $x^* = (x_1^*, \ldots, x_n^*)$ satisfies the system of equations

$$x_i^* \in \underset{x_i \in X^i}{\mathrm{Argmax}}\ E_\theta f_i(\theta, x_1^*, \ldots, x_{i-1}^*, x_i, x_{i+1}^*, \ldots, x_n^*), \quad i = 1, \ldots, n.$$

Assume that the principal reports the set $\Theta' \subset \Theta$ and the agents trust him/her. An equilibrium vector is any action vector meeting the system of equations

$$x_i^* \in \underset{x_i \in X^i}{\mathrm{Argmax}}\ E_{\theta \in \Theta'} f_i(\theta, x_1^*, \ldots, x_{i-1}^*, x_i, x_{i+1}^*, \ldots, x_n^*), \quad i = 1, \ldots, n.$$

Denote by $NE(\Theta')$ the set of all equilibrium vectors (NE – Nash equilibrium).

Suppose that the principal applies informational control based on the partition $S = \{\Theta_\alpha\}, \alpha \in A$. In the sequel, we believe that the sets of Nash equilibria $NE(\Theta_\alpha)$ are nonempty for any $\alpha \in A$.

By analogy to formula (2), let X_α^* be the set of equilibrium action vectors of the agents provided that the principal reports Θ_α: $X_\alpha^* = NE(\Theta_\alpha)$.

Again, similarly to the single-agent case, denote by X^* the set of all equilibrium action vectors (for a certain message of the principal from the partition S): $X^* = \bigcup\limits_{\alpha \in A} X_\alpha^*$.

As a result, for the case of several agents, the concordance condition could be rewritten as

(4) $\forall \alpha \in A \quad \forall \theta \in \Theta_\alpha \quad \forall x^* \in X_\alpha^* \quad \forall x \in X^*$: either $x \in X_\alpha^*$, or $F(\theta, x^*) \geq F(\theta, x)$.

This is done by analogy to the expression (3). Recall that for several agents we have $x = (x_1, \ldots, x_n))$.

Assertion 2.15.2 holds true for several agents and is proved in much the same way. Assertion 2.15.1 is valid if one replaces the coincidence between the goal functions of the principal and agent by the coincidence between the goal functions of the principal and each agent. Apparently, the coincidence condition for the goal functions of all agents is very strong. A possible direction of further research aims at its weakening.

A substantial difference of the case of several agents lies in the feasibility of the following situation. Concordant informational control decreases the average gain of the agents.

Example 2.15.2. Consider a principal and two agents. The uncertain parameter takes values from the set $\Theta = \{1, 2\}$ (the corresponding probabilities are 3/5 for $\theta = 1$ and 2/5 for $\theta = 2$). Gains of the agents are described by the bimatrices in Fig. 2.34; here the rows mean strategies of agent 1, $x_1 \in \{1, 2\}$, while the columns stand for strategies of agent 2, $x_2 \in \{1, 2\}$.

$$\theta = 1 \qquad\qquad \theta = 2$$
$$\begin{pmatrix} (4,4) & (0,0) \\ (0,10) & (6,6) \end{pmatrix} \qquad\qquad \begin{pmatrix} (4,4) & (0,0) \\ (15,0) & (1,1) \end{pmatrix}$$

Figure 2.34 The bimatrices of gains in Example 2.15.2.

The principal's gain $F(\theta, x_1, x_2)$ is defined by

$$F(\theta, x_1, x_2) = \begin{cases} 10, & \text{if } \theta = x_1 = x_2; \\ 0, & \text{otherwise.} \end{cases}$$

In the absence of informational control (equivalently, under trivial informational control), the agents strive for maximizing their expected gains – see Fig. 2.35. This leads to the equilibrium $x_1 = 2$, $x_2 = 1$ and

$$E_\theta f_1(\theta, 2, 1) = E_\theta f_2(\theta, 2, 1) = 6,$$

while the principal's gain equals zero.

$$\begin{pmatrix} (4,4) & (0,0) \\ (6,6) & (4,4) \end{pmatrix}$$

Figure 2.35 The bimatrix of expected gains in Example 2.15.2.

The result of interaction changes if the principal reports to the agents the value θ. Obviously, this control is concordant:

$$S = \{\{1\}, \{2\}\}, \quad X_1^* = \{(1,1)\}, \quad X_2^* = \{(2,2)\};$$
$$F(1, 1, 1) = 10 > 0 = F(1, 2, 2),$$
$$F(2, 2, 2) = 10 > 0 = F(2, 1, 1).$$

In the case of $\theta = 1$, we obtain the outcome $x = (1, 1)$ (the agents gain 4). On the other hand, for $\theta = 2$ we have the outcome $x = (2, 2)$ (the agents gain 1). In both cases, the principal gains 10.

Therefore, the concordant control $\{\{1\}, \{2\}\}$ is optimal; however, the agents' gains have been decreased. •

In this section we have studied the model of concordant informational control, where agents are informed about the fact of control implementation by a principal. Still, the agents rationally trust the messages of the principal. The existence conditions have been established for such control and several properties have been demonstrated. In socioeconomic systems, informational control is applied in combination with other types of control. Therefore, a promising field of further study lies in designing control mechanisms [113] that account for the concordance of the interests pursued by a principal and by an agent (agents) during implementation of informational control.

2.16   REFLEXION IN PLANNING MECHANISMS

Planning mechanisms. Strategy-proofness of *decision-making procedures* may fail due to the following. First, *manipulation* by agents (distortion of preferences reported by them [26, 118, 160]). Second, manipulation of the algorithm adopted to process agents' opinions (the so-called *agenda theory* [101]).

In *planning mechanisms* [113, 135] (i.e., mechanisms of principal's decision-making based on information reported by agents), agents' beliefs often form a common knowledge among them. A planning mechanism is called *strategy-proof*, if each agent benefits from truth-telling irrespective of his/her preferences under any vector of opponents' preferences.

In this section, we discuss the feasibility of weakening the strategy-proofness requirement. Suppose that the principal can perform informational control (i.e., form a certain system of beliefs about opponents' types, the beliefs about beliefs, etc., for agents).

Consider an organizational system with single principal and n agents. The strategy of each agent lies in reporting certain information $s_i \in S_i \subseteq \Re^1, i \in N = \{1, 2, \ldots, n\}$ to the principal. Based on the reported information, the principal assigns *the plans* $x_i = h_i(s) \in X_i \subseteq \Re^1$ to the agents; here $h: S \to X$ is a *planning procedure (mechanism)*, $h_i: S \to X_i, i \in N, s = (s_1, s_2, \ldots, s_n) \in S = \prod_{i \in N} S_i$ means the vector of agents' messages, and $x = (x_1, x_2, \ldots, x_n) \in X = \prod_{i \in N} X_i$ designates the plans vector.

The preference function (goal function) $f_i(x_i, r_i): X_i \times \Re^1 \to \Re^1$ of agent i reflects his/her preferences in planning problems. This function depends on the corresponding component of the plan (assigned by the principal), as well as on a certain parameter $r_i \in \Re^1$ known as agent's *type*.

In models of information revelation, one generally studies planning mechanisms under the following assumption. The preference functions of agents are *single-peaked* [26, 118, 135] ones with the peak points $\{r_i\}_{i \in N}$. This implies that the preference function $f_i(x_i, r_i)$ is a continuous function with several properties. It strictly increases in x_i, reaching a unique maximum point r_i, and strictly decreases then. According to the assumption, the agent's preferences on the set of feasible plans are such that there exists a unique value of the plan, being optimal for the agent (a peak point). For any other plan, the preference level monotonically decreases as one moves off the peak point. Therefore, we will understand the type of an agent as the maximum point of his/her preference function (also referred to as an *ideal point,* a *peak point,* or simply a *"top"*). This is the most beneficial plan to an agent.

The case of a common knowledge ("classic" strategy-proofness). When making decisions, each agent is aware of the following information: the planning procedure, the goal functions and feasible sets of all agents, as well as the types' vector $r = (r_1, r_2, \ldots, r_n) \in \mathfrak{R}^n$. The principal knows the functions $f_i(x_i, \cdot)$ and the sets $\{S_i\}_{i \in N}$ of messages available to the agents; yet, the principal possesses no information on the exact values of the agents' types.

The move sequence is as follows. The principal chooses a planning procedure $h(s) = (h_1(s), h_2(s), \ldots, h_n(s))$ and announces it to the agents. Being aware of the planning procedure, the latter report to the former the source information $\{s_i\}$ for the planning procedure.

A principal's decision (plans assigned to the agents) depends on the information reported by them. Hence, agents can seize the opportunity of exerting an impact on the decision (by reporting information ensuring the most beneficial plans to them). Naturally, in such conditions the information obtained by the principal may appear spurious. Hence, the problem of *strategic behavior* (*data manipulation*) arises immediately.

Suppose that the agents do not cooperate, playing dominant or Nash equilibrium strategies. Let s^* be the vector of equilibrium strategies:

(1) $\forall i \in N, \forall s_i \in S_i: f_i(h_i(s_{-i}^*, s_i^*), r_i) \geq f_i(h_i(s_{-i}^*, s_i), r_i).$

Obviously, the equilibrium point $s^* = s^*(r)$ generally depends on the agents' type profile: $s^* = s^*(r) = (s_1^*(r), s_2^*(r), \ldots, s_n^*(r)).$

If for any agents' preferences $r \in \mathfrak{R}^n$ their truth-telling forms a Nash equilibrium,

(2) $\forall i \in N \quad \forall r \in \mathfrak{R}^n \quad \forall s \in \mathfrak{R}^1: f_i(h_i(r), r_i) \geq f_i(h_i(r_{-i}, s), r_i),$

the corresponding mechanism is said to be *strategy-proof*. In the sequel, this property will be called *strategy-proofness*.

It would seem that the following requirement appears stronger than (2). Truth-telling by each agent forms his/her dominant strategy:

(3) $\forall i \in N \quad \forall r_i, s_i \in \mathfrak{R}^1 \quad \forall s_{-i} \in \mathfrak{R}^{n-1} \quad f_i(h_i(s_{-i}, r_i), r_i) \geq f_i(h_i(s_{-i}, s_i), r_i).$

However, definitions (2) and (3) turn out equivalent [54]. Both definitions of a strategy-proof planning mechanism engage the agents' type vector $r \in \mathfrak{R}^n$ as a "parameter." Hence, strategy-proofness admits a simple interpretation. A planning mechanism is strategy-proof if (whatever actual types of agents) truth-telling represents a dominant strategy for each of them.

Reflexive strategy-proofness. Reject the assumption that the agents' type vector is a common knowledge. Generalize the problem of strategy-proof planning mechanism [26, 118, 135] to this case. Suppose that the awareness of agent i is modeled by a tree

I_i with elements $r_{i\sigma} \in \Re^1$, $\sigma \in \Sigma$. Here r_i is the actual type of agent i (he/she knows it for sure), $i \in N$.

Consequently, agents play a reflexive game, where an informational equilibrium is determined by the following conditions (see Section 2.3):

1 the awareness structure I has a finite complexity;
2 $\forall \lambda, \mu \in \Sigma$: $I_{\lambda i} = I_{\mu i} \Rightarrow x_{\lambda i}^* = x_{\mu i}^*$;
3 $\forall i \in N, \forall \sigma \in \Sigma$:

(4) $s_{\sigma i}^* \in \text{Arg} \max_{s_i \in \Re} f_i(h_i(s_{\sigma i1}^*, \ldots, s_{\sigma i,i-1}^*, s_i, s_{\sigma i,i+1}^*, \ldots, s_{\sigma in}^*), r_{\sigma i})$.

According to (4), the equilibrium message of agent i (a real agent) depends on his/her awareness structure I_i, i.e.,

(5) $s_i^* = s_i^*(I_i)$, $i \in N$.

Under a common knowledge, formula (4) coincides with the definition of a Nash equilibrium.

Denote $s^*(I) = (s_1^*(I_1), s_2^*(I_2), \ldots, s_n^*(I_n))$. The decision x made by the mechanism $h(\cdot)$ depends on the whole awareness structure I:

(6) $x = h(s^*(I))$.

Now, we discuss the concept of strategy-proofness in the absence of a common knowledge.

According to (2) and (3), in the conditions of a common knowledge, a planning mechanism is called strategy-proof if for any actual types of agents truth-telling represents a dominant strategy of each agent.

We endeavor to extend this definition to the case of an informational equilibrium. Require that (regardless of the awareness structure of a real agent) truth-telling is a component of the informational equilibrium (4):

(7) $\forall r \in \Re^n, \forall i \in N, \forall I_i$:

$r_i \in \text{Arg} \max_{s_i \in \Re} f_i(h_i(s_{i1}^*(I_i), \ldots, s_{i,i-1}^*(I_i), s_i, s_{i,i+1}^*(I_i), \ldots, s_{in}^*(I_i)), r_i)$.

Clearly, the condition (3) implies (7). Conversely, the set of all possible awareness structures includes the structure corresponding to a common knowledge. Hence, under (7) a planning mechanism appears strategy-proof (the condition (3) must be met). Thus, the definitions (3) and (7) are absolutely equivalent.

And so, we have arrived at an important conclusion. Suppose that, in a given planning mechanism, each agent has a dominant strategy under any his/her type (in fact, this is an extremely strong requirement – the class of appropriate mechanisms is very small). Then the decisions made by the agent do not depend on the awareness structure.

The strategy-proofness requirement can be weakened. For this, introduce the notion of reflexive strategy-proofness, *viz.*, the existence of awareness substructures

$r_{i\sigma} \in \Re^1, i \in N, \sigma \in \Sigma_+$, such that for any types of real agents their truth-telling is an informational equilibrium.

Formally, the mechanism $h(\cdot)$ is said to be *reflexively strategy-proof*, if there exist *awareness substructures* $r_{i\sigma} \in \Re^1, \sigma \in \Sigma_+$, of real agents ($i \in N$) such that (irrespective of the type of a real agent) truth-telling is a component of an informational equilibrium:

(9) $\forall i \in N \quad \forall r_i \in \Re^1$:

$$r_i \in \underset{s_i \in \Re}{\mathrm{Arg\,max}}\, f_i(h_i(s_{i1}^*, \ldots, s_{i,i-1}^*, s_i, s_{i,i+1}^*, \ldots, s_{in}^*), r_i),$$

(10) $\forall \sigma \in \Sigma \quad \forall j \in N$:

$$s_{i\sigma j}^* \in \underset{s_j \in \Re^1}{\mathrm{Arg\,max}}\, f_j(h_j(s_{i\sigma j1}^*, \ldots, s_{i\sigma j,j-1}^*, s_j, s_{i\sigma j,j+1}^*, \ldots, s_{i\sigma jn}^*), r_{i\sigma j}).$$

Clearly, the set of reflexively strategy-proof mechanisms appears not narrower than the set of strategy-proof mechanisms (indeed, any strategy-proof mechanism is reflexively strategy-proof). Therefore, their characterization represents a relevant problem (including search for appropriate awareness substructures).

Denote by E_I the set of all possible equilibrium combinations of actions chosen by real agents (under all feasible awareness structures), $E_{-i} = Proj_{-i}E_I$, $i \in N$. The definition (9)–(10) can be reformulated as follows. A planning mechanism is reflexively strategy-proof, if for agent i there exists an opponents' type vector $\tilde{r}_{-i} \in E_{-i}$ such that

(11) $\forall r_i, \tilde{r}_i \in \Re^1: f_i(h_i(\tilde{r}_{-i}, r_i), r_i) \geq f_i(h_i(\tilde{r}_{-i}, \tilde{r}_i), r_i), \quad i \in N.$

Let the set of Nash equilibria be designated by

(12) $E_N = \{r \in \Re^n \mid \forall i \in N \quad \forall \tilde{r}_i \in \Re^1: f_i(h_i(r), r_i) \geq f_i(h_i(r_{-i}, \tilde{r}_i), r_i)\},$

(13) $X_i^0 = Proj_i E_N, \quad i \in N.$

In this section, we suppose that the awareness structure of agents is finite and regular.

The following result takes place. As a matter of fact, it states the sufficient condition of reflexive strategy-proofness (the corresponding necessary condition is (11)).

Assertion 2.16.1. A planning mechanism enjoys reflexive strategy-proofness if for any agent i ($i \in N$) there exists a combination of types $\tilde{r} = (\tilde{r}_1, \ldots, \tilde{r}_n) \in E_N$ meeting the condition (11). To construct reflexively strategy-proof mechanisms, one can consider only agents with (at most) reflexion rank 2.

Proof. For agent i, just form an awareness structure I_i such that

(14) $r_{ij\sigma k} = \tilde{r}_k, \quad j \neq i, \quad \text{for any } \sigma \in \Sigma.$

Note that the awareness structures of different agents can be constructed independently. And so, according to agent i, the rest agents believe that a common knowledge does exist (see Fig. 2.36).

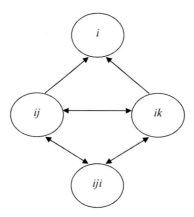

Figure 2.36 The reflexive game graph (from the viewpoint of agent *i*).

Obviously, in the view of agent *I*, truth-telling is an equilibrium in the game of his/her phantom agents. Hence, the condition (10) is valid. By virtue of (11), this yields (9). •

In practice, the beliefs of agent *i* about the opponents' types and their beliefs must be the following. First, the agent must be sure that the opponents' types agree with (11) (it guarantees their truth-telling). Second, the beliefs of opponents (according to this agent) must be such that their truth-telling forms an equilibrium in their game (see (10)). It suffices that a subjective common knowledge takes place at the lower level of the awareness structure.

Finally, we emphasize the constructive character of the above proof. Actually, it suggests the algorithm of forming the awareness structure (14), which guarantees reflexive strategy-proofness.

Now, we give an example of a planning mechanism being manipulatable, but reflexively strategy-proof.

Example 2.16.1. Consider two agents with nonnegative types r_1 and r_2. (Recall that the type of an agent makes up his/her optimal plan.) Analyze the mechanism, which analyzes plans x_1 and x_2 depending on agents' messages s_1 and s_2 by the following rule:

$$(15) \quad x_1 = s_1 - s_2/2, \quad x_2 = s_2 - s_1/2, \quad s_1, s_2 \geq 0.$$

Solving the system of equations

$$\begin{cases} s_1 - \dfrac{s_2}{2} = r_1, \\ s_2 - \dfrac{s_1}{2} = r_2, \end{cases}$$

we obtain the Nash equilibrium strategies (messages) of the agents:

$$s_1^*(r_1, r_2) = \frac{2}{3}(2r_1 + r_2), \quad s_2^*(r_1, r_2) = \frac{2}{3}(2r_2 + r_1).$$

Clearly, the mechanism (15) is manipulable. Under any types of agents (with at least one type being nonzero), at least one agent reports a quantity not coinciding with his/her type. Meanwhile, if $r_1, r_2 \geq 0$, the mechanism becomes reflexively strategy-proof. Indeed, it suffices to convince each agent that (according to his/her opponent) the common knowledge is that both types equal zero. In other words, one should just form the awareness structure $r_{i\sigma} = 0$, $i = 1, 2, \sigma \in \Sigma_+$. Consequently, the unique informational equilibrium consists in the combination

$$s_i^* = r_i, \quad s_{i\sigma}^* = 0, \quad i = 1, 2, \quad \sigma \in \Sigma_+.$$

Therefore, in the equilibrium both agents report their actual types.

Nevertheless, in the case of $r_1, r_2 > 0$, the mechanism (15) looses reflexive strategy-proofness. Regardless of the awareness structure (particularly, its depth), each agent believes that the opponent's message differs from zero. •

Chapter 3

Strategic reflexion and control

In this chapter we study models of strategic reflexion. Section 3.1 deals with the model of strategic reflexion in a two-player game. This allows to solve the *problem of maximal rational rank of strategic reflexion* in bimatrix games (see Section 3.2), as well as to state and analyze the so-called *ranks games*. Next, Section 3.3 is dedicated to the finiteness of reflexion rank, which follows from limited data processing capabilities of humans. Finally, Section 3.4 describes the *reflexive partitions method* for *reflexive control* problems.

3.1 STRATEGIC REFLEXION IN TWO-PLAYER GAMES

Consider reflexive models of decision-making in two-player games (in the ascending order of corresponding awareness).

Reflexion rank 0. Let us analyze the problem of agent's decision-making without any information on the state of nature (the assumption that the goal functions and feasible sets form a common knowledge holds true). On the one hand, it seems reasonable to involve the principle of the maximal guaranteed result. According to it, agent i chooses the guaranteeing strategy (with respect to the state of nature and opponent's action)

$$(1) \quad {}_1x_i^g = \arg \max_{x_i \in X_i} \min_{\theta \in \Theta} \min_{x_{-i} \in X_{-i}} f_i(\theta, x_i, x_{-i}).$$

On the other hand, hypothetically the decision-making principle (1) is not unique. For instance, the agent may expect that his/her opponent chooses not the worst-case action, but the guaranteeing strategy (each agent easily evaluates the opponent's guaranteeing strategy). In this case, the best response makes up

$$(2) \quad {}_2x_i^g = \arg \max_{x_i \in X_i} \min_{\theta \in \Theta} f_i(\theta, x_i, {}_1x_{-i}^g).$$

However, the same line of reasoning can be followed by the opponent. If the agent admits such situation, his/her guaranteeing strategy acquires the form

$$(3) \quad {}_3x_i^g = \arg \max_{x_i \in X_i} \min_{\theta \in \Theta} f_i(\theta, x_i, {}_2x_{-i}^g).$$

Here $_2x^g_{-i}$ results from (2) by substituting index i by index $-i$ (and vice versa). The chain of reflexion rank (the agent's assumptions regarding the opponent's reflexion rank) can be further extended. For this, define recursively the strategies

$$(4) \quad _kx^g_i = \arg \max_{x_i \in X_i} \min_{\theta \in \Theta} f_i(\theta, x_i, {}_{k-1}x^g_{-i}), \quad k = 2, 3, \ldots,$$

where $_1x^g_i$ $(i = 1, 2)$ are evaluated by (1). The set of actions (4) is called the set of *reflexive guaranteeing strategies*.

We provide an illustrative example.

Example 3.1.1. Let the goal functions of the agents be

$$f_1(x_1, x_2) = x_1 - x_1^2/2x_2, \quad f_2(x_1, x_2) = x_2 - x_2^2/2(x_1 + \delta),$$

with $\delta > 0$. Suppose that their feasible sets are $X_1 = X_2 = [\varepsilon; 1]$, $0 < \varepsilon < 1$ $(\varepsilon, \delta << 1)$. The guaranteeing strategies of agents can be found in Table 3.1.

Table 3.1 The guaranteeing strategies of agents in Example 3.1.1.

k	1	2	3	4	5	6	7	
$_kx^g_1$	ε	$\varepsilon + \delta$	$\varepsilon + \delta$	$\varepsilon + 2\delta$	$\varepsilon + 2\delta$	$\varepsilon + 3\delta$	$\varepsilon + 3\delta$	$\ldots$
$_kx^g_2$	$\varepsilon + \delta$	$\varepsilon + \delta$	$\varepsilon + 2\delta$	$\varepsilon + 2\delta$	$\varepsilon + 3\delta$	$\varepsilon + 3\delta$	$\varepsilon + 4\delta$	$\ldots$

First, the values of the guaranteeing strategies appear increasing functions of "reflexion rank." Second, different "reflexion ranks" generally correspond to different guaranteeing strategies. Interestingly, in this example a Nash equilibrium[1] is the vector $(1; 1))$. •

What action should an agent choose? This question remains opened In this context, claim the following. Being informed merely about the set of possible values for the state of nature, agent i can choose among the actions $_kx^g_i$ $(i = 1, 2; k = 1, 2, \ldots)$ determined by (1)–(4).

In the present model, redefine the rational choice of the agent as follows. Imagine the agent is unaware of the opponent's goal function (the assumption that the goal functions and feasible sets form a common knowledge rules such possibility out!). Then his/her unique rational action consists in the choice (1), *viz.*, the classical principle of the MGR. According to the above assumptions, the agent knows the opponent's goal

---

[1]Making a digression, we draw reader's attention to the following. In the current example, suppose that the goal function of agent 2 takes the form $f_2(x_1, x_2) = x_2 + x_2^2/2x_1$. Consequently, there exists a dominant strategy for this agent (in fact, equaling to 1); moreover, the sequence of guaranteeing strategies of agent 1 gets stabilized at the second element: $_1x^g_i = \varepsilon$, $_2x^g_i = 1/2$. If agent 1 can evaluate the dominant strategy of the opponent, then his/her rational choice is the action $_2x^g_i$.

Table 3.2 The gains of agent 1 in Example 3.1.1.

j	1	2	3	4	5	6	7
$2f_1(BR_1(_jx_2^g),_jx_2^g)$	$\varepsilon + \delta$	$\varepsilon + \delta$	$\varepsilon + 2\delta$	$\varepsilon + 2\delta$	$\varepsilon + 3\delta$	$\varepsilon + 3\delta$	$\varepsilon + 4\delta$
$_jx_2^g$	$\varepsilon + \delta$	$\varepsilon + \delta$	$\varepsilon + 2\delta$	$\varepsilon + 2\delta$	$\varepsilon + 3\delta$	$\varepsilon + 3\delta$	$\varepsilon + 4\delta$

function, knows that the opponent is aware of this fact, and so on. Hence, the agent considers the principle of the MGR as irrational. At the very least, he/she should expect that the opponent will adopt the MGR (leading to the choice of $_2x_i^g$). Meanwhile, the goal functions represent a common knowledge; thus, the agent possibly believes that his/her reasoning can be retraced by the opponent (this makes the choice of $_3x_i^g$ rational). And so on (generally, this process turns out infinite). According to this agent, the uncertainty regarding opponent's "reflexion rank" still takes place[2]. Indeed, he/she has no information on the above parameter (otherwise, a certain subjective equilibrium would be implemented based on some beliefs about the uncertain parameter). In this case, it seems rational to apply the maximal guaranteed result with respect to opponent's "reflexion rank":

$$(5) \quad x_i' = \arg \max_{x_i \in X_i} \min_{j=1,2,\dots} \min_{\theta \in \Theta} f_i(\theta, x_i, {}_jx_{-i}^g).$$

Let us point at a couple of features. First, x_i' possibly differs from the classical guaranteeing strategy $_1x_i'$ described by (1). Second, the strategy (5) being adopted, the fact of existing dominant strategy of the opponent will be considered by the agent (see the footnote in Example 3.1.1).

Table 3.2 provides the values of the goal function of agent 1 in Example 3.1.1 (depending on opponent's "reflexion rank"). In addition, this table includes the corresponding actions of the opponent. Obviously, agent i gains $(\varepsilon + \delta)$ by the strategy (5). This result exceeds ε (the gain yielded by the classical MGR).

Therefore, in this model the rational choice of the agent lies in the strategy (4) or (5).

Reflexion rank 1. Now, assume that an agent possesses some information on the state of nature. He/she considers this information as true and knows nothing more for sure.

Within the framework of the existing uncertainty, the principle of determinism suggests two options to the agent performing strategic reflexion. Option 1 is supposing

---

[2]In other words, the original game can be replaced by a game, where agents choose their reflexion ranks. It appears possible to construct reflexive analogs for the new game, and so on (again, this process turns out infinite!). For instance, study several related examples, namely, "Penalty kick" (the Introduction), "Hide-and-seek" and "*Misère* in Preferans" (Section 3.2). To avoid infinite constructions, an agent may use the maximal guaranteed result with respect to opponent's reflexion rank. An alternative approach (being efficient for matrix games) concerns defining the maximal rational rank of agent's reflexion – see Section 3.2.

that his/her opponent has no information. And option 2 is believing that the opponent enjoys the same awareness as he/she does[3].

Yet, an agent may refuse making any assumptions regarding the opponent's awareness and decision principles. Accordingly, he/she has to address the principle of the maximal guaranteed result (MGR). Really, this agent possesses the same information[4] on the opponent (as in the case of reflexion rank 0). And so, the agent expects the worst choice of the opponent from the set of strategies (5). The guaranteeing strategy becomes

$$(6) \quad x_i^g(\theta_i) = \arg \max_{y_i \in A_i} \min_{j=1,2,\ldots} f_i(\theta_i, x_i, {}_jy_{-i}^g).$$

Note that the agent belongs to the informational situation which corresponds to the model under consideration. Meanwhile, evaluating (6), the agent treats his/her opponent as the one belonging to the informational situation of the previous model. This common principle (under certain information available, an agent treats his/her opponent as the one having the same or smaller rank of reflexion) will be involved in other reflexive models of decision-making.

Let agent 1 believe that his/her opponent enjoys the same information as he/she actually does (similar reasoning can be followed by agent 2 – see Assumption A_1 in [136]). Then agent 1 calculates the *subjective equilibrium* (a "Nash equilibrium" for the corresponding subjective description of the game) $E_N(\theta_1) = \{(x_{11}^*(\theta_1), x_{12}^*(\theta_1))\}$ as follows:

$$(7) \quad \forall x_1 \in X_1: f_1(\theta_1, x_{11}^*(\theta_1), x_{12}^*(\theta_1)) \geq f_1(\theta_1, x_1, x_{12}^*(\theta_1)),$$
$$\forall x_2 \in X_2: f_2(\theta_1, x_{11}^*(\theta_1), x_{12}^*(\theta_1)) \geq f_1(\theta_1, x_{11}^*(\theta_1), x_2).$$

In practical interpretation, the presented system of inequalities reflects the evaluation of the corresponding Nash equilibrium by agent 1 and the choice of an appropriate coordinate of this equilibrium. Generally, the agent and his opponent compute different equilibria (their coincidence is possible if $x_{ij}^*(\theta_i) = x_{jj}^*(\theta_j), i, j = 1, 2$).

Thus, in the model with reflexion rank 1, the rational choice of the agent consists in either the reflexive guaranteeing strategy (6), or the subjective equilibrium (7).

The subjective equilibrium (7) defined by agent 1 allows representation in the form of a graph with two nodes, x_1 and x_{12}. They correspond to agent 1 and his/her beliefs[5] about agent 2 (see Fig. 3.1. The subjective equilibrium in the model with rank 1 of strategic reflexion). Incoming arrows reflect the information adopted by an agent about the opponent.

Reflexion rank 2. In this model, agent i has some information on the opponent's beliefs θ_{ij} about the state of nature and his/her own beliefs θ_{ii} about the state of nature. By the axiom of self-awareness, suppose that $\theta_i = \theta_{ii}$.

---

[3]The stated principle (and its generalizations) will be repeatedly employed below to introduce finite awareness structures. Indeed, possessing the awareness structure I_i, agent i may assign to other agents only awareness structures being coordinated with I_i.
[4]No doubt, an agent can expect that the opponent has certain awareness. But this awareness goes beyond the present model, and we do not consider such assumptions.
[5]Recall that agents existing in the minds of other agents are called phantom agents.

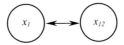

Figure 3.1 The subjective equilibrium in the model with rank 1 of strategic reflexion.

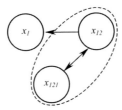

Figure 3.2 The subjective equilibrium in the model with rank 2 of strategic reflexion.

The agent may expect that his/her opponent chooses a guaranteeing strategy (under the awareness of θ_{ij}). Consequently, the best response makes up

$$(8) \quad {}_2x_i^g = \arg \max_{x_i \in X_i} f_i(\theta_i, x_i, x_{-i}^g(\theta_{ij})),$$

where $x_{-i}^g(\theta_{i,-i})$ results from (6).

In addition to the guaranteeing strategy (8), agent 1 can evaluate the *subjective equilibrium*

$$E_N(\theta_1, \theta_{12}) = \{(x_{11}^*(\theta_1, \theta_{12}), x_{12}^*(\theta_1, \theta_{12}))\}$$

of the following form:

$$(9) \quad \forall x_1 \in X_1: f_1(\theta_1, x_{11}^*(\theta_1, \theta_{12}), x_{12}^*(\theta_1, \theta_{12})) \geq f_1(\theta_1, x_1, x_{12}^*(\theta_1, \theta_{12})),$$
$$\forall x_2 \in X_2: f_2(\theta_{12}, x_{121}^*(\theta_1, \theta_{12}), x_{12}^*(\theta_1, \theta_{12})) \geq f_2(\theta_{12}, x_{121}^*(\theta_1, \theta_{12}), x_2),$$
$$\forall x_1 \in X_1: f_1(\theta_{12}, x_{121}^*(\theta_1, \theta_{12}), x_{12}^*(\theta_1, \theta_{12})) \geq f_2(\theta_{12}, x_1, x_{12}^*(\theta_1, \theta_{12})).$$

By analogy to the previous model, generally agent 1 and his/her opponent calculate nonidentical equilibria.

Therefore, in the model with reflexion rank 2, the rational choice of the agent lies in either the reflexive guaranteeing strategy (8), or the subjective equilibrium (9).

Interestingly, the first two systems of inequalities in (9) characterize a Nash equilibrium according to agent 1; on the other part, the second and third systems of inequalities describe a Nash equilibrium to-be-evaluated by agent 2 according to agent 1. The graph in Fig. 3.2. The subjective equilibrium in the model with rank 2 of strategic reflexion presents the "model" of agent 2 adopted by agent 1 in his/her decision-making (see the dashed line).

The analysis of elementary models of strategic reflexion with different ranks testifies to the following. In the case of several agents and their incomplete awareness, the decision processes of agents can be considered independently. Each agent models the behavior of his/her opponents, i.e., strives for constructing his/her own closed model of the game. The differences between the subjective and objective descriptions of a game are discussed in [67]. As a matter of fact, in the case of a common knowledge subjective models do coincide.

Till now, we have studied strategic reflexion of rank 0, 1, and 2. Similarly, one can further increase reflexion rank. All models involve an essential assumption regarding opponent's reflexion rank. Actually, *agent's reflexion rank depends on the reflexion rank assigned by him/her to the opponent* .

For an agent, a priori one would hardly suggest any reasonable recommendations limiting the growth of his/her reflexion rank. Based on this viewpoint, we claim that there exists no universal concept of an equilibrium in games with strategic reflexion. The only remedy here is applying the MGR with respect to reflexion ranks or using a subjective equilibrium (each agent introduces certain assumptions regarding opponent's reflexion rank and chooses an optimal action within the framework of such assumptions).

Thus and so, we concentrate mostly on the cases when reflexion rank demonstrates a bounded growth. There are two reasons explaining finiteness of reflexion rank. First, it may happen that increasing reflexion rank above a certain threshold becomes irrational (further growth of reflexion rank provides no extra gain). Second, data processing capabilities of human beings are naturally limited (infinite reflexion rank represents merely a mathematical abstraction). Therefore, forthcoming sections of this chapter consider models taking the above-mentioned aspects into account. Notably, Section 3.2 deals with bimatrix games, defining the maximal rational rank of strategic reflexion. And the role of informational constraints is analyzed in Section 3.3.

3.2   REFLEXION IN BIMATRIX GAMES AND GAMES OF RANKS

The major idea developed in this section concerns the following. In bimatrix games[6], where there exists no Nash equilibrium (or, under an existing Nash equilibrium, agents choose subjective guaranteeing strategies – see Section 3.1), the gain of each agent depends on his/her reflexion rank and on opponent's reflexion rank. Furthermore, it is demonstrated that infinite growth of the rank of strategic reflexion does not increase their gains [71, 138]. Now, let us pass to formal description.

Consider a bimatrix game[7], where the gains of players 1 and 2 are defined by the matrices $A = ||a_{ij}||$ and $B = ||b_{ij}||$, respectively (both have dimensions of $n \times m$). Denote by[8] $I = \{1, 2, \ldots, n\}$ the set of actions of agent 1 (he/she chooses a row) and by $J = \{1, 2, \ldots, m\}$ the set of actions of agent 2 (he/she chooses a column).

In the above game, agents use the guaranteeing strategies

$$i_0 \in \text{Arg} \max_{i \in I} \min_{j \in J} a_{ij}, \quad j_0 \in \text{Arg} \max_{j \in J} \min_{i \in I} b_{ij}.$$

[6]A bimatrix game is a finite game of two players.
[7]Matrix games (as antagonistic finite games of two players) represent a special case of bimatrix games. Thus, all results derived in this section apply equally to matrix games.
[8]Hopefully, the same (historically established) notation for an awareness structure and the set of actions of agent 1 would cuase no confusion.

Introduce several assumptions. Suppose that the payoff matrices are such that each action of an agent represents the best response to a certain action of the opponent. In addition, we believe that the best response to each action of the opponent appears unique (otherwise, adopt a certain rule redefining the choice of an agent).[9] Hence, determining best responses, we replace the expressions "$i_{...} \in \text{Arg} \max_{i \in I} ...$" and "$j_{...} \in \text{Arg} \max_{j \in J} ...$" with the ones "$i_{...} = \arg \max_{i \in I} ...$" and "$j_{...} = \arg \max_{j \in J} ...$," respectively.

Employ the symbols $a_0 = \max_{i \in I} \min_{j \in J} a_{ij}$ and $b_0 = \max_{j \in J} \min_{i \in I} b_{ij}$ for the maximal guaranteed results (MGR) of agents 1 and 2, respectively.

Define the reflexive bimatrix game MG_{kl} as a bimatrix game with matrices A and B, where agents 1 and 2 have reflexion ranks k and l, respectively ($k, l \in \aleph$, and $\aleph$ stands for the set of natural numbers).

Now, we elucidate the essence of *reflexion rank* (more specifically, the rank of strategic reflexion) in *bimatrix games*. In bimatrix games (and in some other games), an agent may choose actions based on known reflexion ranks of the opponents. For instance, reflexion ranks are defined as follows (see the general model in Section 3.4). An agent possesses reflexion rank 0, if he/she knows only the payoff matrix. An agent has reflexion rank 1, if he/she considers the opponents as the ones possessing reflexion rank 0 (they know only the payoff matrix). Generally, an agent with reflexion rank k believes his/her opponents possess reflexion rank $k - 1$. Such agent follows their line of reasoning (evaluates their actions); accordingly, this agent chooses his/her own strategy based on known payoff matrix (by extrapolating the actions of the opponents). Below we give an illustrative example.

Example 3.2.1 (*Hide-and-seek*) [138]. Two agents play the hide-and-seek game in several rooms with different illumination levels. Agent 1 hides himself/herself in a room, while agent 2 has to choose the room to seek for the former. Illumination levels of all rooms are a common knowledge (a somewhat sophisticated game is described in [82]).

The agents adhere to the following strategies. Under the rest equal conditions, the seeker prefers to look in the rooms having a higher illumination level (finding the hider is easier). The hider understands that the seeker has little chance of success in a room with a lower illumination level. The growing rank of reflexion means that an agent understands that this fact is clear to his/her opponent and so on. Combine the reflexion ranks of agents and corresponding actions in Table 3.3.

Evidently, after reflexion rank 2 the whole set of feasible actions gets exhausted. Moreover, after reflexion rank 3 strategies of the agents are repeated. This fact shows that, in a twoplayer game, increasing reflexion ranks over a certain threshold provides no objectively new opportunities. However, the subjective growth of complexity is still observed.

Reflexion ranks and activity success mismatch in the following sense. Suppose that the hider has reflexion rank 0 (concealing himself/herself in the darkest room).

---

[9]Rejecting the stated assumptions will preserve the results obtained in this section. Indeed, they serve for getting an upper estimate for the maximal rational rank of strategic reflexion.

Table 3.3 The reflexion ranks of agents and corresponding actions.

Reflexion rank	0	1	2	3	4
The room chosen by the hider	The darkest one	Any room except the lightest one	Any room except the darkest one	The lightest one	The darkest one
The room chosen by the seeker	The lightest one	The darkest one	Any room except the lightest one	Any room except the darkest one	The lightest one

If the seeker possesses reflexion rank 1, he/she always wins the game (by seeking in the darkest room). Meanwhile, if the seeker has reflexion rank 3 (and looks for the hider in any room except the darkest one), he/she looses the hider with reflexion rank 0. Indeed, the latter (not troubling about opponent's beliefs) seeks in the darkest room; following the line of reflexive reasoning, the seeker never looks there.

Therefore, we would not claim that a higher reflexion rank is always better than a lower one. A preference for a certain reflexion rank depends on its interaction with the reflexion rank of the opponent. •

The framework of bimatrix games presumes that each agent possesses some belief about opponent's reflexion rank. This allows involving the notion of a subjective guaranteeing strategy. Define *subjective guaranteeing strategies* in the bimatrix game MG_{kl} as follows:

(1) $i_k = \arg \max\limits_{i \in I} a_{ij_{k-1}}, \quad j_l = \arg \max\limits_{j \in J} b_{i_{l-1}j}, \qquad k, l \in \aleph.$

Thus, the game MG_{00} coincides with the original one. The "equilibrium" in the game MG_{kl} is the vector $(a_{i_k j_l}; b_{i_k j_l})$, $k, l \in \aleph$. We observe a couple of interesting facts. First, under $k \geq 1$ and $l \geq 1$, the gain of any agent in the game MG_{kl} may be smaller than the maximal guaranteed one (see the example "*Misère* in Preferans"). Second, each agent assigns to the opponent the reflexion rank which is smaller by unity than his/her own rank. However, this seems somewhat contradictory, as in the game MG_{kl} (under $k \geq 1, l \geq 1$) we obtain simultaneously

$l = k - 1 \quad$ and $\quad k = l - 1,$

(which is impossible!). Hence, the equilibrium in the reflexive game is essentially subjective. The agents do not a priori know the game being played (the reflexion ranks of both agents are not a common knowledge – this would contradict the definition of reflexion rank). And so, studying informational reflexion with respect to reflexion ranks of agents in bimatrix games has obvious prospects.

The self-contradictoriness of strategic reflexion in bimatrix games can be illustrated by the following diagram. Fig. 3.3 shows the subjective description of the game MG_{kl} in terms of the graph of a reflexive game (according to agent 1). And Fig. 3.4 presents the same description, but according to agent 2.

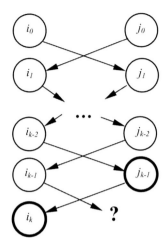

Figure 3.3 The subjective description of the game MG_{kl} (according to agent 1).

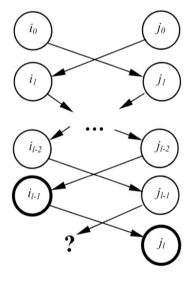

Figure 3.4 The subjective description of the game MG_{kl} (according to agent 2).

In Section 2.4 we have emphasized that the graph of a reflexive game enjoys a relevant property. Notably, for each node, the number of incoming arcs is by unity smaller than the number of agents. In the case of bimatrix games, we obtain one incoming arc for each node. Subjectively equilibrium actions are in bold type. They lead to the "equilibrium" (i_k, j_l). The corresponding subjective descriptions of the game do not engage the actions i_{k-1} (for agent 1) and the actions j_{l-1} (for agent 2). See the schemes in Fig. 3.3 and Fig. 3.4. This means that each subjective description is not interconsistent.

We have briefly discussed the self-contradictoriness of strategic reflexion in bimatrix games. Now, revert to the relationship between a subjective equilibrium, agents' gains and the ranks of their reflexion.

Denote $I_K = \bigcup\limits_{k=0,1,\ldots,K} i_k$, $J_L = \bigcup\limits_{l=0,1,\ldots,L} j_l$, $K = 0, 1, 2, \ldots, L = 0, 1, 2, \ldots$. Here I_∞ and J_∞ are the unions over all reflexion ranks (from 0 to infinity).

Assume that an agent (or both agents) does not know the reflexion rank of the opponent. Then it appears reasonable to consider the game $MG_{\infty\infty}$, where each agent evaluates the guaranteed result with respect to the reflexion rank of the opponent. Introduce guaranteeing strategies that correspond to the complete uncertainty about opponent's reflexion rank:

$$(2) \quad i_\infty = \arg\max_{i\in I} \min_{j\in J_\infty} a_{ij}, \quad j_\infty = \arg\max_{j\in J} \min_{i\in I_\infty} b_{ij}.$$

Similarly, define guaranteeing strategies provided that opponent's reflexion rank does not exceed a known level (agent 1 (agent 2) believes that opponent's reflexion rank is not larger than L (K, respectively):

$$(3) \quad i^L = \arg\max_{i\in I} \min_{l\in J_L} a_{ij_l}, \quad j^K = \arg\max_{j\in J} \min_{k\in I_K} b_{i_k j}.$$

In contrast to (1), in formula (3) the strategy of each agent turns out independent of his/her reflexion rank. It is fully defined by the information on opponent's reflexion rank.

The expressions (1)–(3) do not exhaust the whole variety of possible situations. For instance, agent 1 may expect agent 2 to choose j_∞; consequently, his/her best response equals $\arg\max\limits_{i\in I} a_{ij_\infty}$, and so on. Only "strong" agents can increase reflexion ranks; nevertheless, the following seems intuitively clear. As reflexion rank grows (which extends the line of reasoning "I think that he/she thinks that I think . . ."), one faces the danger of "overcomplicating the things." A strong agent with a high reflexion rank overestimates the opponent (believing that the latter possesses a high reflexion rank, as well). Meanwhile, if opponent's reflexion rank is actually small, a strong agent looses to a pushover (see the examples "Hide-and-seek" and "*Misère* in Preferans"). Hence, one should systematically analyze the connection between agents' gains and the type of the game played. We do this below.

The following aspects are essential here: the presence/absence of a Nash equilibrium and the choice of guaranteeing strategies/actions being Nash equilibria by agents (including the usage of such actions in subjective equilibria construction). We have four possible variants as follows.

Variant 1 (there exists a pure strategies' Nash equilibrium, and agents adopt Nash equilibrium actions).

Denote by $(i^*; j^*)$ the numbers of pure strategies representing Nash equilibria. By analogy to (1), suppose that in the reflexive game each agent chooses the best response to opponent's choice of a corresponding component of the equilibrium. In this case, we obtain

$$(4) \quad i_k = \arg\max_{i\in I} a_{ij^*}, \quad j_l = \arg\max_{j\in J} b_{i^*j}, \quad k, l \in \aleph$$

By definition of a Nash equilibrium, formula (4) implies that $i_k = i^*$, $j_l = j^*$, k, $l \in \aleph$. Hence, in Variant 1 strategic reflexion becomes pointless[10] (possibly, except the case when best responses are such that agents choose the components of different Nash equilibria (if any)).

Variant 2 (a pure strategies' Nash equilibrium exists, but agents choose the guaranteeing strategies (1)).

If the guaranteeing strategies form a Nash equilibrium (see the antagonistic games with a saddle point), we obtain Variant 1. Hence, strategic reflexion makes sense provided that (within the framework of Variant 2) a Nash equilibrium does not coincide with the guaranteeing strategies' equilibrium (i_0, j_0).

Variant 3 (a pure strategies' Nash equilibrium does not exist, and agents apply mixed strategies' Nash equilibrium[11]).

Suppose that agents expect the opponent to choose a mixed strategies' Nash equilibrium (when they define the best responses by analogy to (4)). Evidently, the maximal expected gain of each agent corresponds to his/her mixed strategies' Nash equilibrium. Thus, in Variant 3 any equilibrium coincides with a mixed strategies' Nash equilibrium. In other words, strategic reflexion becomes meaningless.

Variant 4 (a pure strategies' Nash equilibrium does not exist, and agents choose the guaranteeing strategies (1)).

Here analysis of reflexion seems appropriate.

We have briefly considered four possible variants of agents' behavior. This leads to the following result.

Assertion 3.2.1. In bimatrix games, strategic reflexion makes sense, if agents use the subjective guaranteeing strategies (1) which are not Nash equilibria.

Introduce the quantities

(5) $K_{\min} = \min\{K \in \aleph | I_K = I_\infty\}$,
(6) $L_{\min} = \min\{L \in \aleph | J_L = J_\infty\}$.

In practical interpretation, $K_{\min}$ and $L_{\min}$ stand for the minimal reflexion ranks of agents 1 and 2, respectively, such that the sets of their subjective equilibrium actions coincide with maximally possible sets of subjective guaranteeing strategies (in the game under consideration).

By definition we have $\forall K, L \in \aleph$: $I_K \subseteq I_{K+1}, J_L \subseteq J_{L+1}$. Consequently, $\forall K \geq K_{\min}$: $I_K = I_\infty$, $\forall L \geq L_{\min}$: $J_L = J_\infty$.

Assume that the reflexion ranks of agents 1 and 2 do not exceed K and L, respectively. In this case, the sets of subjective guaranteeing strategies of agents 1 and 2 (according to the opponent's viewpoint) equal I_{L-1} and J_{K-1}, respectively.

---

[10]In bimatrix games, strategic reflexion is pointless, if the equilibrium in a reflexive game with any combination of nonzero reflexion ranks of agents coincides with the equilibrium in the original game.

[11]In bimatrix games, a mixed strategies' Nash equilibrium surely exists.

And so, increasing the reflexion ranks may enlarge the set of subjective guaranteeing strategies if

(7) $L - 1 < K_{min}$,
(8) $K - 1 < L_{min}$.

Note that the *maximal rational rank of reflexion*[12] of agent 1 depends on the properties of subjective guaranteeing strategies of agent 2 (see (8)) and vice versa.

On the other hand, an agent would not increase his/her reflexion rank if the set of his/her feasible subjective equilibrium actions has been "exhausted." From this view, further increase in reflexion ranks may extend the set of subjective guaranteeing strategies under the condition

(9) $K < K_{min}$,
(10) $L < L_{min}$.

Combine formulas (8) and (9), as well as (7) and (8) to establish the following result. Agent 1 has no sense to increase his/her reflexion rank above the threshold

(11) $K_{max} = \min\{K_{min}, L_{min} + 1\}$.

Accordingly, agent 2 benefits nothing by increasing his/her reflexion rank above the threshold

(12) $L_{max} = \min\{L_{min}, K_{min} + 1\}$.

Denote

(13) $R_{max} = \max\{K_{max}, L_{max}\}$.

Therefore, we have proved.

Assertion 3.2.2. In a bimatrix game, agents benefit nothing[13] by making the ranks of their strategic reflexion higher than the quantities (11) and (12).

In a concrete case (i.e., for a specific game), Assertion 3.2.2 assists each agent in evaluating the maximal rational ranks of strategic reflexion of both agents.

The quantities (11)–(13) depend on the game, *viz.*, the payoff matrices. And so, we obtain the estimated relationships between these quantities and the dimensionality of the payoff matrices (clearly, $|I_\infty| \leq |I| = n$, $|J_\infty| \leq |J| = m$; for two-player games, one has a better estimate – see Assertion 3.2.3). For this, introduce the graph of best responses.

---

[12] We understand the maximal rational rank of agent's reflexion as follows. This is a reflexion rank such that the agent generates no new subjective equilibria (according to his/her view) by further increase of the rank.
[13] For any reflexion rank exceeding the above estimates, there exists a reflexion rank meeting the same estimates and leading to the same subjective equilibrium.

The graph of best responses $G = (V, E)$ is a finite two-partite directed graph, where the set of nodes $V = I \cup J$, and arcs come from each node (corresponding to the action of an agent) to the best response of the opponent. We describe the properties of this graph:

1   For each node in the set I, there is an outgoing arc to the node in the set J (agent 2 possesses a best response to any action of agent 1); similarly, for each node in the set J, there is an outgoing arc to the node in the set I (agent 1 possesses a best response to any action of agent 2).

2   Each node in the set V has exactly one incoming arc (each action of an agent represents the best response to a certain action of the opponent).

3   Imagine that the same path passed the same node twice; by definition of best responses, a part of this path is a loop (and new nodes do not appear in this path).

4   Consider paths starting in node i_0. The maximal number of pairwise different actions of agent 1 constitutes min $(n; m + 1)$.

5   Consider paths starting in node i_0. The maximal number of pairwise different actions of agent 2 makes up min $(n; m)$.

6   Consider paths starting in node j_0. The maximal number of pairwise different actions of agent 1 constitutes min $(n; m)$.

7   Consider paths starting in node j_0. The maximal number of pairwise different actions of agent 2 makes up min $(n + 1; m)$.

The established properties of the graph of best responses lead to some upper estimates for the rational ranks of strategic reflexion in bimatrix games.

Assertion 3.2.3. Suppose that a bimatrix game 2×2 admits no Nash equilibria. Consequently, $I_\infty = I, J_\infty = J$.

Proof.   Take an arbitrary bimatrix game 2×2 such that a Nash equilibrium does not exist. Set $X_1 = \{x_1, x_2\}$, $X_2 = \{y_1, y_2\}$. Evaluate the guaranteeing strategies i_0 and j_0. For definiteness, let $x_1 = i_0, y_1 = j_0$.

Two mutually exclusive options are possible: $j_1 = y_1$ and $j_1 = y_2$.

If $j_1 = y_1$, then $i_1 = i_2 = x_2$ (otherwise, (x_1, y_1) is a Nash equilibrium). In this case, $j_2 = j_3 = y_2$ (otherwise, (x_2, y_1) is a Nash equilibrium). Hence, $i_3 = i_4 = x_1$ (otherwise, (x_2, y_2) is a Nash equilibrium). Thus, for option 1 we obtain $I_\infty = I, J_\infty = J$.

If $j_1 = y_2$, then $i_2 = x_2$ (otherwise, (x_1, y_2) is a Nash equilibrium). In this case, $j_3 = y_1$ (otherwise, (x_2, y_2) is a Nash equilibrium). Hence, $i_4 = x_1$ (otherwise, (x_2, y_1) is a Nash equilibrium). Therefore, for option 2 we again have $I_\infty = I, J_\infty = J$. •

In practice, Assertion 3.2.3 means the following. In bimatrix games 2×2 without Nash equilibria, any outcome can be implemented as a subjective equilibrium.

A promising direction of further research lies in subjective equilibria analysis in basic *ordinary finite two-player games* 2×2 (recall that there are 78 structurally different ordinary games, ie., games with two agents, where each agent has two feasible actions and can sort his/her gains in the strictly descending order from the best gain to the worst one [20, 148]).

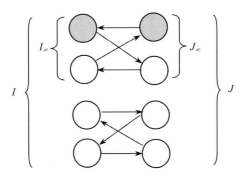

Figure 3.5 An example of the graph of best responses in a bimatrix game 4 × 4, where $I_\infty \subset I, J_\infty \subset J$.

Assertion 3.2.3 suggests the idea that (probably) all bimatrix games without Nash equilibria satisfy the estimates $I_\infty = I$, $J_\infty = J$. However, a counterexample can be found in Fig. 3.5 (the graph of best responses in a game 4 × 4, where nodes i_0 and j_0 are shaded).

We have rough upper estimates ($|I_\infty| \leq n$, $|J_\infty| \leq m$) for the "dimensions" of the sets I_∞ and J_∞. To proceed, analyze how fast these sets are "covered" by appropriate subjective equilibria (What are the corresponding minimal ranks of strategic reflexion?).

Property 3 above implies the following. In a bimatrix game, a rational increase in the rank of strategic reflexion (starting from step 2) surely modifies the set of strategies that must be subjectively guaranteeing ones under smaller (or the same) reflexion ranks.

The sets of feasible actions appear finite in bimatrix games. Hence, the sets I_∞ and J_∞ are finite either. By virtue of Properties 4–7, the quantities L_{min} and K_{min} possess finite values Notably, **in bimatrix games, the unbounded growth of reflexion rank is *a fortiori* irrational**. Again, the finiteness of feasible sets enables easy evaluation of the quantities (11) and (12) (the maximal rational ranks of reflexion) for a specific bimatrix game. The properties of the graph of best responses lead to upper estimates of the maximal rational ranks of strategic reflexion.

In a bimatrix game $n \times m$, the guaranteed estimates[14] of the quantities (11)–(13) naturally depend on the dimensions of the payoff matrices: $K_{min} = K_{min}(n)$, $L_{min} = L_{min}(m)$. Hence,

(14) $K_{max}(n,m) = \min\{K_{min}(n), L_{min}(m) + 1\}$,
(15) $L_{max}(n,m) = \min\{L_{min}(m), K_{min}(n) + 1\}$.

The expression (13) acquires the form

(16) $R_{max}(n,m) = \max\{K_{max}(n,m), L_{max}(n,m)\}$.

---

[14]We understand a guaranteed estimate as an upper estimate, i.e., the maximal possible quantity in a given class of games.

Properties 4–7 and formulas (14)–(16) bring us to

Assertion 3.2.4. In bimatrix games $n \times m$, the maximal rational ranks of strategic reflexion of agents 1 and 2 meet the inequalities

(17) $K_{\max}(n, m) \leq \min\{n, m + 1\}$,
(18) $L_{\max}(n, m) \leq \min\{m, n + 1\}$,
(19) $R_{\max}(n, m) \leq \max\{\min\{n, m + 1\}, \min\{m, n + 1\}\}$.

Corollary 1. In a bimatrix game $n \times n$ $(n \geq 2)$, the maximal rational rank of strategic reflexion of any agent[15] satisfies

$R_{\max}(n, n) \leq n$.

Finally, we formulate the analogous statement in the case of two feasible actions (this situation is common in applied models).

Corollary 2. In a bimatrix game 2×2, the maximal rational rank of strategic reflexion does not exceed 2.

Again, we emphasize that formulas (17)–(19) give the upper estimates. The existence of several best responses to the same action, the presence of a Nash equilibrium (or dominated strategies) in the original model possibly decreases the maximal rational rank of strategic reflexion.

Assertion 3.2.4 can be reformulated as follows. **In a bimatrix game, the maximal rational rank of strategic reflexion exceeds the minimal number of feasible strategies of agents (at most) by unity.**

Below we give some examples illustrating the derived theoretical results for strategic reflexion in bimatrix games (also, see Example 3.2.1).

Example 3.2.2 (*Prisoners' dilemma*) [121, 148]. Consider a classic bimatrix game ("Prisoners' Dilemma"). Two prisoners are accomplices. They choose between two actions, *viz.*, "N" – "not admitting a crime" and "A" – "admitting a crime." If both prisoners admit the crime, they sustain a punishment (their gain makes up the vector (1; 1)). Yet, if accomplice 1 admits the crime and accomplice 2 does not, the former is set at liberty and the latter sustains a major punishment (the gain vector is (6; 0)). We have similar situation when accomplice 2 admits the crime and accomplice 1 does not. Finally, when both do not admit the crime, they sustain a small punishment. Notably, each prisoner then gains 5 (less than in the case of setting at liberty). The payoff matrix is provided by Fig. 3.6 (the action "A" dominates the action "N" for both agents).

---

[15]Obviously, reflexion becomes meaningless in a game, where an agent has just one feasible action.

Actions	N	A
N	(5; 5)	(0; 6)
A	(6; 0)	(1; 1)

Figure 3.6 The payoff matrix in the "Prisoners' dilemma" game.

The only Nash equilibrium in this game forms the vector ("A"; "A"), which consists of the guaranteeing strategies of agents and yields the gains $a_0 = 1$ and $b_0 = 1$ to them. Hence, $i_0 = i_1 = i_2 = \cdots = i_\infty = $ "A," $j = j_1 = j_2 = \cdots = j_\infty = $ "A," and reflexion seems meaningless in the given game (at least, none of the definitions (1)–(3) leads to a "new" equilibrium, i.e., an equilibrium differing from the Nash equilibrium; moreover, none of the definitions substantiates stability of the Pareto-efficient outcome ("N"; "N"), which is a test problem in game theory). •

Example 3.2.3 (*Battle of sexes*) [121]. Consider another classic bimatrix game ("Battle of Sexes") played by spouses. A husband prefers watching soccer ("S"), while a wife favors going to theater ("T"). Each of them would rather spend time with the partner (than being alone). The payoff matrix can be found in Fig. 3.7.

Actions	S	T
S	(3; 1)	(0; 0)
T	(0; 0)	(1; 3)

Figure 3.7 The payoff matrix in the "battle of sexes" game.

The game in question has two pure strategies' Nash equilibria, namely, ("S"; "S") and ("T"; "T"). In the sense of reflexion, each agent benefits from repeating the opponent's choice. However, the choice of any feasible action represents a guaranteeing strategy. Hence, separating out a specific outcome in this reflexive game appears impossible.

In addition to two pure strategies' Nash equilibria, the game considered admits a mixed strategies' Nash equilibrium. Let $p \in [0; 1]$ ($q \in [0; 1]$) be the probability that the husband (wife) chooses watching soccer (going to theater, respectively). Then the equilibrium makes up $p = q = 3/4$ (ie, the mixed strategies' equilibria become (3/4, 1/4) and (1/4, 3/4), yielding the expected gains (3/4; 3/4) to the agents)

Suppose that the husband believes he knows the mixed strategy of the wife. Accordingly, the husband possibly chooses $p_1 = \arg \max_{p \in [0;1]} [3p/4 + 3(1-p)/4]$. Clearly, the expected gain of the husband is independent of his mixed strategy and constitutes 3/4. Similar conclusion applies to the wife's strategy q_1 (as well as to all other reflexive mixed strategies of both agents). In other words, any mixed strategies' reflexive equilibrium is the mixed strategies Nash equilibrium (reflexion turns out pointless). •

Example 3.2.4 *"Misère" in Preferans*). This example is a special case of the "hide-and-seek" game. Imagine a Preferans play, where a player (agent 1) bids "miser" (no-tricks game); we will call him the declarer. The rest players (the defenders) represent agent 2.

Assume that agent 1 chooses between standard "miser" strategy ("S") and non-standard "miser" strategy ("N"). His opponent (agent 2 who whists) has two possible actions, "S"–whisting against standard "miser" strategy and "N"– whisting against nonstandard "miser" strategy. If agent 1 follows standard "miser" strategy and agent 2 whists against nonstandard "miser" strategy, then agent 1 wins the play (the gain vector is $(5; 0)$). The gains $(5; 1)$ correspond to the situation when agent 1 selects nonstandard "miser" strategy, whereas agent 2 whists against standard "miser" strategy. In fact, standard "miser" strategy is easier to whist; and so, we suppose that the outcomes ("S"; "S") and ("N"; "N") correspond to the gains $(2, 3)$ and $(3, 2)$. Therefore, the payoff matrix takes the form presented in Fig. 50.

Actions	N	S
N	$(3; 2)$	$(5; 1)$
S	$(5; 0)$	$(2; 3)$

Figure 3.8 "Miser" in Preferans: the payoff matrix.

This example admits no pure strategies' Nash equilibrium. The guaranteeing strategies are defined by $i_0 = $ "N," $j_0 = $ "S." According to (1),

$i_1 = $ "N," $j_1 = $ "N,"

$i_2 = $ "S," $j_2 = $ "N,"

$i_3 = $ "S," $j_3 = $ "S,"

$i_4 = $ "N," $j_4 = $ "N,"

$\ldots$

Evidently, level 4 of identical reflexion ranks repeats level 1 (and subjective guaranteeing strategies are periodically repeated). Furthermore, $I_K = I$ under $K = 2$, and $J_L = J$ under $L = 1$. This means that first two ranks of reflexion exhaust the sets of feasible actions of the agents; in addition, first three ranks of reflexion exhaust all combinations of pure strategies. In other words, $I_\infty = I$, $J_\infty = J$ and $i_\infty = i_0$, $j_\infty = j_0$.

Agent 1 benefits from the following games (the following combinations of reflexion ranks): MG_{00}, MG_{03}, MG_{10}, MG_{13}, MG_{21}, MG_{22}, and MG_{32}. And five times out of seven he/she possesses reflexion rank not smaller than the opponent does.

Agent 2 benefits from the following games: MG_{01}, MG_{02}, MG_{11}, MG_{12}, MG_{23}, and MG_{33}. And in all games he/she has reflexion rank not smaller than the opponent does.

We have underlined an important aspect above. The gain of an agent may be less than his/her MGR. For instance, the MGR of agent 1 (agent 2) in the current game equals 3 (1, respectively). In the games MG_{20}, MG_{23}, and MG_{33}, agent 1 obtains 2 (which is strictly smaller than his/her MGR $a_0 = 3$). In the games MG_{22}, MG_{31}, and MG_{32}, agent 2 obtains nothing (which is strictly smaller than his/her MGR $b_0 = 1$).

Within the framework of the game considered, let us evaluate a mixed strategies' equilibrium. Denote by p the probability of nonstandard "miser" strategy by agent 1. Set q as the probability of whisting against nonstandard "miser" strategy by agent 2. Thus, we have $p = 3/4$ and $q = 3/5$. Notably, the mixed strategies' equilibrium takes the form $(3/4, 1/4)$ and $(3/5, 2/5)$ (leading to the expected gains $(19/5; 3/2)$ to the agents).

Imagine that agent 1 believes he/she knows the mixed strategy of agent 2. Then he/she can choose $p_1 = \arg \max\limits_{p \in [0;1]} [p(9/5 + 2) + (1 - p)(3 + 4/5)]$. Evidently, the expected gain of agent 1 is independent of his/her mixed strategy and makes up 19/5. Similar conclusion can be drawn for the strategy q_1 of agent 2, as well as for all other reflexive mixed strategies of both agents. Notably, any reflexive mixed strategies' equilibrium is the mixed strategies' Nash equilibrium (reflexion becomes meaningless). •

Once again, recall the following. In real two-player games (including bimatrix games), none of the agents surely knows the type of the game being played (the reflexion ranks k and l of both agents). Therefore, their decision-making should be treated not as a game, but as reflexive decision-making, which consists of two stages. First, accepting the assumption about opponent's reflexion rank. Second, choosing the best response in accordance with the above rank. In this context, the results derived in Section 3.2 interconnect the cardinal numbers of the sets of agents' strategies and the maximal ranks of strategic reflexion (which are rational to consider).

Discussing the strategic reflexion of agents in bimatrix games, one can claim the following for convenience. Instead of selecting their actions (a row and a column in the payoff matrix), agents pass to the choice of their reflexion ranks. The corresponding game is called the game of ranks [71].

Game of ranks. Introduce the following assumptions. Let the payoff matrices be such that each agent has a unique best response to any action of the opponent:

$$(20) \quad \forall j \in J \left| \operatorname*{Arg\,max}_{i \in I} a_{ij} \right| = 1, \quad \forall i \in I \left| \operatorname*{Arg\,max}_{j \in J} b_{ij} \right| = 1.$$

Here and in the sequel, $|M|$ indicates the number of elements in the set M.

Moreover, suppose that the maximal guaranteed result of each agent is attained exactly on one action:

$$(21) \quad \left| \operatorname*{Arg\,max}_{i \in I} \operatorname*{min}_{j \in J} a_{ij} \right| = \left| \operatorname*{Arg\,max}_{j \in J} \operatorname*{min}_{i \in I} b_{ij} \right| = 1.$$

The conditions (20)–(21) ensuring the univocal correspondence between agent's reflexion rank and his/her action are presumed valid.

Each agent can choose a finite rank of his/her reflexion; this results in a corresponding action. Having zero reflexion rank, agent 1 chooses the guaranteeing strategy – the action $i_0 = \arg \max\limits_{i \in I} \min\limits_{j \in J} a_{ij}$. And possessing reflexion rank $k \geq 1$, this agent chooses the action $i_k = \arg \max\limits_{i \in I} a_{ij_{k-1}}$. The analogous formulas apply to the actions of agent 2: $j_0 = \arg \max\limits_{j \in J} \min\limits_{i \in I} b_{ij}$ (under zero reflexion rank) and $j_k = \arg \max\limits_{j \in J} b_{i_{k-1}j}$ (under reflexion rank $k \geq 1$).

Assertion 3.2.4 implies that the set of feasible actions (the choice of reflexion ranks) is finite. Thus, we pass from the original game to the **game of ranks**, where agent's strategy consists in choosing the rank of strategic reflexion (see Table 3.4).

Table 3.4 Reflexion ranks and actions of the agents.

Rank k	0	1	...	R
The action of agent 1	i_0	i_1	...	i_R
The action of agent 2	j_0	j_1	...	j_R

The upper estimate for the number of possible pairwise distinguishable strategies makes up $|I| \times |J| = m \times n$. Then the original bimatrix game can be transformed into a bimatrix game $R \times R$.

Obviously, some rows and columns in the new matrix can coincide (i.e., the choice of different ranks by agents leads to the same action in the original game). By identifying such rows and columns, we get the matrix of the new game called the *choice game for the rank of strategic reflexion* (or simply the *game of ranks*).

Since $i_k \in I$ and $j_k \in J$, all actions of the agents in the game of ranks correspond to their actions in the original game. Hence, we have

Assertion 3.2.5. In the game of ranks, the payoff matrix represents a submatrix for the matrix of the original bimatrix game.

Assertion 3.2.5 suggests that transition to the game of ranks possibly eliminates some equilibria (i.e., they do not appear in the new matrix).

Example 3.2.5. Take the bimatrix game given by $\begin{pmatrix} (2,3) & (0,0) & (3,2) \\ (0,0) & (4,4) & (0,1) \\ (3,2) & (1,0) & (2,3) \end{pmatrix}$.

To construct the matrix of the game of ranks, we analyze the choice of agents under different reflexion ranks (see Table 3.5).

Table 3.5 Reflexion ranks and actions of the agents in Example 3.2.5.

Rank k	0	1	2	3	4	...
The action of agent 1	3	1	1	3	3	...
The action of agent 2	3	3	1	1	3	...

And so, this matrix acquires the form $\begin{pmatrix} (2,3) & (3,2) \\ (3,2) & (2,3) \end{pmatrix}$. Evidently, the equilibrium pair of gains $(4,4)$ has disappeared after transition to the game of ranks. •

The following question seems natural. Will transition to the game of ranks yield new equilibria? Actually, the answer is negative.

Assertion 3.2.6. For an arbitrary bimatrix game, transition to the game of ranks yields no new equilibria.

Proof. Suppose that I is the action set of agent 1, and J is the action set of agent 2. Next, let $I' \subseteq I$ and $J' \subseteq J$ be the action sets of agent 1 and agent 2, respectively, in the game of ranks.

Consider the pair of actions (i_u, j_v), $i_u \in I'$, $j_v \in J'$, which represents the equilibrium in the game of ranks.

First, demonstrate that j_v is the best response of agent 2 to the action i_u of agent 1 (in the original game). Indeed, the best response on the set J enters J' (by the construction procedure of the game of ranks). Thus, the best response on the set J' coincides with that on the set J. By the definition of an equilibrium, the best response on the set J' makes up j_v.

By analogy, i_u constitutes the best response of player 1 to the strategy j_v of player 2 (in the original game). Therefore, the pair of actions (i_u, j_v) forms an equilibrium in the original game.

Recall that the choice of the equilibrium pair has been arbitrary. Hence, any equilibrium in the game of ranks turns out an equilibrium in the original game (i.e., new equilibria do not appear). •

Thus, transition to the game of ranks does not generate new equilibria (see Assertion 3.2.6). Moreover, the existing equilibria may even disappear (Example 3.2.5). The number of equilibria in the game of ranks satisfies the following estimate (based on formulas (20) and (21)).

Assertion 3.2.7. Under the conditions (20) and (21), the game of ranks admits at most two equilibria.

Proof. Suppose that the game of ranks has three different equilibria: (i_u, j_v), $(i_{u'}, j_{v'})$ and $(i_{u''}, j_{v''})$. By Assertion 3.2.6, they represent equilibria in the original game. Consequently, according to (20), we obtain $i_u \neq i_{u'} \neq i_{u''}$. Without loss of generality, assume that $u = \max\{u; u'; u''\}$. In an equilibrium, the action of an agent provides the best response to the opponent's action. And so, the following expressions hold true: $i_u = i_{v+1} = i_{u+2} = i_{v+3} = i_{u+4} = \cdots$; $j_v = j_{u+1} = j_{v+2} = \cdots$ Similar equalities take place for $i_{u'}$, $i_{u''}$. Hence, $i_{u+1} = i_{u'}$ and $i_{u+1} = i_{u''}$. However, in this case, $i_{u'} = i_{u''}$. This contradiction concludes the proof of Assertion 3.2.7. •

Interestingly, sometimes any outcome of the agents' game leads to a better result than an equilibrium.

Example 3.2.6. Consider the bimatrix game defined by $\begin{pmatrix} (6,10) & (0,0) & (10,6) \\ (0,0) & (5,5) & (0,1) \\ (10,6) & (1,0) & (6,10) \end{pmatrix}$.

Here the equilibrium brings to the pair of gains $(5,5)$. Obviously, this is worse than any outcome without duplicated strategies (for both agents): $\begin{pmatrix} (6,10) & (10,6) \\ (10,6) & (6,10) \end{pmatrix}$. •

3.3 BOUNDEDNESS OF REFLEXION RANKS

Throughout this book, we study the normative aspect[16] of reflexive interaction. Notably, which is the reflexion rank of an agent (the maximal rational rank of his/her reflexion), maximizing his/her gain in a reflexive game? In other words, exceeding the maximal rational rank of strategic reflexion provides no extra gain. Research aims at deriving the corresponding constraints within the framework of a game-theoretic model.

In many cases, the normative model of decision-making dictates that an agent should increase the rank of his/her strategic reflexion infinitely. On the other hand, there exist definite *informational constraints*, i.e., data processing capabilities of real agents are naturally limited (the model may explicitly account for the "cognitive costs" – see [2]). Thus, infinite reflexion is none other than a mathematical abstraction. The present section serves for qualitative discussion of the relationship between informational constraints and reflexion rank. All assertions possess nonrigorous character, since the underlying "axioms" are merely debatable assumptions.

As an informational constraint, we choose the Miller's "magic number 7 plus and minus 2" generally accepted in psychology [7, 115]. It reflects the maximal number of objects (attributes, alternatives, etc.) that can be simultaneously operated by a human.

In Section 3.2 we have emphasized the following. The decision process of reflexing agents represents reflexive decision-making (rather than a game), which incorporates two stages. First, introducing some assumptions regarding possible values of opponent's reflexion rank. And, second, choosing the corresponding best responses.

Consider the decision process of agent i (more specifically, an elementary act of his/her analysis of a current game situation). Agent i may adopt the following reasoning, "Suppose that I'm going to choose an action $x_i \in X_i$, whereas my opponent (i.e., the set of all agents[17] except agent i) intends to choose an action $x_{-i} \in X_{-i}$ (zero reflexion rank). If this fact is reflected by the opponent (reflexion rank 1), he/she can choose $BR_{-i}(x_i)$, and I will accordingly choose $BR_i(x_{-i})$. Next, if this fact is again reflected by the opponent (reflexion rank 2), and we can choose $BR_{-i}(BR_i(x_{-i}))$ and $BR_i(BR_{-i}(x_i))$, respectively." Such construction chain for the graph of best responses continues until (a) the whole set of feasible actions is completely exhausted (see Sections 2.6 and 3.2) or (b) new actions do not appear in the chain (see Section 3.2), if we neglect informational constraints. Let w be the reflexion rank. The number of actions (real and phantom agents) to-be-considered by an agent (under arbitrary $x_i \in X_i$ and $x_{-i} \in X_{-i}$) makes up $2(w + 1)$. Accordingly, the number of relations between them is $(w + 1)!$ (an agent believes that his/her opponent possesses the same level of rationality as he/she actually does).

---

[16]See Chapter 4 for special models considered within the descriptive aspect.

[17]In this section, we presume that (in a multi-agent system) each agent treats all opponents as one agent. By rejecting this assumption (believing that each agent analyzes possible behavior of each opponent), one decreases the estimates of the maximal rational rank of strategic reflexion. Indeed, in this case, an agent has to consider n real agents and $n!$ relations among them. This exceeds the Miller's number even for a system with four agents ("no space" for reflexion).

Informational constraints being accounted, one arrives at the inequality $2(w + 1) \leq 7 \pm 2$ or $(w + 1)! \leq 7 \pm 2$. The first inequality yields $w \in \{0; 1; 2; 3\}$, and the second one admits the solution $w \in \{0; 1; 2\}$. If the Miller's number equals 7, **the maximal rational rank of strategic reflexion (due to informational constraints) is 2.**

The finiteness of an awareness structure (*ergo*, the boundedness of the rational rank of strategic reflexion) has an alternative explanation. The amount of information in any message of a finite length is finite. Really, the hierarchy of beliefs (an awareness structure) forms as agents acquire some information. If the latter is limited, the corresponding awareness structure turns out finite. A relevant issue of future investigations concerns establishing a relation between the information acquired by agents and the resulting awareness structure.

Therefore, in Sections 3.1–3.3 we have considered some approaches to strategic reflexion in two-player games. Now, address the general case, *viz.*, the reflexive partitions method. It allows describing strategic reflexion of any finite number of interacting agents (the concept of a reflexive structure – see Section 3.4). Furthermore, this method enables posing and solving reflexive control problems (purposeful formation of required reflexive structures).

3.4   REFLEXIVE STRUCTURES AND REFLEXIVE CONTROL

Reflexive partitions. The hypothesis of indicator behavior discussed in Section 1.3 indirectly implies the following. Choosing his/her actions by the procedure (1.3.2), an agent does not ponder over that the rest agents act similarly. Otherwise, an agent would perform reflexion and (making decisions at a time instant t) seek for the best response to the actions of the rest agents, forecasted according to (1.3.2). In this case, the state of goal is no more defined by formula (1.3.1). Instead, we obtain

$$(1) \qquad w_i(x^t_{-i}) = \arg \max_{y \in \Re^1} F_i(y, x^t_{-i}).$$

Here x^t_{-i} satisfies (1.3.1). We will believe that a reflexing agent of rank 1 considers the rest agents as non-reflexing. This agrees with a tradition established in decision theory (see Sections 3.1–3.2) – an agent having a certain rank of strategic reflexion treats the rest agents as possessing a smaller rank (by unity) [2]. Alternative suppositions regarding agents' beliefs about the reflexion ranks of their opponents will be studied below (e.g., assume that intelligent agents of the same reflexion rank play a Nash equilibrium [162]).

Similarly, it is possible to consider agents with higher reflexion ranks[18]. For this, define $\aleph = \{N_0, N_1, \ldots, N_m\}$ as a partition of the agents' set N, where N_i is the set of agents with reflexion rank i, $i = \overline{0, m}$, and m specifies the *maximal reflexion rank*, $n_i = |N_i|$, $i \in N$, $\sum_{i=0}^{m} n_i = n$. We will call $\aleph$ a *reflexive partition* [92].

---

[18]The term "an agent of reflexion rank k" possesses many synonyms (a step k player, a level k player, a k-level player, a smart k-player [104, 156], see the overview in [127]).

Suppose that an agent with reflexion rank k exactly knows the sets (shares) of the agents with ranks $k' < k - 1$. Moreover, assume that he/she considers all agents as having reflexion rank $k - 1$. In other words, this agent does not concede the existence of agents with the same (or even higher) reflexion rank than his/her rank. In addition, the agent in question may incorrectly estimate the sets of agents possessing reflexion ranks $k - 1, k, \ldots$.

Consider a given initial action vector x^0 of the agents. Let us study the following dynamic reflexive model of their decision-making. The corresponding expressions for the one-step "game" model represent a special case, when decisions are made one-time under $\gamma_i^1 \equiv 1$, $i \in N$.

Reflexion rank 0. Take agents with reflexion rank 0 (belonging to the set N_0). Assume that they choose actions, thinking that the rest agents act similarly to the previous period. Formula (1.3.1) yields

$$(2) \quad x_i^t = x_i^{t-1} + \gamma_i^t \, [w_i(x_{-i}^{t-1}) - x_i^{t-1}], \quad i \in N_0, t = 1, 2, \ldots.$$

In the case $N_0 = N$ (no reflexing agents), all agents observe the *real trajectory* $(x^0, \ldots, x^t, \ldots)$ of the agents' action vectors, see (2).

Reflexion rank 1. Agent j with reflexion rank 1 ($j \in N_1$) considers the rest agents as having reflexion rank 0. According to formula (2), he/she "forecasts" their choice. Hence, his/her choice $x1_j^t$ represents the best response on the outcome expected by this agent:

$$(3) \quad x1_j^t = x1_j^{t-1} + \gamma_j^t \, [w_j(x_{-j}^t) - x1_j^{t-1}], \quad j \in N_1.$$

For agent $j \in N_1$, the *forecasted trajectory* is defined by $(x^0, \ldots, (x1_j^t, x_{-j}^t), \ldots)$; however, actually the trajectory $(x^0, \ldots, (x1_{j \in N_1}^t, x_{i \in N_0}^t), \ldots)$ is realized. This means that the real trajectory may differ from the forecasted trajectories of agents with reflexion ranks 0 and 1. We will discuss possible noncoincidence of these trajectories (and consequences) later.

Let us make a couple of remarks. This book proceeds from the assumption that agents with any reflexion rank are sufficiently "intelligent" (they choose actions by maximizing their goal functions). It is possible to study less intelligent agents (known as *imitating agents*); they *de bene esse* possess reflexion rank -1. Their actions are defined by a known function of the current and previous actions of the rest agents (e.g., the arithmetical mean of the actions chosen by other agents or by the agents related to a given one; copying the action of a specific agent). Perhaps, such models provide an adequate description for some phenomena (such as innovation diffusion, etc.).

Reflexion rank 2. Suppose that each agent j with reflexion rank 2 ($j \in N_2$) exactly knows the set N_0; moreover, he/she considers all agents from the set $N_1 \cup N_2 \backslash \{j\}$ as having reflexion rank 1. In the general case of several agents with reflexion rank 2, this agent wrongly assigns rank 1 to them. Consequently, he/she can "forecast" the behavior of the opponents. Therefore, his/her choice is the best response to the expected outcome:

$$(4) \quad x2_j^t = x2_j^{t-1} + \gamma_j^t \, [w_j(x_{i \in N_0}^t, x1_{l \in N_1 \cup N_2 \backslash \{j\}}^t) - x2_j^{t-1}], \quad j \in N_2.$$

For agent $j \in N_2$, the *forecasted trajectory* is given by $(x^0, \ldots, (x2^t_j, x1^t_{l \in N_1 \cup N_2 \setminus \{j\}}, x^t_{i \in N_0}), \ldots)$, while actually the trajectory $(x^0, \ldots, (x2^t_{j \in N_2}, x1^t_{l \in N_1}, x^t_{l \in N_0}), \ldots)$ is realized.

Reflexion rank k $(k \leq m)$. The behavior of agents with reflexion rank k is described by analogy to the three cases above (reflexion ranks 0, 1 and 2). This is done on the basis of the following *awareness structure* of the agents. For agent j with reflexion rank k, denote by $\aleph_{jk}$ the *subjective reflexive partition* (the beliefs of the agent about the partitions of all agents):

(5) $\aleph_{jk} = (\underbrace{N_0, N_1, \ldots, N_{k-2}, N_{k-1} \cup N_k \cup \ldots \cup N_m \setminus \{j\}}_{k}, \{j\}, \underbrace{\emptyset, \ldots, \emptyset}_{m-k-1}), \quad j \in N_k.$

An agent with reflexion rank k chooses actions by the procedure

(6) $xk^t_j = xk^{t-1}_j + \gamma^t_j [w_j(x^t_{l \in N_0}, x1^t_{l \in N_1}, \ldots, x[k-1]^t_{l \in N_{k-1} \cup N_k \cup \ldots \cup N_m / \{j\}}) - xk^{t-1}_j],$
$j \in N_k.$

In the "static" case, this agent selects the action

(7) $xk^*_j(\aleph_{jk}) = \arg \max_{y \in \mathfrak{R}^1} F_j(y, x^1_{l \in N_0}, x1^1_{l \in N_1}, \ldots, x[k-1]^1_{l \in N_{k-1} \cup N_k \cup \ldots \cup N_m / \{j\}}), \quad j \in N_k.$

Therefore, a **reflexive structure** represents the set of subjective reflexive partitions of all agents. Assume that agents' beliefs about the reflexion ranks of each other satisfy (5). Then the awareness structure is uniquely defined by the reflexive partition $\aleph$.

Interestingly, the reflexive structure (5) introduced within the framework of strategic reflexion resembles awareness structures used in models of informational reflexion (see Section 2.2).

The vector of agents' actions

(7) $x^*(\aleph) = \{xk^*_j(\aleph_{jk})\}_{j \in N_k, k=\overline{0,m}}$

is said to be a *reflexive equilibrium* of the game $\Gamma_\aleph = \{N, F_i(\cdot)_{i \in N}, \aleph\}$ [92, 132, 133]. In other words, a **reflexive equilibrium** forms the set of agents' actions being the best responses to opponents' actions (according to an existing reflexive structure). By virtue of the assumptions regarding the existence and uniqueness of best responses, a reflexive equilibrium always exists. Furthermore, a reflexive equilibrium seems rather exotic. Generally, the actions of agents are not the best responses to opponents' actions.

By proper variation of reflexive partitions, one can change the actions of agents, i.e., perform **reflexive control**.

Recall the following assumptions adopted for defining a reflexive equilibrium:

–   the initial actions vector x^0 of agents is fixed and represents a common knowledge;
–   an agent with reflexion rank k has correct beliefs about the reflexion ranks of all agents with smaller ranks;
–   an agent with reflexion rank k considers all agents with the same or higher ranks as the ones possessing rank $k-1$.

The agents' beliefs about the reflexion ranks of their opponents can be stated differently (see the discussion below). In this case, the expressions for the best response, current goal and reflexive equilibrium (7) would be altered accordingly. However, the construction procedure of a reflexive equilibrium gets retained.

The general reflexive model of *collective behavior* introduced above would hardly lead to general analytical derivations. Nevertheless, it may provide a basis for developing particular analytical models or general simulation models (e.g., according to the classification suggested in [129]). Such models serve for describing and forecasting collective behavior (human beings, mobile robots, program agents) in various situations. For instance, we refer an interested reader to reflexive simulation models of evacuation, reflexive models of transport flows. In addition, see the examples in subsection 4.26.

Models of strategic reflexion: A classification. Denote by n_{ijl} the belief of agent i with reflexion rank j about the reflexion rank of agent l. Homogeneous agents being analyzed, let q_{ijk} designate the beliefs of agent i with reflexion rank j about the share of agents possessing reflexion rank k. Finally, set $q_k = n_k/n$ as the "true" share of agents with reflexion rank k.

A common **postulate** (accepted almost in all models of collective reflexive behavior) declares the following. An agent with some reflexion rank is "unaware" of other existing agents with the same or higher ranks, i.e., $\forall k > j$: $q_{ijk} = 0$, $q_{ijj} = 1/n$.

Classification bases for models of strategic reflexion [92].

1) The set of possible agent's actions (finite or infinite, e.g., a segment in $\Re^1$).
2) The principle of actions choice by agents with reflexion rank 0:

   - fixed (a priori given) actions, e.g., a focal point;
   - best response to some fixed (a priori given) actions (e.g., the results of a past period);
   - random actions according to a given (uniform) distribution.

3) Agents are identical (homogeneous, i.e., differing only in their reflexion ranks) or nonidentical (differing also in their goal functions).
4) The objective distribution of agents by the reflexion ranks is:

   - arbitrary fixed;
   - random (the Poisson distribution $q_k = e^\tau \tau^k/k!$, $\tau > 0$ is a parameter).

5) The awareness of an agent with reflexion rank k about the set of agents:

   - he/she knows for sure the set N and believes this information is a common knowledge;
   - he/she possesses individual beliefs about the set of agents.

   Almost all models of collective reflexive behavior rely on the former assumption.
6) The awareness of an agent with reflexion rank k about the agents having lower ranks (from 0 to $k - 1$ inclusive):

   - he/she knows them for sure (or with a certain error; in the case of $m \leq 2$, the corresponding model is studied in [164]);

- he/she assumes that these agents are distributed by the reflexion ranks from 0 to $k - 1$ inclusive according to a normalized random distribution ($\forall k < j$: $q_{ijk} = \frac{n_j}{\sum_{l=0}^{j-1} n_l}(n - 1)$). Generally, Poisson distributions are involved, see [30];

- he/she believes that the rest (!) agents have reflexion rank $k - 1$ (e.g., see the models of k-level reasoning [123, 162], as well as further development and experimental study of these models in [51, 52]).

7) The awareness of an agent with reflexion rank k about other agents having the same or higher ranks:

  - he/she believes all of them belong to rank 0;
  - he/she believes all of them belong to rank $k - 1$;
  - he/she assumes that these agents belong to reflexion ranks from 0 to $k - 1$ inclusive according to a random (Poisson) distribution;
  - he/she knows their reflexion ranks (in this case, the above "postulate" fails); by choosing his/her actions, an agent eliminates the uncertainty regarding their behavior (expecting them to choose the worst-case actions).

The following aspect has been emphasized earlier. For any attribute values in this classification system, a reflexive equilibrium is constructed according to the general procedure presented above.

All existing models of strategic reflexion (developed in Russia and overseas) agree with this classification system. Let us glance over them.

An overview of theoretical investigations. Publications on strategic reflexion models, the so-called level k models, appeared in the mid-1990s [52, 123, 164]. In 2004, they were generalized by the *cognitive hierarchies model* (CHM) [30].

The survey [177] identified four basic approaches to the construction and study of strategic reflexion models within the framework of game theory and experimental economics. We cite fundamental works only (references to later research can be found in [177]). Notably, the four basic approaches are:

- the level k approach [51];
- the approach of quantal best response equilibria [112];
- the quantal level k approach [163];
- the approach of cognitive hierarchies [30].

Now, consider these approaches in detail. All models possess the following features. First, agents are identical (have the same goal functions and sets of feasible actions), differing only in their reflexion ranks. Second, an agent with rank k knows nothing (regards as inconceivable) about other existing agents with the same or higher reflexion ranks. Third, agents with rank 0 choose their actions randomly according to the uniform distribution on the set of their feasible actions. Some experimental works [45, 64, 123] analyze the structure of agents' interaction (information about opponents' actions being available to agents).

Level k models. Take each agent $i \in N$ and assign a certain reflexion rank (i.e., the number $k_i \in \{0; 1; 2; \ldots\}$) to him/her. An agent with rank k treats all other agents (!)

as having rank $k - 1$. Thus, he/she chooses the best response to their actions. If such response is not unique, the agent chooses any equiprobably.

A modification of the level k model represents the so-called L_k-*model*. It employs agents having reflexion ranks 0, 1, and 2 only. Here, each agent of rank k is characterized by the probability ε_k of "being mistaken" (choosing equiprobably any nonoptimal action differing from the best response to the actions of agents with reflexion rank $k - 1$, see [177]). Therefore, the L_k-model is described by four parameters, *viz.*, the shares q_1 and q_2 of agents having reflexion ranks 1 and 2, respectively, and by their errors ε_1 and ε_2.

Quantal best response equilibrium. Let us consider the case of finite sets of feasible agents' actions. The *quantal best response* $QBR_i(p_{-i}; \lambda)$ of agent i is defined as his/her mixed strategy with the probabilities $p_i(x_i) = \frac{\exp(\lambda F_i(x_i, p_{-i}(\cdot)))}{\sum_{y_i} \exp(\lambda F_i(y_i, p_{-i}(\cdot)))}$, where the parameter λ characterizes agents' sensitivity to variations of their goal functions [46, 112]. The concept of quantal best response assists in determining a *quantal response equilibrium* [112] (QRE) as a profile of mixed strategies $p^* = (p_1^*, \ldots, p_n^*)$ such that $\forall i \in N: p_i^* = QBR_i(p_{-i}^*(\cdot), \lambda)$. Any normal-form game with a nonnegative parameter λ admits a quantal response equilibrium [112].

Quantal level k models (QL_k) are "at the junction" of the QRE-based models and L_k-models [163]. In contrast to the L_k-model, each agent believes that agents with lower hierarchical levels "make no mistakes." Accordingly, these models are described by five parameters: $q_1, q_2, \varepsilon_1, \varepsilon_2$, and λ.

CHM-models. An agent chooses his/her best response depending on the rank distribution of all agents with lower reflexion ranks. Suppose that reflexion ranks obey the Poisson distribution with a parameter τ. Consequently, the Poisson-CHM model is characterized by the parameter τ.

The four classes of models were compared in [177] using the experimental data provided by the authors of corresponding models. The comparative analysis drew the following conclusion. The assumption regarding the adequacy of Poisson-like distributions to describe reflexion ranks of agents is incorrect. In addition, higher efficiency of the QL_k-model was substantiated in statistical terms.

An overview of some experimental investigations. A standard scheme of an experimental study is as follows. Select a sufficiently simple game and a corresponding theoretical model which describes the expected behavior of participants. Subsequently, organize a series of plays, where participants are researchers themselves (or their under- and postgraduates). Finally, process the experimental results by statistical methods to identify the model parameters (see the survey [53]).

Most researchers acknowledge that such experiments essentially depend on the type of audience. For instance, students interested in game theory demonstrate higher reflexion ranks [5, 34, 47]. For various groups of agents, the results of the beauty contest game are presented in [33].

The beauty contest game provides a classic example illustrating both the model of cognitive hierarchies [30] and other models of strategic reflexion [119, 123, 177]. An up-to-date review of foreign experimental research on CHM can be found in [24].

Consider a finite number n of agents. Let each agent choose a number from 0 to 100. Denote by $x_i \in \{0, \ldots, 100\}$ the action of agent i. The winner is the agent choosing a number closest to a given share $\alpha \in (0; 1]$ of the mean action $\frac{1}{n}\sum_{i=1}^{n} x_i$ of all agents.

The winner receives a prize, whereas the rest agents get nothing. Imagine there are several "winners." Then the prize is divided among them in equal parts.

In the case of $\alpha = 1$ and more than two agents, any vector of identical agents' actions forms a Nash equilibrium. The situation gets complicated if $\alpha < 1$ (the unique Nash equilibrium lies in zero actions of all agents).

The following reasoning is suggested to determine an "equilibrium" outcome in the models of cognitive hierarchies [30]. Let the actions of agents with reflexion rank 0 be uniformly distributed and $\alpha < 1$. Consequently, their mean action makes up 50. Alternatively, assume that agents with reflexion rank 0 choose action 50 as the focal point [154]. Or else, suppose that (in a repetitive game) agents with reflexion rank 0 choose their actions based on agents' actions in the past period. The best response (see (1)) of agents with reflexion rank 1 is the action 50α. For those having reflexion rank 2 (reflexion rank 3), we obtain the best response in the form of $50\alpha^2$ ($50\alpha^3$, respectively), and so on. The results of various experiments corroborating the existence of such characteristic points can be found in [24, 30, 33]. In addition, different models are analyzed and compared in [177] (including their compliance with experimental data).

The "11–20" game. It involves two agents choosing simultaneously and independently a number between 11 and 20. Each agent gains the amount of money equal to the chosen number. Additionally, if an agent chooses a number smaller by unity than the opponent's choice, he/she receives 20 extra units of money [5].

A reasonable choice of agents with reflexion rank 0 lies in 20. The best response to this choice (i.e., the action of players with rank 1) is 19. Similarly, the best response of players having reflexion rank 2 constitutes 18, and so on.

The game conditions (payments depending on the combinations of the agents' actions) may be modified somehow. Moreover, similar games are easily constructed by analogy (e.g., the "91–100" game) [5].

Other "experimental" games:

-   "the centipede game" engaging players with reflexion ranks 0, 1, and 2 [19, 111];
-   "the market entry game" [31, 172].

Generally, Poisson distributions are used in CHM. The matter has obvious grounds. First, this distribution appears sufficiently simple in the sense of computations. Second, the share of corresponding agents decreases rapidly (cognitive constraints) with reflexion rank (the rate q_k/q_{k-1} decreases simultaneously with $1/k$). And finally, Poisson distributions match experimental data. Indeed, the paper [30] demonstrated that, as $\tau \to +\infty$, the agents' action vector in the Poisson-CHM model converges to the Nash equilibrium resulting from successive elimination of the weakly dominating actions. In the general case (non-Poisson distributions), this result takes no place.

In the discrete Poisson distribution $q_k = e^\tau \tau^k / k!$, the parameter τ serves as mean and variance; in addition, $\frac{q_{k-2}}{q_{k-1}} = \frac{\tau}{k-1}$. Therefore, its "mean" experimental value can be interpreted as the "mean rank" of agents' reflexion. The lion's share of experiments testifies to the following. This value lies between 1 and 2, usually, about 1.5 (see [30, 177] for experimental results in the beauty contest game).

The following distribution of agents by reflexion ranks was obtained in [5] for the "11–20" game with 108 participants.

Reflexion rank	0	1	2	3	4	5	6	7	8	9
Action	20	19	18	17	16	15	14	13	12	11
The share of agents (in %)	6	12	30	32	6	1	6	3	0	4

In [167] one can find experimental results indicating an important fact. From 40% to 60% of agents have nonzero reflexion rank (i.e., choose actions differing from (a) agents with reflexion rank 0 and (b) a Nash equilibrium). The following summary of agent distributions by reflexion ranks was presented and discussed in [32]:

Reflexion rank	The share of agents [164]	The share of agents [49]	The share of agents [51]
0	0.25	0.42	0.21
1	0.12	0.44	0.21
2	0.12	0.11	0.27
3	0.12	0.03	0.19
4	0.12	0.01	0.09
5	0.12	0	0.03
6	0.12	0	0.01
7 and higher	0	0	0

Finally, we should underline an important aspect. The experiment-based conclusions regarding the presence (and shares) of agents with reflexion ranks 5, 6, 7 and higher seem doubtful. Actually, they contradict to *sensus communis*. Generally speaking, the experimental distributions of agents by reflexion ranks are different and (probably) strongly dependent both on the game itself and the experimental conditions, staff of participants, and so on.

Reflexive control problem [92]. Consider reflexive partition as a control parameter. It is possible to formulate *controllability problem*, as follows. Under a given set $\Im$ of feasible reflexive partitions, find the set of agents' action vectors $X(\Im) = \bigcup_{\aleph \in \Im} x(\aleph)$ that can be realized by reflexive control. The inverse problem lies in obtaining the "minimal" set of feasible reflexive partitions (in a certain sense), allowing to realize a given agents' action vector.

Now, let us address the control problem. Suppose that the preferences of a control subject (a *principal*) are described by his/her real-valued goal function $F_0(Q(x^*))$ defined on the set of *aggregated outcomes* ($Q: \Re^n \to \Re^1$), i.e., $F_0(\cdot) : \Re^1 \to \Re^1$. Using the expression (12), the *efficiency of the reflexive partition* $\aleph$ can be characterized by $K(\aleph) = F_0(Q(x^*(\aleph)))$.

Consequently, the *problem of reflexive control* (in terms of reflexive partitions) can be formally stated as [92]

(8) $\quad K(\aleph) \to \max_{\aleph \in \Im}.$

Let K_m be the maximal value of the efficiency criterion in the problem (8) under a fixed maximal reflexion rank m. Recall the technique adopted for the models of strategic reflexion (Sections 2.6 and 3.2). By analogy, pose the problem of the *maximal rational rank of reflexion* (a rank being pointless to exceed for the principal in the sense of controllability or/and efficiency of reflexive control):

$$m^* = \min\{m | m \in \text{Arg} \max_{w=0,1,2,\dots} K_w\}.$$

To proceed, we discuss conformity of subjective reflexive partitions of the agents. Suppose that each agent observes merely the aggregated outcome. Trajectories forecasted by the agents may differ from the real trajectory (see the general reflexive model of collective behavior). This motivates the agents to doubt the correctness of their subjective reflexive partitions. Imagine that the agents observe just the aggregated outcome of the game (in addition to their own actions). By analogy to the condition of stable informational control (see Section 2.7), one can introduce the *condition of stable reflexive partition*. Notably, require that the aggregated outcome for the real trajectory coincides with the forecasted aggregated outcomes for all agents. Stability of reflexive partitions is closely associated with learning in games [62]. Observing the behavior of opponents (which differs from the forecasted behavior), agents may modify their beliefs about the reflexion ranks of the opponents or pass to higher levels of reflexion. For instance, see the experiments described in [109] (these phenomena are manifested in dynamic games; we refer to dynamic level k games [85]).

Under a fixed reflexive partition $\aleph \in \mathfrak{I}$, we have realization of the action vector (7). And the aggregated outcome $Q(x^*(\aleph))$ is realized.

According to agent j with reflexion rank k, the following vector is realized:

$$\tilde{x}_{jk}(\aleph_{jk}) = (x_{l \in N_0}, x1_{l \in N_1}, x2_{l \in N_2}, \dots, x[k-1]_{l \in N_{k-1} \cup N_k \cup \dots \cup N_m \setminus \{j\}}, xk_j),$$
$$j \in N_k, k = \overline{0, m}.$$

The condition of stable reflexive partition $\aleph \in \mathfrak{I}$ takes the form

$$Q(\tilde{x}_{jk}(\aleph_{jk})) = Q(x^*(\aleph)), \quad j \in N_k, k = \overline{0, m}.$$

The problem of reflexive control $(\aleph)$ can be stated on the set of stable reflexive controls (if nonempty). In practice, this means that the principal forms an optimal partition of the agents into reflexion ranks. In such partition, the agents do not doubt the correctness of their beliefs about reflexion ranks of the opponents (based on observing the results of the "game").

Concluding this section, let us outline how the principal can control agents' partition into reflexion ranks. Today, scientific literature suggests two approaches on the subject. The first approach stems from that the agents trust the principal, interpreting his/her information as true (regardless of their initial beliefs). In this case, by appropriate messages to different collectives of agents, the principal forms various (but not arbitrary – see Section 2.12!) awareness structures and corresponding reflexive structures. The second approach lies in that the agents not just substitute their beliefs by the principal's messages, but reduce the agents' uncertainty (narrow the set of feasible "worlds" according to the agents' view, see Sections 2.13–2.14). Generally speaking, designing the models of agents' awareness structures under incoming information represents a promising direction for further research.

Therefore, the method of reflexive partitions of the set of rational agents into subsets of agents possessing different ranks of strategic reflexion allows the following:

- from the decision theory viewpoint, extending the class of collective behavior models for intelligent agents performing a joint activity under incomplete awareness and missed common knowledge;
- from the descriptive viewpoint, enlarging the set of outcomes that can be "explained" (within the framework of the model) as stable results of agents' interaction; accordingly, extending the controllability domain (for control problems);
- from the normative viewpoint, posing/solving the problems of collective behavior by choosing a proper structure of agents' awareness.

Unfortunately, the modern level of investigations in the field of strategic reflexion models suggests no universal "tools" for deriving analytical solutions in wide classes of reflexive control problems. All fruitful results cover just special cases and simple models (see Section 4.27). Apparently, a breakthrough can be made in two directions. First, studying agents' decision-making experimentally and establishing general laws based on analysis of experimental results. Second (a theoretical direction), developing a descriptive language of reflexive models for simple uniform formulation and solution of different reflexive control problems.

Moreover, reflexive control problems, the problem of the maximal rational rank of strategic reflexion and others might and should be posed within alternative modifications of reflexive models of collective behavior (differing from the suggested one). This represents a subject for future-oriented investigations. We mean the problems of *active forecast* (see [136] and Section 4.23), where agents involve the principal's information about the future state of a system to "retrieve" the current state and make decisions using this new information. Here reflexive partitions appear to have definite prospects.

Almost all researchers of strategic reflexion models in collectives of agents adhere to the following opinion. A promising field of applied usage of these models concerns *multiagent systems*. On the one hand, such systems are artificial; hence, assigning certain reflexion ranks to different agents is not ambiguous (in contrast to reflexive control of collectives of people – see a brief description of informational impact methods above) and can be performed at the stage of multiagent system design. And "control" problem consists in choosing an optimal share of reflexing agents (increasing their "intelligent level" requires additional costs for computational resources, communication resources and other resources [91, 92]). On the other hand, scientific publications on multiagent systems witness the following. The usage of artificial autonomous agents with capacity for reflexion has yet not become widespread. Evacuation models and some particular models of collective behavior are an exception (see Section 4.26).

Chapter 4

Applied models of informational and reflexive control

Sections 4.1–4.4 consider a series of applied models of reflexive games (implicit control, informational control by mass media, as well as reflexion in psychology and art works). They illustrate the above general models of decision-making and game-theoretical interaction under different awareness. A common feature of these models lies in the following. In most cases, there exist two agents[1] whose reflexion rank (within the scope of the confidence principle) makes up 0, 1 or 2. According to the confidence principle, the receiver of a message has no doubts about its truth. Numerous references provide detailed information on the capabilities of this approach in various applied problems.

Sections 4.5–4.26 are dedicated to applied models of informational or/and reflexive control in economic (Sections 4.6–4.9, 4.11, 4.16, 4.18, 4.21, 4.22, 4.26.3, 4.26.7), social (Sections 4.13, 4.14, 4.19, 4.20, 4.23–4.25), and organizational (Sections 4.10, 4.12, 4.15, 4.17, 4.26.5, 4.26.6) systems, military problems (Sections 4.5, 4.26.1, 4.26.2, 4.26.4) and other fields.

This book is naturally limited. Many well-studied applications of reflexion in control problems have not been discussed here. For instance, the following reflexive models are beyond consideration:

– control of reputations and norms of activity (i.e., institutional control) [135];
– planning mechanisms (resource allocation, expertise, etc.); see stability analysis for mutual beliefs of agents in [113];
– threshold collective behavior [22, 23].

In this context, we also mention models of informational control of opinions, reputation and confidence levels of agents in social networks [60, 72, 73, 74].

4.1 IMPLICIT CONTROL

Nowadays, psychology, neurolinguistic programming and other fields of science (to say nothing of information warfare) include a considerable number of research works [12, 13, 17, 90, 120, 143, 178] focused on the so-called *implicit control* of people

---

[1]In many models, an agent acts as the principal (performs informational control). Nevertheless, an informational equilibrium takes place (as a subjective equilibrium of a controlled agent).

(and resistance to such control). Generally speaking, implicit control is understood as a veiled control action causing no objections of a controlled subject. A particular case of implicit control consists in *manipulation* – implicit control of a human being (against his/her own free will). By manipulation, an initiator gains certain advantages. Analysis of existing publications leads to the following formal *model of implicit control* (to-be-stated below). Actually, such control is described in terms of reflexive games (as a form of reflexive control). Bold type emphasizes the basic conclusions of this analysis. The conclusions are intensively used below as assumptions in model construction.

First, implicit control aims at the information adopted by a controlled subject in his/her decision-making. Hence, according to the classification system proposed in [136], **implicit control represents informational control** .

Second, in most famous situations of implicit control, **just two subjects interact** (possibly, collective ones). In a rough approximation, it suffices to identify a control subject (*active agent*) and a controlled subject (*passive agent*).

Third, an **active agent has an adequate awareness about a passive agent.** Hence, the former can forecast the latter's behavior in any informational situation. Furthermore, (a) **an active agent knows the actual value of an uncertain parameter** and (b) **a passive agent** often **believes that an active agent possesses an adequate awareness about him/her.** The 2-subjective adequate awareness of a passive agent about an active agent is immediate from Assertion 2.2.3.

Suppose that an active agent can change (influence, modify, etc.) the awareness structure of a passive agent. Consequently, the reflexive control by an active agent imposes a certain awareness structure on a passive agent, such that the resulting decision of the latter is most beneficial to an active agent. Thus, **the efficiency criterion of informational impact (implicit control) is defined by the payoff gained by an active agent.**

And so, implicit control is described in terms of a reflexive game as follows. There exist two agents (active and passive ones) with known goal functions, sets of feasible actions and the following awareness structure. The active agent has an adequate awareness about the passive one; he/she also knows the true value of an uncertain parameter. The passive agent believes that the active one is adequately aware of him/her. Special stipulations can be made in the model, as well.

Consider a given set of feasible awareness structures of the passive agent. *The problem of optimal informational impact* (more specifically, an impact on awareness) is finding a feasible awareness structure of the passive agent such that the corresponding informational equilibrium appears most beneficial to the active agent (maximizes his/her goal function). Recall that, according to the introduced assumptions, the awareness structure of the active agent is fixed.

The following fact is extremely important, since it simplifies the model appreciably. In all common models of implicit control, **reflexion rank does not exceed 2** (i.e., the maximal length of the essential index sequence describing an element of the awareness structure makes up 3).

Solution of the posed implicit control problem (in terms of a reflexive game) defines the awareness structure to-be-imposed on the passive agent by the active one (from the normative viewpoint). However, *the technology of implicit control* **lies beyond the formal models discussed below.** We understand such technology as a sequence of methods, tricks, stages, etc. leading to solution of the formulated problem

[127]. Applied models of informational control are studied and adopted in social psychology, neurolinguistic programming, psychotherapy, etc. [12, 13, 36, 90, 95, 120, 142].

Notably, the framework of the reflexive game-theoretic model recommends an active agent to form a certain awareness structure for a passive agent, but declares nothing about *how* this must be done. At the same time, numerous ways of informational impact (see the cited works) concern the technology of forming a certain awareness structure of a passive agent by an active agent (under given goals of such impact). Hence, formal models answer the question *what* an active agent should do under informational control. Social psychology and other humanities [16, 35, 36, 120, 150, 156] accumulate and generalize the empirical knowledge on efficient ways of achieving these goals.

The number of real situations admitting informational control is huge and these situations are extremely heterogeneous. Consequently, no universal recipes have been found to date; **the technology of implicit control represents an art.** (The matter is slightly simplified by the fact that, in existing publications, a passive agent trusts the information reported by an active agent; i.e., the trust problem (see the confidence principle in [136]) is almost not studied). Making the technology of implicit control a science (by systematization and development of the normative models of informational impact, their analysis, identification, etc.) forms a promising and pressing problem. Yet, it exceeds the limits of the book. We focus on the analysis of normative models and study synthesis problems for optimal informational impacts.

Let us adopt the terminology of hierarchical games and call an active agent by agent 1 for convenience. Accordingly, a passive agent will be assigned index 2. Recall that agent 1 knows the actual value of the uncertain parameter; then it follows that $\theta_1 = \theta$. Moreover, he/she possesses an adequate awareness about the passive agent, i.e., $I_{12} = I_2$. Next, the passive agent believes that the active one has an adequate awareness about him/her: $I_{212} = I_2$. Reflexion rank does not exceed 2. And so, the active agent can influence $\theta_2, \theta_{21}, \theta_{212}$ and their combinations.

As a result, we obtain *seven variants*:

- the reflexion rank of the passive agent is 0, and informational impact aims at θ_2 (*Problem A*);
- the reflexion rank of the passive agent is 1, and informational impact aims at θ_{21} (*Problem B*);
- the reflexion rank of the passive agent is 2, and informational impact aims at θ_{212} (*Problem C*);
- the reflexion rank of the passive agent is 1, and informational impact aims at θ_2 and θ_{21} (*Problem AB*);
- the reflexion rank of the passive agent is 2, and informational impact aims at θ_2 and θ_{212} (*Problem AC*);
- the reflexion rank of the passive agent is 2, and informational impact aims at θ_{21} and θ_{212} (*Problem BC*);
- the reflexion rank of the passive agent is 2, and informational impact aims at θ_2, θ_{21} and θ_{212} (*Problem ABC*).

Consider the listed variants. Without loss of generality, suppose that the sets of feasible awareness structures of the passive agent (i.e., feasible values of θ_2, θ_{21} and θ_{212})

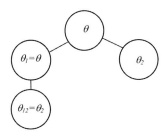

Figure 4.1 The awareness structure in Problem A.

equal Θ and do not depend on reflexion rank. As practical illustrations, we refer to the numbers of *stratagems* (strategies used in politics and military art – see their description in [155][2]).

Problem A. Here we believe that the goal function of the passive agent, $f_2(\cdot)$, is independent of the actions of the active agent or incorporates a fixed action of the active agent. Fig. 4.1 demonstrates essential components[3] of the awareness structure.

Informational impact aims at θ_2. According to the classification suggested in [136], Problem A is an informational regulation problem and no reflexive control takes place. The rest problems from the above list do have reflexive control.

The above reflexive game represents an hierarchical game of the type Γ_1 [67, 121] (a Stackelberg game), where the active agent (making the first move) has to forecast the response of the passive agent. Denote by $X_2(\theta_2) = \mathrm{Arg}\max_{x_2 \in X_2} f_2(\theta_2, x_2)$ the action set of the passive agent (the set of "best responses"), maximizing his/her goal function under the awareness $\theta_2 \in \Theta$. Within the framework of the existing awareness structure, this set can be computed by the active agent. Hence, the latter strives for choosing a "message" $\theta_2^* \in \Theta$ which would assure the following. The passive agent selects an action from the set of best responses (according to the active agent's viewpoint). Let us compare the sets of best responses by the guaranteed value of the goal function of the active agent. *The solution to Problem A takes the form*

$$\theta_2^*(\theta) = \arg\max_{w \in \Theta} \max_{y_1 \in A_1} \min_{x_2 \in X_2(w)} f_1(\theta, x_1, x_2).$$

For this problem, the graph of reflexive game can be found in Fig. 4.2 (the arrow from x_1 to x_2 is absent; indeed, the goal function of the passive agent does not depend on the actions of the active agent).

---

[2]Most "classic" stratagems (they are 36 totally) include no informational control (stratagems No. 2, 4, 5, 9, 11, 12, 15–26, 28–31, 33, 35, 36). They can be treated as popular time-proved technologies of implicit or institutional control (e.g., stratagems No. 2, 4, 5 and others), i.e., control of activity conditions for an active and/or passive agent.
[3]The left-hand branch reflects the awareness of agent 1 $(\theta_1, \theta_{12}, \theta_{121}, \ldots)$. And the right-hand branch corresponds to the awareness of agent 2 $(\theta_2, \theta_{21}, \theta_{212}, \ldots)$.

Figure 4.2 The graph of reflexive game in Problem A.

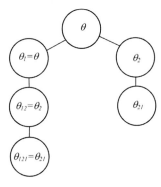

Figure 4.3 The awareness structure in Problems B and AB.

Problem A arises in stratagems No. 7 ("extracting something from nothing" – a modeling strategy for reality which does not actually exist) and No. 32 ("the stratagem of an open city gate" – representing the false as the true and vice versa).

Problems B and AB. They reflect the widespread situation of implicit control, *viz.*, the impact on θ_{21}. Notably, the passive agent possesses specific beliefs θ_2 about the state of nature (and the active agent knows them for sure). Fig. 4.3 illustrates the corresponding awareness structure.

The active agent evaluates the subjective equilibrium of the passive one:

$$X_2(\theta_2, \theta_{21}) = \{x_2 \in X_2, x_{21} \in X_1 \mid \forall y_2 \in X_2 \quad f_2(\theta_2, x_{21}, x_2) \geq f_2(\theta_2, x_{21}, y_2),$$
$$\forall y_1 \in X_1 \quad f_1(\theta_{21}, x_{21}, x_2) \geq f_1(\theta_{21}, y_1, x_{21})\}.$$

By virtue of $I_{212} = I_2$, we have $x_{212} = x_2$. In other words, defining his/her subjective equilibrium, the passive agent believes that his/her equilibrium action x_2 can be "computed" by the active agent.

Suppose that the value of θ_2 is fixed and known to both agents for sure. For the active agent, *the reflexive control problem* lies in the following. Find a value of θ_{21} such that the action chosen by the passive agent from the set $X_2(\theta_2, \theta_{21})$ maximizes the goal function of the active agent:

$$\theta_{21}^*(\theta) = \arg\max_{w \in \Theta} \max_{x_1 \in X_1} \min_{x_2 \in \text{Proj}_2(X_2(\theta_2, w))} f_1(\theta, x_1, x_2).$$

Imagine that, in Problem B, the active agent has an opportunity to influence θ_2. In this case, we obtain Problem AB, where the goal function of the active agent

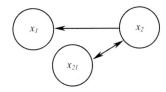

Figure 4.4 The graph of reflexive game in Problems B and AB.

is maximized also with respect to the parameter θ_2. Consequently, *the solution to Problem AB* acquires the form

$$(\theta_2^*(\theta), \theta_{21}^*(\theta)) = \arg \max_{u,w \in \Theta} \max_{x_1 \in X_1} \min_{x_2 \in \mathrm{Proj}_2(X_2(u,w))} f_1(\theta, x_1, x_2).$$

In the problem under consideration, the graph of reflexive game can be observed in Fig. 4.4.

Problem B occurs in the following stratagems:

- No. 1 – "deceiving an emperor for sailing across a sea" (certain actions are performed openly to mask a real target; thus, a specific template of perception is dictated on an opponent);
- No. 3 – "killing by somebody else's knife" (an opponent is destroyed or weakened by proxy);
- No. 6 – "making a noise on the East, and attacking on the West" (the actual direction of an attack or aggression is concealed);
- No. 8 – "under the pretence of repairing bridges, setting out to Chen Tsan" (an opponent is pursued in correct expectations of one's own plans, and a victory is achieved by a nonstandard maneuver);
- No. 10 – "hiding a dagger by a smile" (hostile attitude, disaffection and aggressive plans are masked by friendly greeting);
- No. 14 – "borrowing a corpse for regaining a soul" (well-known methods, ideas, means and leaders are used to solve new problems);
- No. 27 – "making reckless gestures, still preserving the balance" (an opponent is underestimating one's own abilities, intelligence, and awareness).

A detailed description of these stratagems (including their interpretations) can be found in [155].

Stratagem No. 3 is remarkable for the following. Here a conflict explicitly engages *the third party* (besides the active and passive agent). In other words, the active agent uses the third party (pursuing his/her own ends). Actually, the active agent convinces that the third party acts in its own interests; in addition, the active agent forms a corresponding belief for the passive agent. Thus, the impact aims at θ_{23} or θ_{31}, θ_{32}, etc.

Problem B reflects the most common type of implicit control. Notably, the active agent imposes a certain pattern of actions to the passive agent (by forming required beliefs about the pattern of actions of the passive agent). Manipulation appears efficient only if the fact of its existence is not comprehended by the passive agent (see the assumption regarding the awareness structure above).

Consider the following illustrative example. The book [36] describes a psychological experiment conducted by the owner of a beef-importing company in the USA. "The company's customers – buyers for supermarkets and other retail food outlets – were called on the phone as usual by a salesperson and asked for a purchase in one of three ways. One set of customers heard a standard sales presentation before being asked for their orders. Another set of customers heard the standard sales presentation plus information that the supply of imported beef was likely to be scarce in the upcoming months. A third group received the standard sales presentation and the information about a scarce supply of beef, too; however, they also learned that the scarce supply news was not generally available information – it had come, they were told, from certain exclusive contacts that the company had. . . . Compared to the customers who got only the standard sales appeal, those who were also told about the future scarcity of beef bought more than twice as much . . . The customers who heard of the impending scarcity via "exclusive" information . . . purchased six times the amount that the customers who received only the standard sales pitch did." This example incorporates Problem AB: some customers were informed about the state of nature θ_i, others were also reported the beliefs θ_{ij} possessed by the rest agents.

The term "*pattern*" fits the practical interpretations of Problem B. It corresponds to typical connections among situations and control actions most efficient in such situations (see models of *situational control* [131]).

The first possible interpretation is the following. The active agent convinces the passive one that the former adopts a certain pattern. In response, the passive agent chooses some pattern. We obtain a "patterns' equilibrium." Next, the active agent demonstrates a nonstandard behavior (out of his/her pattern). If the passive agent still applies the same pattern, the active one possibly gains a certain payoff. A similar effect occurs when the passive agent has a preset pattern known to the active agent.

Interestingly, the inefficiency of the pattern-based activity of the passive agent is often seeming. For instance, his/her pattern may be efficient "on the average" (as the *unified response* to the behavior of different opponents in alternating external conditions). Formally, patterns can be described as best responses of agents to opponents' actions.

Problem C. It characterizes an infrequent case of implicit control (the impact on θ_{212}). In such situation, the passive agent has specific beliefs θ_2 about the state of nature and the belief about the state of nature of the active agent (i.e., θ_{21}). Both beliefs are known to the active agent. The passive agent *does not* consider the active agent as being adequately aware of him/her (otherwise, $I_{212} = I_2$, and the impact on θ_{212} turns out pointless). The corresponding awareness structure is shown in Fig. 4.5.

The active agent computes the subjective equilibrium of the passive agent:

$$X_2(\theta_2, \theta_{21}, \theta_{212}) = \{x_2 \in X_2, x_{21} \in X_1, x_{212} \in X_2 \mid$$
$$\forall y_2 \in X_2 \quad f_2(\theta_2, x_{21}, x_2) \geq f_2(\theta_2, x_{21}, y_2),$$
$$\forall y_1 \in X_1 \quad f_1(\theta_{21}, x_{21}, x_{212}) \geq f_1(\theta_{21}, y_1, x_{212}),$$
$$\forall y_2 \in X_2 \quad f_2(\theta_{212}, x_{21}, x_{212}) \geq f_2(\theta_{212}, x_{21}, y_2)\}.$$

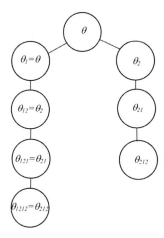

Figure 4.5 The awareness structure in Problems C, AC, BC and ABC.

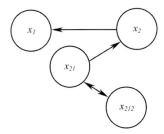

Figure 4.6 The graph of reflexive game in Problems C, AC, BC and ABC.

Suppose that the values of θ_2, θ_{21} are fixed and both agents know them for sure. For the active agent, *the informational control problem* consists in the following. Find a value of θ_{212} such that the action of the passive agent chosen from the set $X_2(\theta_2, \theta_{21}, \theta_{212})$ maximizes the goal function of the active agent:

$$\theta^*_{212}(\theta) = \arg\max_{w \in \Theta} \max_{x_1 \in X_1} \min_{x_2 \in \mathrm{Proj}_2(X_2(\theta_2, \theta_{21}, w))} f_1(\theta, x_1, x_2).$$

Imagine that the active agent can influence the parameter θ_2 and/or θ_{21} in Problem C. Consequently, we derive Problem AC, BC or ABC, where the goal function of the active agent is maximized also with respect to θ_2. That is, *the solution to Problem ABC* takes the form

$$(\theta^*_2(\theta), \theta^*_{21}(\theta), \theta^*_{212}(\theta)) = \arg\max_{u,w,h \in \Theta} \max_{x_1 \in X_1} \min_{x_2 \in \mathrm{Proj}_2(X_2(u,w,h))} f_1(\theta, x_1, x_2).$$

Fig. 4.6 illustrates the graph of reflexive game in this problem.

Problem C arises in stratagem No. 13 ("touching grass to frighten a snake away"), i.e., a display action provokes an opponent to demonstrate his/her actual intentions. In addition, this problem appears in stratagem No. 34 ("the stratagem of persons with a self-inflicted wound"), i.e., some contradictions are demonstrated to exist in one own's camp.

Thus, **the basic problems of implicit control (among the listed seven ones) are Problems A, AB and ABC, differing by unity in reflexion ranks.** The results of Section 4.1 serve for posing and solving these problems in concrete cases.

4.2 THE MASS MEDIA AND INFORMATIONAL CONTROL

Let us adopt the framework of awareness structures to describe informational control (implicit control, manipulations, etc.) implemented by the mass media. We involve advertizing and the hustings as the corresponding examples. This type of informational impact can be viewed as implicit control; however, due to the wide spread nature of implicit control we discuss the latter in a separate section.

Suppose there is an agent representing the object of informational impact. The impact should form a required attitude of the agent to a specific object or subject.

In the case of advertizing, the agent is a customer, while a certain product or a service acts as the object [77, 153]. It is required that the customer purchases the product or service in question.

In the case of the hustings, a voter serves as the agent, and the subject is provided by a candidate. It is required that the voter casts an affirmative vote for the given candidate.

Consider agent i. Combine the rest agents in one agent (in the sequel, we will use subscript j for him/her). Denote by $\theta \in \Theta$ an objective characteristics of the object (being unknown to all agents). For instance, the characteristics could be customer-related properties of the products, personal properties of the candidates, etc.

Let $\theta_i \in \Theta$ be the beliefs of agent i about the object, $\theta_{ij} \in \Theta$ be his/her beliefs about the beliefs of agent j about the object, and so on.

For simplicity, make the following assumptions. First, the set of feasible actions of the agent consists of two actions, *viz*, $X_i = X_j = \{a; r\}$; the action a ("accept") means purchasing the product (service) or voting for the candidate, while the action r ("reject") corresponds to not purchasing the product (service) or voting for alternative candidates. Second, the set Θ is composed of two elements describing the object's properties: g ("good") and b ("bad"), i.e., $\Theta = \{g; b\}$.

Below we study several models of agent's behavior (according to the growing level of their complexity).

Model 0 (reflexion rank 0). Suppose that the behavior of the agent is described by a mapping $B_i(\cdot)$ of the set Θ (object's properties) into the set X_i (agent's actions); notably, we have $B_i: \Theta \to X_i$. Here is an example of such mapping: $B_i(g) = a, B_i(b) = r$. In other words, if the agent believes the product (candidate) is good, then he/she purchases the product (votes for this candidate); if not, he/she rejects the product (or the candidate).

Within the given model, informational control forms specific beliefs of the agent about the object, leading to the required choice (informational regulation [136]). In the example above, the agent purchases the product (votes for the required candidate) if the following beliefs have been a priori formed: $\theta_i = g$. This book does not discuss any technologies of informational impact (i.e., the ways to form specific beliefs).

Model 1 (reflexion rank 1). Suppose that the behavior of the agent is described by a mapping $B_i(\cdot)$ of the sets $\Theta \ni \theta_i$ (object's properties) and $\Theta \ni \theta_{ij}$ (the beliefs of the agent about the beliefs of the rest agents) into the set X_i of his/her actions, notably, B_i: $\Theta \times \Theta \rightarrow X_i$. The following mappings are possible examples:

$$B_i(g,g) = a, \quad B_i(g,b) = a, \quad B_i(b,g) = r, \quad B_i(b,b) = r$$

and

$$B_i(g,g) = a, \quad B_i(g,b) = r, \quad B_i(b,g) = a, \quad B_i(b,b) = r.$$

In the first case, the agent follows his/her personal opinion, while in the second case he/she acts according to the "public opinion" (the opinions of the rest agents).

In fact, for the stated model, informational impact is reflexive control and informational regulation. It serves to form agent's beliefs about the object and about the beliefs of the rest agents, leading to the required choice. In our example, the agent purchases the product (votes for the required candidate) if the following beliefs have been a priori formed: $\theta_i = g$ with arbitrary θ_{ij} (case 1) and $\theta_{ij} = g$ with arbitrary θ_i (case 2).

Moreover, informational impact by the mass media not always aims to form θ_{ij} directly; in the majority of situations, the impact is exerted indirectly when the beliefs about behavior (chosen actions) of the rest agents are formed for the agent in question. Consequently, the latter may adopt these data to estimate their actual beliefs. Examples of indirect formation of the beliefs θ_{ij} could be provided by famous advertizing slogans like "Pepsi: The Choice of a New Generation," "IPod: Everybody Touch." In addition, this could be addressing the opinion of competent people or revealing information that (according to a public opinion survey) the majority of voters are going to support a given candidate, etc.

Model 2 (reflexion rank 2). Suppose that the behavior of the agent is described by a mapping $B_i(\cdot)$ of the sets $\Theta \ni \theta_i$ (object's properties), $\Theta \ni \theta_{ij}$ (the beliefs of the agent about the beliefs of the rest agents) and $\Theta \ni \theta_{iji}$ (the beliefs of the agent about the beliefs of the rest agents about his/her individual interests) into the set X_i of his/her actions, i.e., B_i: $\Theta \times \Theta \times \Theta \rightarrow X_i$. A corresponding example is the following:

$$\forall \theta \in \Theta \quad B_i(\theta, \theta, g) = a, \quad B_i(\theta, \theta, b) = r.$$

It demonstrates some properties being uncommon for Models 1–2. In this case, the agent acts according to his/her "social role" and makes the choice expected by the others (this model resembles ethical models described in Section 4.3.4).

For this model, informational impact is reflexive control; it consists in formation of agent's beliefs about the beliefs of the rest agents about his/her individual beliefs (leading to the required choice). In the example above, the agent purchases the product (votes for the required candidate) if the following beliefs have been a priori formed: $\theta_{iji} = g$.

Interestingly, informational impact not always aims at forming θ_{iji} directly. In many situations, the impact is exerted indirectly when the beliefs about expectations (actions expected from the agent) of the rest agents are formed for the agent in question. The matter concerns the so-called social impact; numerous examples of this phenomenon are discussed in the textbooks on social psychology [120, 156]. Indirect formation of the beliefs θ_{iji} could be illustrated by the slogans like "Do you ... Yahoo!?", "It is. Are you??", "What the well-dressed man is wearing this year" and similar ones. Another example is revelation of information that, according to a public opinion survey, the majority of members in a social group (the agent belongs to or is associated with) would support a given candidate, etc.

Therefore, we have analyzed elementary models of informational control by the mass media. The models have been formulated in terms of reflexive models of decision-making and awareness structures. In all these models[4], reflexion has the maximal rank of 2. One would hardly imagine real situations when informational impact aims at the components of the awareness structure that have a greater depth. Thus, a promising direction of further investigations lies in studying formal models of informational control (and the corresponding procedures) for the agents performing collective decision-making under the conditions of interconnected awareness.

The formal model of informational impact by the mass media is considered in Section 4.19 (using the example of product advertizing).

4.3   REFLEXION IN PSYCHOLOGY

This section adopts the terminology of reflexive games to describe some phenomena and processes studied by psychologists, namely, playing chess, transactional analysis, the Johari window, and ethical choice. The authors do not claim to obtain new results in the field of psychology. Instead, they aim at illustrating the feasibility and reasonability of the framework of reflexive games in the humanities.

4.3.1   Playing chess

A classical example of reflexive analysis concerns *chess*. Here, two aspects of agents' interaction can be discussed. First, choosing a move, each player analyzes the tree of the game – his/her possible moves, possible responses of the opponent, responses to this response, etc. Second, judging a certain position or a set of positions (corresponding to some tree of the game), a player should view it "in opponent's eyes." Studying possible best responses, a player has to forecast the opponent's beliefs about his/her judgement, and so on. For details, we refer to the description of strategic and informational reflexion above. This subsection does not deal with the reflexive aspects of repeated games. Instead, most attention belongs to the reflexive analysis of positions by agents.

Consider the following model. Denote by θ an objective position on the board. Next, let θ_1 be the judgement of this position by agent 1 (for definiteness, suppose that agent 1 is white), θ_2 mean the judgement of this position by agent 2 (black),

---

[4]Probably, the exception is a rare situation when informational impact aims to form the whole awareness structure (e.g., by thrusting a "common opinion" like "Vote by your heart!", "This is our choice!" and so on).

θ_{12} indicate the beliefs of agent 1 about the judgement of the position by agent 2, and so on.

Now, analyze which components of the awareness structure do appear in the judgements. How does an opponent affect the judgement of a position? Such influence is sometimes called a *psychological impact*. We list the major factors of a psychological impact[5]:

- moves on the board;
- the opponent's behavior;
- the pace of a game (the speed of opponent's decisions);
- the opinion regarding the skill level of the opponent;
- the agent's information on the opponent's opinion about the skill levels of both players;
- tournament standings of the opponents, etc.

The following situation provides a striking example of the influence exerted by player's behavior on a game. A Soviet chess grandmaster, M. Tal, played against R. Fischer (another chess virtuoso from the USA) at a World Championship. During a game, M. Tal was heavily attacked, and his position became critical. At that moment, young R. Fischer wrote the next move (the strongest one for that position!) and almost submitted the notebook to the opponent. Later, M. Tal shared his thoughts: "Fischer asks for my approval. But how should I respond? Frowning is not allowed. Smiling can reveal my thoughts." According to M. Tal, he selected the only admissible option – standed up and calmly walked along the stage. R. Fischer was totally amazed by the imperturbable and self-confident behavior of the opponent. He doubted the correctness of the intended plan. As a result, the American player made another (weaker) move, committed one more serious mistake and lost the game[6].

Therefore, the awareness structure includes the components θ_i ($i = 1, 2$), as reflecting the beliefs of players about the position. As a matter of fact, most guides on chess treat the problems of "correct" judgement of positions. Players learn to perform adequate judgements. Recall the list of major factors of a psychological impact. The influence on θ_i corresponds to moves on the board, the opinion regarding the skill level of the opponent, and tournament standings of the opponents.

Textbooks on chess theory focus less attention on the components θ_{ij} ($i, j = 1, 2$). The latter can be influenced by the opponent's behavior and the pace of a game.

Finally, the components θ_{ijk} ($i, j, k = 1, 2$) corresponding to reflexion rank 2 are almost not considered (e.g., the agent's information on the opponent's opinion about the skill levels of both players).

---

[5]The usage of certain factors of psychological impact is not the content but the technology of informational control.

[6]In the example considered, Fischer and Tal exchanged the roles of active agents. Just imagine that the former did not initiated informational impact. Apparently, the game would had evolved differently.

Thus, we naturally arrive at an important conclusion. In chess description[7] involving informational control (in terms of a reflexive game), researchers often adopt the following framework:

1   The number of agents equals 2.
2   Reflexion rank does not exceed 2 (the maximal number of relevant indexes is 3).
3   Most situations of reflexive control in chess are modeled by Problem B (reflexion rank makes up 1).

There exist numerous books, where the technology of reflexive control in chess is described in a vivid, figurative and breathtaking manner. (However, writers generally have no idea that they portray the process and result of informational control). On the other hand, players, psychologists and mathematicians acknowledge that chess represent a convenient "base" for constructing and verifying different models of conflict interaction. Hence, the systematic treatment of formal models for reflexive phenomena arising in chess seems promising.

4.3.2 Transactional analysis

In fact, *transactional analysis* was founded by E. Berne, an outstanding American psychotherapist and psychoanalyst [17]. The essence of his approach consists in the following. At each instant, a human being in a social group recognizes one of his ego-states, namely, *Parent*, *Adult* or *Child*. An agent staying in one of these states (*situations*) creates a transaction *stimulus* to another agent. The latter reacts by a transaction *response*, which (by-turn) represents a stimulus, and so on.

A "stimulus-response" pair forms a *transaction*. Imagine that the source of a stimulus and the addressee of a reaction, as well as the addressee of a stimulus and the source of a reaction coincide. In this case, we obtain a *complementary (reciprocal) transaction* (otherwise, a *crossed transaction* takes place). According to Berne, a game represents "a series of transactions that is complementary (reciprocal), ulterior, and proceeds towards a predictable outcome" [17].

A game engages several agents (probably, collective ones). However, a **transaction is always between two agents.**

We describe an arbitrary transaction in terms of a reflexive game. For convenience, suppose that agent 1 is the initiator. Any transaction between two agents can be defined by $(a \to b)$, $(c \to d)$, where $a, b, c, d \in \Theta = \{$Parent, Adult, Child$\}$ stand for the situations of stimulus source, stimulus addressee, reaction source, and reaction addressee, respectively. A transaction corresponds to one of the finite awareness structures in Fig. 4.7–Fig. 4.8[8], where $\theta_1 = a$ $(\theta_{121} = a), \theta_{12} = b, \theta_2 = c$ $(\theta_{212} = c), \theta_{21} = d$.

Thus, a transaction is rewritten as $(\theta_1 \to \theta_{12})$, $(\theta_2 \to \theta_{21})$ or $(\theta_{121} \to \theta_{12})$, $(\theta_{212} \to \theta_{21})$.

---

[7]In contrast to implicit control, one would hardly distinguish active and passive agents in chess. Indeed, the distribution of roles is a priori unknown and depends on personal qualities of players.
[8]The initiator of a transaction considers himself/herself independently $(a = \theta_1)$ or also adopts the available beliefs of the opponent about himself/herself $(a = \theta_{121})$.

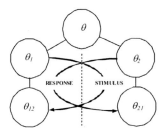

Figure 4.7 Awareness structure I of an transaction.

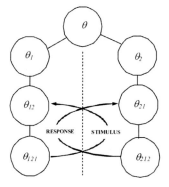

Figure 4.8 Awareness structure 2 of an transaction.

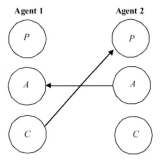

Figure 4.9 An example of a crossed transaction.

The graph of reflexive game has 6 nodes (2 real and 4 phantom agents). In transactional analysis, the specifics of an informational equilibrium lie in the following. Making decisions (occupying the positions of Parent, Adult, or Child), agents 1 and 2 possess different information. Agent 1 models the desired situation, whereas agent 2 states it (fixes the distribution of roles under known choice of agent 1).

Study an example of a transaction illustrated by Fig. 4.9.

Here agent 1 "addresses" agent 2 from the position of Child (i.e., suggests agent 2 to treat him/her as Child: θ_{121} = "Child"). Agent 1 considers agent 2 as Parent (θ_{12} = "Parent"). Agent 2 "responds" to agent 1 as Parent to Parent (i.e., suggests

Others I	What others know about myself	What others don't know about myself
What I know about myself	I	II
What I don't know about myself	III	IV

Figure 4.10 The Johari window.

agent 1 to treat him/her as parent: $\theta_{212} =$ "Parent"). Still, agent 2 considers agent 1 as Parent: $\theta_{21} =$ "Parent."

Finally, note that the maximal rank of agents' reflexion in transactional analysis constitutes 2.

4.3.3 The Johari window

In social psychology, a common method of analyzing the potential variability of personal characteristics in a social environment employs the *Johari method* (see Fig. 4.10). It was named after Joseph Luft and Harrington Ingham, the originators [103, 104].

This method expresses an agent in two dimensions, *viz.*, "I" and "Others." It serves for training of employees that (due to their activity) should understand the following. How and why do others have a different opinion of these employees (varying from their own opinion)?

Quadrant I (see Fig. 4.10) is the *Open quadrant* (it contains information on an agent, which appears known to him/her and others).

Quadrant II is the *Hidden quadrant* (it contains information on an agent, which appears known to him/her and unknown to others).

Quadrant III is the *Blind Spot quadrant* (it contains information on an agent, which appears known to others and unknown to him/her).

And finally, Quadrant IV is the *Unknown quadrant* (it contains information on an agent, which appears known to others and him/her).

Within the framework of reflexive games studied in this book, denote by θ the objective information on an agent, by θ_1 the agent's subjective information on himself/herself, and by θ_2 the information of other agents on this agent. Without loss of generality, suppose that all informational components (θ, θ_1, and θ_2) are subsets of a universal set Θ.

Evidently, the Open quadrant (quadrant I in Fig. 4.10) corresponds to the information $\theta_1 \cap \theta_2$, the Hidden quadrant (quadrant II) corresponds to the information $\theta_1 \cap (\theta \backslash \theta_2)$, the Blind Spot quadrant (quadrant III) corresponds to the information $\theta_2 \cap (\theta \backslash \theta_1)$, and the Unknown quadrant (quadrant IV) corresponds to the information $(\theta \backslash \theta_1) \cap (\theta \backslash \theta_2)$. The sum of all four sets yields the universal set θ.

Here description uses a certain external (objective) viewpoint. Notably, we implicitly believe that $\theta_1 \subseteq \theta, \theta_2 \subseteq \theta$. Reflexion rank equals 0. Reject the objective picture and consider the situation from an agent's viewpoint (e.g., take the agent engaged in a Johari window). Consequently, reflexion rank makes up 1. If there are two agents only ("I" and "others"), reflexion rank does not exceed 2. Furthermore, it is possible to construct Johari windows existing in the minds of phantom agents (τ-agents, where $|\tau| \geq 2$). In other words, the awareness structure of a two-player reflexive game can be interpreted as the set of τ-subjective Johari windows.

4.3.4 Ethical choice

We have repeatedly underlined that this book deals with second-type reflexion (related to the beliefs of a subject about the beliefs of other subjects) rather than self-reflexion. In economic models with rational individuals, self-reflexion gains nothing. Indeed, "I know that I know that I know ..." does not provide new information in comparison with "I know that ..." This fact matches the axiom of self-awareness (see Section 2.2).

Meanwhile, the situation changes in psychology. This science considers a human being as a complicated integral object.

V.A. Lefebvre [96–99] and other researchers [169] analyzed the reflexive model of decision-making by a human being obeying some system of cultural and ethical norms. In particular, the book [96] describes a model of game interaction between two agents. Each of them chooses between two alternatives (the action is the choice probability for alternative 1) under the utilitarian aspect and the ethical aspect (see the Introduction). The utilitarian aspect corresponds to maximization of one's own payoff, whereas the final choice takes into account the ethical "load" of alternatives. Following the cited work, let us present the *model of ethical choice* in terms of the concept of an informational equilibrium (developed in our book).

For each of two agents (denote them by indexes 1 and 2), it is possible to separate out the utilitarian aspect of their choice in the form of "agents" 1_u and 2_u, respectively. And the process of final choice is organized as follows (see Fig. 4.11).

First, "agents" 1_u and 2_u "play" a game with a common knowledge. They evaluate the equilibrium strategies $x_{1_u}^*$ and $x_{2_u}^*$ (interpreted as the choice probabilities for alternative 1). Subsequently, agents 1 and 2 play another game for the final choice (the pair $(x_{1_u}^*, x_{2_u}^*)$ is a common knowledge for them). The agents search for the action vector meeting the system

$$
(1) \quad \begin{cases} x_1 \in BR_1(x_2, x_{1_u}^*, x_{2_u}^*), \\ x_2 \in BR_2(x_1, x_{1_u}^*, x_{1_u}^*). \end{cases}
$$

The solution to the system (1) gives the numbers x_1^* and x_2^* (the final choice probabilities of alternative 1).

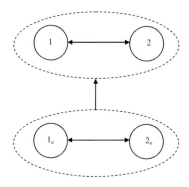

Figure 4.11 The model of ethical choice.

4.4   REFLEXION IN BÉLLES-LÉTTRES

The following aspect has been emphasized earlier. The graph of a reflexive game can be constructed without concretizing the goal functions of agents. In this case, it reflects the qualitative relationship between the awarenesses of reflexing agents. We give several examples from art works.

Scientific literature contains a subjective description of an objective reality and strives for maximal objectivization. Contrariwise, imaginative literature (also known as *bélles-léttres*) naturally has reflexion – any fiction portrays a reflexive reality, i.e., results from author's reflexion.

Moreover, *plots of many literary works proceed from the noncoincidence of an objective reality and/or reflexive realities of heroes.* We elucidate this below.

Set aside trivial "reflexion" such as "There was a pope with a dog ..." (from *The Tale of the Pope and of His Workman* by A.S. Pushkin). Each literary work includes a group of characters – people playing some roles[9] (situational, communicative, social roles, etc.). The roles depend on a surrounding environment and a hero interacts with them.

Interestingly, the same hero may act in different roles (he/she corresponds to different phantom agents). And the perception of this hero by himself/herself possibly differs from that by other heroes.

Some plots involve the change or exchange of roles. The humor (or even tragedy) consists in a mismatch among roles of the same character (see non-compliance with a pattern in Section 4.1).

Describe formally possible cases of the mutual awareness of two subjects, "*Hero*" and "*Environment.*" Denote them by "H" and "E," respectively.

In addition, we introduce the following system of symbols: H – nonreflexive beliefs of a hero about an objective reality (all except the hero and his/her environment); E – nonreflexive beliefs on the environment about an objective reality. The hero and his/her environment represent real agents, whereas the following agents are phantom: HH – the beliefs of the hero about himself/herself; HE – the beliefs of the hero about the environment; HEH – the beliefs of the hero about the beliefs of the environment about him/her; EE – the beliefs of the environment about itself; EH – the beliefs of the environment about the hero; EHE – the beliefs of the environment about the beliefs of the hero about the environment[10].

Finite awareness structures (reflexion rank equal 2) of the hero and his/her environment can be found in Fig. 4.12. Here, the components of an awareness of the hero (or about the hero) are drawn by thick lines.

We discuss the meaning of nodes in the graph demonstrated by Fig. 4.12. According to the hero, the node HH corresponds to self-reflexion (the perception of his/her own "role"), the node HE corresponds to the role of the environment, and the node

---

[9]The following question is separate and goes beyond the scope of our research (still, it is traditionally raised in discussions about artistry). How deep should an actor get the feel of his/her role? There exists no definite answer even among professionals. Some experts believe in complete internal identification of a character by an actor. Others declare that an actor should control the difference between himself/herself and a character.

[10]Practical interpretations of HHE, HHH, HEE, EEE, EEH, EHH seem difficult.

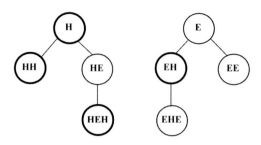

Figure 4.12 The awareness structures of the hero and his/her environment.

HEH corresponds to the hero's role in the eyes of the environment (to the hero's role by the normative viewpoint). By analogy, one easily characterizes nodes for the environment.

Recall that many plots (e.g., "situation comedy") employ the mismatch (a conflict) among the roles of the same subject (probably, a collective one). The awareness structure in Fig. 4.12 enables listing possible conflicts[11].

First of all, distinguish between *internal conflicts* and *external conflicts*. The former are realized by a subject within his/her awareness (i.e., incorporate conflicts arising between components of his/her own awareness structure). The latter occur between components of the awareness structures of the subject and his/her environment. And so, four components of awareness (see the shaded ones in Fig. 4.12) admit six types of conflicts. Below we mention them and provide examples[12].

There exist three types of internal conflicts:

1   *The mismatch between H and HH* – an internal conflict between a hero and his/her beliefs about himself/herself. For instance, consider almost all literary works created by F.M. Dostoevsky (the master and wizard of self-reflexion effects); *Childhood*, *Boyhood*, and *Youth* by L.N. Tolstoy; all works belonging to the genre of confession, including St. Augustine Aurelius, A. Musset, J.-J. Rousseau, L.N. Tolstoy, N.A. Berdyaev, and others.

2   *The mismatch between H and HEH* – an internal conflict between a hero and his/her beliefs about his/her role (in the eyes of an environment). The examples are *The Hero of Our Time* by M.Yu. Lermontov (the character of Pechorin); *The Adolescent* by F.M. Dostoevsky, *Father Gorio* by H. de Balzaq (the character of E. de Rastignac), and others.

3   *The mismatch between HH and HEH* – an internal conflict between the values of an environment and a hero (according to the latter). Here the examples are *Crime and Punishment*, *Notes from Underground*, *The Gambler*, *Demons* by F.M. Dostoevsky; *Princess Mary* by M.Yu. Lermontov; *Lady Macbeth of Mtsensk* by N.S. Leskov; *Lost Illussions* by H. de Balzaq; most authors of marginal literature, namely, F. de Sade, H. Miller, V. Erofeev, and others.

---

[11]Many literary works include conflicts of different types simultaneously.
[12]The authors are grateful to Prof. E.V. Zharinov for valuable remarks and assistance in selection of the examples of reflexive conflicts in bélles-léttres.

In addition, there exist *three types of external conflicts*:

4   *The mismatch between H and EH* – an external conflict between a hero and
    the beliefs (requirements) of an environment about him/her. For instance, *Woe
    from Wit* by A.S. Griboedov (the character of Chatsky), *Notre Dame de Paris* by
    V. Hugo; many classic literary works of Russian and foreign realism: L.N. Tolstoy
    (e.g., *Three Deaths, The Death of Ivan Ilych*, etc.), I.S. Turgenev, M.E. Saltykov-
    Shchedrin, J.B. Molière, P. Corneille, J. Racine, and others.
5   *The mismatch between HH and EH* – an external conflict between the beliefs of a
    hero about himself/herself and the beliefs about him/her by an environment. Here
    we mention *Eugene Onegin* by A.S. Pushkin; *The Hero of Our Time* by M. Yu.
    Lermontov (the characters of Grushnitsky and Pechorin); *Rudin* by I.S. Turgenev,
    Beltov by A.I. Herzen, and others.
6   *The mismatch between EH and HEH* – an external conflict between the beliefs
    of an environment about a hero and the vision of these beliefs by the hero. See
    The Inspector-General by N.V. Gogol (the character of Хлестаков); *The Little
    Tragedies* by A.S. Pushkin; *Gobseck* by H. de Balzaq, and others.

These six types of conflicts are typical in classical literature. The matter turns out
somewhat simpler in modern literature. Most plots belong to one of the following
types: "Detective story," "Spy story," "Eternal triangle (or polygon)." Let us provide
examples of the corresponding graphs of reflexive game.

Example 4.4.1 ("Detective story"). Consider an investigation officer and an
offender. Denote them by 1 and 2, respectively. Consequently, the procedure of crime
detection is described by the graph of the reflexive game in the form $2 \leftarrow 1 \leftrightarrow 12$ (here
the component 12 corresponds to that the offender strives for convincing the investigation officer in his own innocence). The fact of crime detection is described by the
graph $1 \leftrightarrow 2$.

More sophisticated cases of awareness are also possible. For instance, Smerdyakov
and Ivan Fedorovich (*The Brothers Karamazov* by F.M. Dostoevsky) possess nonidentical awareness about the murder of the father and the attitude of each other to this
fact. In the eyes of Smerdyakov, the situation (the graph of reflexive game) has the
form "Smerdyakov" ← "Ivan Fedorovich wishes father's death" ↔ "Smerdyakov
is the murderer." According to Ivan Fedorovich, the situation appears as "Ivan
Fedorovich" ← "Smerdyakov is innocent" ↔ "Ivan Fedorovich does not wish father's
death."

A similar situation takes place in *Crime and Punishment*. Raskol'nikov does not
know that the investigation officer knows he is the murderer. Denote them by 1 and
2, respectively. In the mind of Raskol'nikov, the graph becomes $1 \leftarrow 12 \leftrightarrow 121$. On
the other hand, the complete graph of reflexive game acquires the form presented by
Fig. 4.13. •

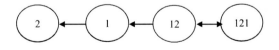

Figure 4.13 The graph of reflexive game in "Detective story".

Example 4.4.2 ("Spy story-1"). Suppose that two states (A and B) and an agent play the following game. The agent represents a high-level official of state A and (simultaneously) an intelligencer of state B; this fact is unknown to state A. The graph of reflexive game in such situation[13] can be found in Fig. 4.14. The nodes of the graph indicate the following (real and phantom) agents: 1 – state A; 2 – state B; 3 – the agent; 12 – state B perceiving the agent as a faithful official of state A; 13 – the agent as a faithful official of state A.

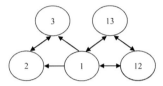

Figure 4.14 The graph of reflexive game in "Spy story1".

An uncertain parameter consists in the belief about the actual status of the official, i.e., $\Theta = \{f, i\}$ (a faithful official, an intelligencer). These beliefs are reflected by horizontal (f) and diagonal (i) shading, see Fig. 4.15.

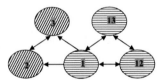

Figure 4.15 The graph of reflexive game in "Spy story1" (including uncertainty).

Does the mechanism of simple messages (see Section 2.12) lead the above awareness structure? The graph shows the following. The basis of the awareness structure contains no agents from the same world with identical last indexes. This is good, since the awareness structures of such agents do coincide under a simple mechanism. Let us verify the premise of Assertion 2.12.9: $\theta_{21} = \theta_1 = f = \theta_{12}$; $\theta_{31} = \theta_1 = f = \theta_{13}$; $\theta_{23} = \theta_3 = i = \theta_2 = \theta_{32}$ (in each chain of equalities, we are concerned with left and right terms). By virtue of the assertion, we obtain the affirmative answer to the question. For instance, take the following sequence of messages: $s_1 = \{1, 2, 3\}[f]$, $s_2 = \{2, 3\}[i]$, where s_2 means the recruitment of the official of state A by state B. •

Next, study a slightly complicated modification of the previous plot.

Example 4.4.3 ("Spy story-2"). The situation resembles the one described in Example 4.4.2. The difference is that the agent actually works for state A (and sends specially made information to state B). In this case, the graph of reflexive game is demonstrated by Fig. 4.16.

---

[13]Clearly, a similar awareness is natural for "Eternal triangle."

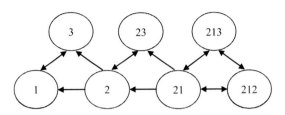

Figure 4.16 The graph of reflexive game in "Spy story-2".

The nodes of this graph correspond to the following (real and phantom) agents: 1 – state A; 2 – state B; 3 – the agent; 21 – state A believing wrongly that the agent represents its official having no contacts with state B; 23 – the agent working for state B; 212 – state B having no contacts with the agent as a high-level official of state A; 213 – the agent being a faithful official of state A, having no contacts with state B. As a result, $\Theta = \{f, i, d\}$ (a faithful official, an intelligencer, a double agent). By analogy, the graph in Fig. 4.17 uses shading for the beliefs of players: f – horizontal shading, i – diagonal shading, d – vertical shading.

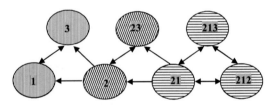

Figure 4.17 The graph of reflexive game in "Spy story-2" (including uncertainty).

Unfortunately, even such a slightly complicated situation is impossible in the mechanism of simple messages. The basis of the awareness structure comprises agents from the same world with identical last indexes. Notably, $I_{12} = I_2 \neq I_{212}$ (agents 12 and 212 belong to the same world). Alternatively, apply Assertion 2.12.9, since $\theta_{12} = \theta_2 = i \neq f = \theta_{212}$. •

For all examples discussed in this section, the maximal rank of reflexion equals 2 (and the length of the maximal sequence of indexes makes up 1). In literary works, higher reflexion ranks appear "once in a blue moon." Still, some examples do exist.

Example 4.4.4. *The Emperor and the Assassin* (1998), a movie directed by Kaige Chen, describes the interaction of a Chinese emperor and an assassin. The latter is sent to the former as an ambassador of a neighboring state. Meanwhile, the emperor knows that the ambassador is an assassin. And the assassin knows this, as well.

The graph of reflexive game in this situation is illustrated by Fig. 4.18.

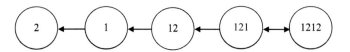

Figure 4.18 The graph of reflexive game in the movie *The Emperor and the Assassin*.

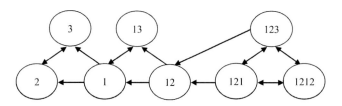

Figure 4.19 The role of emperor's wife in the movie. The Emperor and the Assassin.

The nodes of the graph stand for the following (real and phantom) agents: 1 – the emperor; 2 – the assassin; 12 – the assassin believing that the emperor knows nothing about him; 121 – the emperor believing that the visitor is an ambassador of a neighboring state; 1212 – an ambassador of a neighboring state.

The role of emperor's wife in the movie's intrigue can be observed in the graph of reflexive game (see agent 3 in Fig. 4.19). •

Concluding this section, we emphasize the following. In recent years, the existence of several reflexive (virtual, probably, embedded) realities underlies the plots of many movies. In this context, we mention *The Matrix*, *The Thirteenth Floor*, *Vanilla Sky*, *Avalon*, *The Truman Show*, and others. Interested readers would easily draw the corresponding graphs of reflexive games (using the approach suggested here).

Therefore, the language of graphs of reflexive game is a convenient tool for uniform description of reflexion effects in literary works.

Summarizing the discussion of applied models of reflexive games, we arrive at the following conclusion. Formal models serve for systematic unified formulation and solution (с единых методологических позиций) of analysis and synthesis problems for efficient informational impacts in different situations of collective activity. Nowadays, a shortcoming of such approach relates to its normative character. Notably, there is no opportunity to design efficient technology of informational impact (still, this is the matter of craftsmanship). And a promising subject of future[14] investigations concerns mathematical modeling in this field.

---

[14]Being honest with readers, we acknowledge the following. There exist no systematic and complete results of experimental (not retrospective!) studies in this field. And so, one should not expect a breakthrough here in the coming years.

4.5   REFLEXIVE SEARCH GAMES

This section focuses on the interaction of two players representing "moving points" in a "corridor" with a closed side (i.e., a half-strip). Notably, we have an evading player (e.g., a submarine) and a searching player (e.g., a helicopter or an anti-submarine ship). The evading player chooses a point of hiding, whereas the searching player chooses his/her speed. Each player possesses some beliefs about the maximal distance of detection. Moreover, each player has certain beliefs about opponent's beliefs, the beliefs about the beliefs, and so on. It appears that the maximal rational rank of reflexion of the searching player (the evading player) makes up 3 (2, respectively). Depending on the awareness structure, the evading player chooses between two equilibrium strategies, namely, "standing still to postpone the instant of detection" and "attempting to move away as soon as possible."

Search game: the case of common knowledge. Searching for an active evading object is a typical example of conflict interaction. The methods of game theory assist in its study. Elementary game-theoretic problems of search were analyzed in the late 1960s (in particular, see the classical monograph [1]). The general theory of search games has not been developed to date; a series of results are obtained in [43, 44, 145, 146].

Consider a *search game*, where two players control point objects, *viz.*, the searching object (1) and the evading object (2). In the sequel, we identify players and objects controlled by them. The search area is the rectangle $\{(x, y) \mid 0 \le x \le D, 0 \le y \le 2L)\}$, $D, L > 0$, on the Euclidean plane; the objects move within this rectangle. The searching object starts from point (D, L) and moves along the segment $y = L$ with a chosen speed α $(0 \le \alpha \le A, A > 0)$. At a certain instant, it detects object 2, if the distance between the objects has reached the quantity $l = \theta - k\alpha, k > 0$ (at this instant). In practical interpretation, l means the level of "sight" of the searching object (the maximal distance of detection). Of course, the level of sight decreases as the searching object accelerates. As its location, the evading object can choose a point inside the rectangle. Evidently, this point lies on the segment $y = 0$ or $y = L$. Thus, suppose that the choice of object 2 is described by a number δ $(0 \le \delta \le D)$, i.e., the abscissa of the point used for hiding.

The searching object strives for detecting the evading one (at the minimal time). Assume that detection takes no place. Then the searching object benefits, if the evading object is located as close to it as possible (the distance between them is characterized by δ).

According to the aforesaid, the goal function of the searching object is defined by

$$(1) \quad f(\alpha, \delta) = \begin{cases} M - \frac{D - \delta}{\alpha}, & \theta \ge L + k\alpha, \\ -\frac{D}{\alpha} - c(\delta), & \theta < L + k\alpha, \end{cases}$$

where $M > 0$ specifies detection "bonus" (detection occurs if $l \ge L$), and $c(\delta)$ is an increasing function such that $c(0) = 0$; it reflects the losses of player 1 in the case of no detection. This game is antagonistic – the interests of player 1 are opposite to the interests of player 2.

Therefore, the game is fully described by the goal function $f(\alpha, \delta)$, the positive parameters D, L, A, M, k, θ and the function $c(\delta)$; they all form a common knowledge for players 1–2.

To find the saddle point in the game (1), evaluate $\max\limits_{\alpha} \min\limits_{\delta} f(\alpha, \delta)$:

$$\min_{\delta} f(\alpha, \delta) = \begin{cases} M - \frac{D}{\alpha}, & \theta \geq L + k\alpha, \\ -\frac{D}{\alpha} - c(D), & \theta < L + k\alpha, \end{cases}$$

$$\max_{\alpha} \min_{\delta} f(\alpha, \delta) = \max\{M - \frac{D}{\alpha^0}, -\frac{D}{A} - c(D)\},$$

where $\alpha^0 = (\theta - L)/k$ indicates the speed such that $l = L$. And so, detection occurs in most "economical" way.

Suppose validity of the following conditions:

$$(2) \quad \alpha^0 \leq A, \quad M - \frac{D}{\alpha^0} \geq -\frac{D}{A}.$$

The first condition in (2) states that the speed α^0 is admissible for object 1; the second condition shows that object 1 yields an appreciable extra gain by detection. We note that the first condition has purely technical character (it simplifies further exposition). Yet, the second condition is fundamental – it ensures the existence of an equilibrium in the game (1).

Under the conditions (2), we have

$$\max_{\alpha} \min_{\delta} f(\alpha, \delta) = M - \frac{D}{\alpha^0} = f(\alpha^0, 0).$$

Similarly,

$$\max_{\alpha}(\alpha, \delta) = \max\left\{M - \frac{D - \delta}{\alpha^0}, -\frac{D}{A} - c(\delta)\right\}.$$

By virtue of (2), write down the chain of inequalities

$$M - \frac{D - \delta}{\alpha^0} \geq M - \frac{D}{\alpha^0} \geq -\frac{D}{A} \geq -\frac{D}{A} - c(\delta),$$

Therefore, $\max\limits_{\alpha} f(\alpha, \delta) = M - \frac{D-\delta}{\alpha^0}$, and

$$\min_{\delta} \max_{\alpha} f(\alpha, \delta) = M - \frac{D}{\alpha^0} = f(\alpha^0, 0).$$

Under the condition (2), the game (1) admits a unique equilibrium $(\alpha^0, 0)$.

Informational reflexion. In the current subsection, we consider informational reflexion in the search game (1). Imagine that the function $f(\alpha, \delta)$ and the parameters D, L, A, M, k form a common knowledge. Players 1 and 2 possess point-type regular awareness structures regarding the parameter θ ($I_1 = (\theta_1, \theta_{12}, \theta_{121}, \ldots)$ and $I_2 = (\theta_2, \theta_{21}, \theta_{212}, \ldots)$), respectively). Regular and point-type awareness structures are discussed in Chapter 2. In addition, suppose that both players know about the validity of the conditions (2). Rewrite them as the two-sided inequality

$$(3) \quad L + \frac{k}{MD^{-1} + A^{-1}} \leq \theta \leq L + kA.$$

Concerning the function $c(\delta)$, it suffices that both players are aware of its increase and the fact that $c(0) = 0$.

Consider agent's decision-making (in the ascending order of their reflexion ranks, starting from rank 2).

a) Let the beliefs of player 1 be characterized by the graph $1 \leftarrow 12 \leftrightarrow 121$ (reflexion rank 2). Then the 12-player (player 2 in the mind of player 1) chooses the action $\delta_{12} = 0$. The best response of player 1 lies in choosing the speed

$$(4) \quad \alpha_1^0 = \frac{\theta_1 - L}{k}.$$

b) Let the beliefs of player 2 be characterized by the graph $2 \leftarrow 21 \leftrightarrow 212$ (reflexion rank 2). Then the 21-player chooses the speed $\alpha_{21}^0 = \frac{\theta_{21} - L}{k}$. The action of player 2 depends on the occurrence of detection (according to his/her view). Assume that detection takes place (the condition $\theta_2 - k\alpha_{21}^0 \geq L$, or, equivalently, $\theta_2 \geq \theta_{21}$ holds true). Then the best response is $\delta_2 = 0$. Otherwise (under $\theta_2 < \theta_{21}$), the best response makes up $\delta_2 = D$ (index 2 marks the action of real player 2).

c) Let the beliefs of player 1 be characterized by the graph $1 \leftarrow 12 \leftarrow 121 \leftrightarrow 1212$ (reflexion rank 3). The case b) implies that two situations are feasible: $\theta_{12} \geq \theta_{121}$, $\delta_{12} = 0$ and $\theta_{12} < \theta_{121}, \delta_{12} = D$. In situation 1, the best response is α_1^0 (see (4)). In situation 2, the best response consists in any value of $\alpha_1 \in [0, \alpha_1^0]$.

As a result, we obtain

$$(5) \quad \alpha_1 \begin{cases} = \alpha_1^0 = \frac{\theta_1 - L}{k}, & \theta_{12} \geq \theta_{121}, \\ \in [0, \alpha_1^0], & \theta_{12} < \theta_{121}. \end{cases}$$

d) Let the beliefs of player 2 be characterized by the graph $2 \leftarrow 21 \leftarrow 212 \leftrightarrow 2121$ (reflexion rank 3). By analogy, we have

$$\alpha_{2121} = \alpha_{2121}^0, \quad \delta_{212} = 0, \quad \alpha_{21} = \alpha_{21}^0 = \frac{\theta_{21} - L}{k},$$

$$(6) \quad \delta_2 = \begin{cases} 0, & \theta_2 \geq \theta_{21}, \\ D, & \theta_2 < \theta_{21}. \end{cases}$$

This is similar to the case b).

e) Let the beliefs of player 1 be characterized by the graph $1 \leftarrow 12 \leftarrow 121 \leftarrow 1212 \leftrightarrow 12121$ (reflexion rank 4). As in the case c), two situations are possible: $\theta_{12} \geq \theta_{121}, \delta_{12} = 0, \alpha_1 = \alpha_1^0 = \frac{\theta_1 - L}{k}$ and $\theta_{12} < \theta_{121}, \delta_{12} = D, \alpha_1 \in [0, \alpha_1^0]$.

f)   Let the beliefs of player 2 be characterized by the graph $2 \leftarrow 21 \leftarrow 212 \leftarrow 2121 \leftrightarrow 21212$ (reflexion rank 4). According to the case *c)*, we have two situations: $\theta_{212} \geq \theta_{2121}$, $\delta_{212} = 0$, $\alpha_{21} = \alpha_{21}^0 = \frac{\theta_{21} - L}{k}$ and $\theta_{212} < \theta_{2121}$, $\delta_{212} = D$, $\alpha_{21} \in [0, \alpha_{21}^0]$. In situation 1, the best response is defined by (6).

Situation 2 seems slightly complicated for analysis. Indeed, player 2 may expect any action from the segment $[0, \alpha_{21}^0]$ from player 1. We deal with interval-type uncertainty. A common way of eliminating such uncertainty lies in evaluation of the maximal guaranteed result. Obviously,

$$\max_{\alpha \in [0, \alpha_{21}^0]} f(\alpha, \delta) = \begin{cases} M - \frac{D - \delta}{\alpha_{21}^0}, & \theta_2 \geq \theta_{21}, \\ -\frac{D}{\alpha_{21}^0} - c(\delta), & \theta_2 < \theta_{21}. \end{cases}$$

Thus, the guaranteeing action $\delta_2 = \arg\min_{\delta} \max_{\alpha \in [0, \alpha_{21}^0]} f(\alpha, \delta)$ is again determined by the expressions (6).

As the rank of players' reflexion grows, the set of their subjective equilibrium actions is not enlarged (as compared with reflexion rank 2 for player 2 and reflexion rank 3 for player 1). Rank 3 for player 1 and rank 2 for player 2 indicate the *maximal rational* ranks of their reflexion. We formulate this idea as

Assertion 4.5.1. Consider the reflexive search game (1). The maximal rational ranks of reflexion equal 3 (the searching player) and 2 (the evading player).

Proof (by induction).   Basis of induction: if the rank of player 1 makes up 3, then his/her action is defined by formulas (5); if the rank of player 2 constitutes 2 or 3, then his/her action is defined by formulas (6). These cases have been studied above.

Now, consider decision-making by player 1 with reflexion rank n ($n \geq 4$). The rank of the 12-player equals $n - 1$, and his/her action is determined by (6) (by supposition). Consequently, the best response of player 1 (see the case *c)*) satisfies the expressions (5). Again, the expressions (5) correspond to the best response of player 1 with reflexion rank 3. And part I of the assertion is immediate.

Assume that player 2 with reflexion rank n makes a decision. By supposition, the action of the 21-player with reflexion rank $n - 1$ is defined by formulas (5). The best response of player 2 to these actions is described by (6), see the case *f)*. And the expressions (6) characterize the best response of player 2 with reflexion rank 2 or 3. And part II of the assertion follows. ●

In practice, Assertion 4.5.1 means the following. Player 2 either stands as farther as possible from player 1 (by choosing $\delta_2 = 0$, if player 2 believes in detection), or attempts to move away as soon as possible (by choosing $\delta_2 = D$). A strategy $0 < \delta_2 < D$ is not an equilibrium strategy for any awareness structures.

The equilibrium strategy of player 1 represents $\alpha_1 = \alpha_1^0 = \frac{\theta_1 - L}{k}$ (if $\delta_{12} = 0$) or any value from the segment $[0, \alpha_1^0]$ (if $\delta_{12} = D$).

As a result, the capabilities of informational control by player 2 are exhausted; both equilibria turn out attainable under reflexion rank 2. Assume that the principal can affect the beliefs of player 1 about his/her capabilities (reflected by the parameter θ).

Then all depends on whether player 1 overestimates them or not. If he/she does, detection occurs (otherwise, no detection takes place).

Suppose that, as the game is played, both players observe the fact of detection (or undetection) and the choice of player 2 (δ_2) in the case of detection. Consequently, the informational equilibrium enjoys stability if $\theta_2 \geq \theta_{21}$, $\theta_1 \leq \theta$. This has the following interpretation:

1) player 2 believes that player 1 does not overestimates his/her capabilities;
2) this is actually the case (player 2 does not necessarily has an adequate estimate for the capabilities of player 1).

Some generalizations. Finally, we discuss possible generalizations of the results (to the case of more sophisticated search situations). Search often consists in "combing" a search set. Imagine that the evading player can move during the game (instead of simple choice of its location at the beginning). Under appropriate conditions imposed on the speeds of players and the parameters of a search set, detection is possible by systematic "combing" (the so-called second-type search problems). For several search sets, the strategies (paths) of "combing" and evasion were proposed in [43, 44]. Construction of such paths uses the properties of variables of informational sets characterizing the awareness of the searching player about the location of the hiding one.

If the searching player adopts these strategies, the evading player again has 2 alternatives ("standing still" and "moving away"). Increasing the rank of reflexion yields no "intermediate" equilibrium strategies.

4.6   MANUFACTURERS AND INTERMEDIATE SELLERS

In this section, we discuss a model which involves an agent manufacturing a certain product and a principal acting as an intermediate seller between the agent and a market. The intermediate seller is supposed to know the exact market price (actually, the manufacturer does not know it). The manufacturer and intermediate seller a priori negotiate their shares in the income. Afterwards, the intermediate seller reports to the manufacturer (not necessarily true) information on the market price. Finally, the manufacturer chooses an output (the amount of products to-be-manufactured). The intermediate seller's message with the market price can be viewed as informational control. Stable informational control guarantees that the real income of the manufacturer coincides with the expected one (on the basis of the message reported by the intermediate seller). It appears that (by a proper choice of informational control) the intermediate seller ensures his/her maximal income irrespective of the shares (in other words, the intermediate seller can agree with any offer of the manufacturer, since he/she would be able to gain almost any payoff). Interestingly, sometimes the manufacturer makes higher profits than in the case of truth-telling by the intermediate seller.

Consider the interaction of an agent manufacturing a certain product and a principal being an intermediate seller of the product. They interact as follows:

1) negotiate the shares λ and $(1 - \lambda)$ of the manufacturer and intermediate seller, respectively, in the income ($\lambda \in (0; 1)$);

2)   the intermediate seller reports to the manufacturer the estimate $\tilde{\theta}$ of the *market price* θ;
3)   the manufacturer makes a certain *amount of the product* $y \geq 0$ and passes it to the intermediate seller;
4)   the intermediate seller vends the product at the market price and gives the income $\lambda\,\theta\,y$ to the manufacturer (the intermediate seller keeps the income $(1-\lambda)\theta y$).

The model presumes that the intermediate seller knows the exact market price (in contrast, the manufacturer possesses no a priori information about the price).

The manufacturer is characterized by the cost function $c(y)$, relating the product output and the corresponding manufacturing costs. Suppose there exist no output constraints – any amount of the product can be manufactured.

In the stated situation, one would identify three key parameters, notably, the share λ, the price θ and the product output y. Both sides negotiate the share in advance, the price is reported by the intermediate seller, while the manufacturer chooses the product output.

Now, analyze the behavior of the participants as soon as the shares λ and $(1-\lambda)$ have been settled. Striving to maximize his/her profits, the manufacturer chooses the product output y^* depending on his/her cost function, his/her share and the market price reported by the intermediate seller. Assume that the manufacturer trusts the intermediate seller, and the former has no opportunity to verify truth-telling of the latter. In this case, the intermediate seller may generally report a certain value $\tilde{\theta}$ not coinciding with the actual value θ of the market price. The message $\tilde{\theta}$ by the intermediate seller can be treated as informational control.

Finally, imagine that the intermediate seller adheres to stable informational control; he/she endeavors to guarantee to the manufacturer the income expected by the latter on the basis of the value $\tilde{\theta}$.

According to the above assumptions, the goal functions of the intermediate seller and manufacturer make up $f_0(y, \tilde{\theta}) = \theta y - \tilde{\theta}\,\lambda y$ and $f(y, \tilde{\theta}) = \tilde{\theta}\lambda y - c(y)$. The goal functions are rewritten by accounting the stabilization effect (income reallocation by the principal – the intermediate seller). This is done for control stability, see Section 2.7.

Require the cost function to be such that the manufacturer's profits (the difference between the income and costs) attains a unique maximum at a point $y^* = y^*(\tilde{\theta}) > 0$. It suffices that the function is twice differentiable and satisfies the conditions

$$c(0) = c'(0) = 0, \; c'(y) > 0, \; c''(y) > 0 \quad \text{for } y > 0,$$

$$c'(y) \to \infty \quad \text{as } y \to \infty.$$

Moreover, suppose that the following properties take place: $(yc'(y))'$ is a continuous increasing function tending to infinity as $y \to \infty$.

Then the following result is immediate.

Assertion 4.6.1.
1)   By choosing an optimal value $\tilde{\theta}$ (according to his/her viewpoint), the intermediate seller can ensure the maximal value of his/her goal function irrespective of λ.

2) There exist a quantity $\lambda^* = \lambda^*(\theta)$ such that
   a) if $\lambda = \lambda^*$, then truth-telling is optimal for the intermediate seller (i.e., $\tilde{\theta} = \theta$);
   b) if $\lambda < \lambda^*$ ($\lambda > \lambda^*$), then the manufacturer gains larger (smaller) profits as against the profits ensured in the case of $\tilde{\theta} = \theta$ (under truth-telling by the intermediate seller).
3) The value of λ^* is constant and independent of the price θ if the cost functions are power-type $c(y) = ky^\alpha$ ($k > 0$, $\alpha > 1$). In this case, $\lambda^* = 1/\alpha$.

Proof. Under the message $\tilde{\theta}$ reported by the intermediate seller, the manufacturer maximizes his/her goal function by choosing the product output $\tilde{y} = \arg\max_{y \in A} f(y, \tilde{\theta})$ using the condition $c'(\tilde{y}) = \tilde{\theta}\lambda$. Substitute $\tilde{y}$ in the goal function of the intermediate seller. Applying the first-order necessary optimality conditions and the expression $\frac{d\tilde{y}}{d\tilde{\theta}} = \frac{\lambda}{c''(\tilde{y})}$, one obtains the equation

(1) $c'(\tilde{y}) + \tilde{y}c''(\tilde{y}) = \theta.$

This equation admits a unique solution $\tilde{y}$ (the product output $\tilde{y}$ depends only on the actual market price θ), and the corresponding optimal message of the intermediate seller is

(2) $\tilde{\theta} = \dfrac{c'(\tilde{y})}{\lambda}.$

Clearly, the utility function of the manufacturer

$$f_0(\tilde{y}, \tilde{\theta}) = \tilde{y}(\theta - c'(\tilde{y}))$$

appears independent of the share λ. Furthermore, the manufacturer's profits do not depend on λ:

(3) $f(\tilde{y}, \tilde{\theta}) = \tilde{y}c'(\tilde{y}) - c(\tilde{y}).$

Evaluate λ^* in the following way:

(4) $\lambda^* = \dfrac{c'(\tilde{y})}{\theta}.$

Compare (2) and (4) to observe that, if $\lambda = \lambda^*$, the message $\tilde{\theta} = \theta$ is optimal for the intermediate seller.

Now, let $\lambda < \lambda^*$. Then formulas (2) and (4) imply that the optimal message of the intermediate seller constitutes

(5) $\tilde{\theta} = \dfrac{\lambda^*}{\lambda}\theta > \theta.$

The intermediate seller reporting θ, the manufacturer chooses y^* by solving the equation

(6) $c'(y^*) = \theta\lambda.$

Therefore, the manufacturer gains the profits

(7) $f(y^*, \tilde{\theta}) = y^* c'(y^*) - c(y^*).$

Compare (2), (5) and (6) to see that $\tilde{y} > y^*$ (indeed, $c'(y)$ is an increasing function). Next, the function $y c'(y) - c(y)$ increases. By virtue of formulas (3) and (7), the message θ reduces the profits of the manufacturer (in comparison with the message $\tilde{\theta}$).

Similarly, one can prove that, if $\lambda > \lambda^*$, the converse statement holds true (the manufacturer's profits are higher under the message θ than under the message $\tilde{\theta}$). Point 2 of Assertion 4.6.1 is demonstrated.

Let us establish the condition to-be-imposed on the cost function c (y) for making the right-hand side of (4) independent of θ. It follows from (1) that one suffices to select

(a) $\dfrac{c'(y)}{\theta} = k_1$ and   (b) $\dfrac{y c''(y)}{\theta} = 1 - k_1$ (k_1 is a constant).

Divide (b) by (a) to derive the differential equation

(8) $y c''(t) - k_2 c'(y) = 0,$

where $k_2 = (1 - k_1)/k_1$ stands for an arbitrary constant. Solve equation (8) to obtain that $c(y) = k y^\alpha$ ($k > 0, \alpha > 1$). Use formulas (1) and (4) to check that $\lambda^* = 1/\alpha$. •

4.7 THE SCARCITY PRINCIPLE

This section studies a model which elucidates the case of seeming irrationality in the behavior of economic agents [36]. Agents (customers of a beef-importing company) were divided into three categories. Subsequently, each category was reported "specific" information. As a result, the actions of different categories of agents varied. Interestingly, the actions of agents can be explained by assuming that different informational impacts form different awareness structures of agents.

The book by American psychologist R. Cialdini [36] deals with the description and classification of stereotypes often followed by people in their behavior (when they make certain decisions). These stereotypes represent a kind of "programs" being "launched" under specific circumstances and determine human actions, including obviously irrational actions. In particular, R. Cialdini identifies six "fundamental psychological principles that direct human behavior." The principles are "reciprocation, consistency, social proof, liking, authority, and scarcity". Let us discuss the last principle.

The idea of the scarcity principle is the following. "Opportunities seem more valuable to us when they are less available." In particular, this is the case for scarce information, and "exclusive information is more persuasive information." The following experiment is described in corroboration of these words; it was conducted by a successful businessman interested in psychology, the owner of a beef-importing company in the USA.

"The company's customers – buyers for supermarkets and other retail food outlets – were called on the phone as usual by a salesperson and asked for a purchase in one of three ways. One set of customers heard a standard sales presentation before being asked for their orders. Another set of customers heard the standard sales presentation plus information that the supply of imported beef was likely to be scarce in the upcoming months. A third group received the standard sales presentation and the information about a scarce supply of beef, too; however, they also learned that the scarce supply news was not generally available information – it had come, they were told, from certain exclusive contacts that the company had.

... Compared to the customers who got only the standard sales appeal, those who were also told about the future scarcity of beef bought more than twice as much ... The customers who heard of the impending scarcity via "exclusive" information ... purchased six times the amount that the customers who received only the standard sales pitch did. Apparently, the fact that the news about the scarcity information was itself scarce made it especially persuasive."

Not doubting the correctness of R. Cialdini's conclusions, we endeavor to take a different view of the situation. Notably, we explain the actions of company's customers based on a game-theoretic model.

And so, there exist n customers (called agents) of the company that make decisions regarding the amount of beef purchase. Suppose the number n of the agents is sufficiently large, while all agents are identical and compete in the Cournot framework. The price linearly depends on the supply; i.e., the goal functions of the agents have the form

$$f_i(x_1, \ldots, x_n) = \left(Q - \sum_{j \in N} x_j\right) x_i - cx_i,$$

where $x_i \geq 0, i \in N = \{1, \ldots, n\}, c \geq 0$. These functions have the following interpretation: x_i means the sales volume of the agent during the period considered, $(Q - \sum_{j \in N} x_j)$ is the corresponding market price, and c denotes a wholesale purchase price. Then the first term of the goal function is the revenue (as the price multiplied by the sales volume), and the second one expresses the product purchase costs.

Next, we involve the first-order necessary optimality conditions to evaluate the following equilibrium actions of the agents (under the conditions of a common knowledge):

$$(1) \quad x_i = \frac{Q - c}{n + 1}, \quad i \in N.$$

Recall all agents are identical according to the introduced assumption; therefore, they have the same equilibrium actions. This is the case in the absence of informational impact. Having received standard offers, the first group of agents purchase the product in the volume (1); the agents expect selling it during the given period.

Now, study the behavior of the second group (these agents are informed of the coming reduction in beef deliveries). It seems possible to assume that the agents consider this information as a common knowledge (see Section 2.12). In this case, a rational action of the agents is purchasing a doubled volume of the product; the agents would aim to sell the product in the same equilibrium amount (1) during the next period (simultaneously, they would search for alternative suppliers).

Finally, consider the behavior of the third group. These agents have been informed of the coming reduction in beef deliveries (and that this information is available to a few agents only). Probably, it would be rational to adopt the following assumption for such agents. There are two types of agents, notably, uninformed and informed ones (outsiders and insiders, respectively). Evidently, the agents of the third group believe they are insiders. During the given period, uninformed agents will sell the product in the volume (1); being short of the product, they will not participate in the game in the next period. Thus, the number of players during the next period (in fact, coinciding with the number of insiders) goes down from n to kn ($k < 1$ indicates the percentage of insiders). Hence, during the next period, the equilibrium action will be defined by

$$(2) \quad x_i' = \frac{Q - c}{kn + 1}.$$

Compare formulas (1) and (2) to observe an important fact. Under a sufficiently large n, we have

$$\frac{x_i'}{x_i} = \frac{n + 1}{kn + 1} \approx \frac{1}{k}.$$

This is why the agents of the third group purchase the product in the volume $(x_i + x_i')$, i.e., by $\left(\frac{1}{k} + 1\right)$ times greater than the ones belonging to the first group. For instance, suppose that the share of outsiders (the agents of the third group) is equal to 20%. In other words, $k = 1/5$ and this is a common knowledge). One obtains

$$x_i + x_i' = 6x_i.$$

Then it appears rational for the third group to purchase a 6-times higher volume as compared with the first group. Thus, our assumptions lead to the result described in the book [36].

4.8 JOINT PRODUCTION

Let us consider a model, where several agents manufacture the same product sold at an external market. The total income is allocated among agents according to fixed shares. Each agent seeks to maximize the profit (the difference between the income and costs). On the other hand, the principal (a regulator) strives for maximizing the total profit of the agents. In this case, an uncertain parameter consists in the market price of the product. Different ways of forming the awareness structure of the game are actually discussed (namely, informational regulation, active forecast, and reflexive control). What is relevant, these methods demonstrate the same efficiency: the principal gains no extra efficiency by forming more complicated awareness structures of agents (in comparison with informational regulation and active forecast). Interestingly, the gain of each agent is smaller than the expected one (but larger than the gain in the case of principal's truth-telling).

Take a multi-agent two-level organizational system composed of the principal and n agents. The strategy of each agent lies in choosing an action, whereas the principal's strategy concerns the choice of a message for agents.

Denote by $x_i \in X_i = \Re^1_+$ the action of agent i ($i \in N$), by $N = \{1, 2, \ldots, n\}$ the set of agents, by $x = (x_1, x_2, \ldots, x_n) \in X' = \prod_{i \in N} X_i$ the agents' action vector, and by $x_{-i} = (x_1, x_2, \ldots, x_{i-1}, x_{i+1}, \ldots, x_n) \in X_{-i} = \prod_{j \neq i} X_j$ the opponents' action profile for agent i.

The preferences of participants (the principal and agents) are expressed by their goal functions. The goal function of agent i, $f_i(x, r_i)$, represents the difference between the income $h_i(x)$ from joint activity and the costs $c_i(x, r_i)$; here r_i stands for the efficiency parameter (*type*) of the agent. In other words, we have $f_i(x, r_i) = h_i(x) - c_i(x, r_i)$, $i \in N$.

Assume the following form of income functions and cost functions:

(1) $h_i(x) = \lambda_i \theta X$, $i \in N$,

(2) $c_i(x, r_i) = \dfrac{x_i^2}{2(r_i \pm \beta_i \sum\limits_{j \neq i} x_j)}$, $i \in N$,

where $X = \sum_{i \in N} x_i$, $\sum_{i \in N} \lambda_i = 1$ Suppose that $\sum_{j \neq i} x_j < \frac{r_i}{\beta_i}$ in the case of minus mark in the denominator (see (2)).

In practice, the set of agents has an interpretation as a certain firm, whose departments (agents) manufacture the same product sold at the price θ. The total income θX is allocated among agents according to fixed shares $\{\lambda_i\}$. For an agent, the costs are an increasing function of his/her actions. And the efficiency of activity (the denominator in (2)) depends on the agent's type. The interaction of agents is modeled by the relationship between the costs (the efficiency of activity) of each agent and the actions of all other agents. The plus mark in the denominator of the expression (2) corresponds to the efficient interaction of agents (decreasing costs for greater actions of agents, i.e., positive externalities' effects). Notably, the higher are the actions of other agents, the smaller are the resulting costs (and the greater is the efficiency of activity) of a given agent. In applications, this agrees with decreasing specific fixed costs, experience exchange, technology exchange, and so on. The minus mark in the denominator of (2) corresponds to the inefficient interaction of agents (increasing costs for greater actions of agents, i.e., negative externalities' effects). Notably, the higher are the actions of other agents, the larger are the resulting costs (and the lower is the efficiency of activity) of a given agent. In applications, this agrees with basic assets shortage, constraints imposed on secondary indicators (e.g., environmental pollution), etc. The coefficients $\{\beta_i \geq 0\}$ reflect the level of agents' interdependence.

Assume that all system participants know the market price θ. Take the agents' goal functions and apply the first-order necessary conditions of optimality to obtain

$$x_i = \lambda_i \theta \left(r_i \pm \beta_i \sum_{j \neq i} x_j \right), \quad i \in N.$$

And the total actions represent the following function of the parameter θ:

$$X(\theta) = \frac{\sum\limits_{i \in N} \dfrac{\lambda_i \theta r_i}{1 \pm \lambda_i \theta \beta_i}}{1 \mp \sum\limits_{i \in N} \dfrac{\lambda_i \theta \beta_i}{1 \pm \lambda_i \theta \beta_i}}.$$

Set $n=2, \lambda_i = \beta_i = \frac{1}{2}$ $(i=1,2)$; then the total action and the Nash-equilibrium actions of the agents are defined by

(3) $X(\theta) = 2\theta R/(4 \mp \theta)$,

(4) $x_i^*(\theta) = \dfrac{2\theta}{16 - \theta^2}(4r_i \pm \theta\, r_{-i})$, $i=1,2.$

The functions $X^-(\theta)$ and $X^+(\theta)$ are demonstrated by Fig. 4.20 and Fig. 4.21, respectively (the superscript indicates the mark in the denominator of (2)). In the case of $X^-(\theta)$, we believe that $\theta < 4$ (otherwise, there exists no Nash equilibrium).

The expression (3) leads to the relationship between the parameter θ and the total actions X:

(5) $\theta = \Theta(X) = \dfrac{4X}{2R \pm X}.$

To proceed, introduce control into the system. Establish the following sequence of moves. At the moment of decision-making, the principal and agents (the former chooses control, and the latter choose actions) know the goal functions and feasible sets of all system participants. The principal enjoys the right of first move. He/she chooses some values of control variables and reports them to the agents. Subsequently, the agents choose certain actions (under known control).

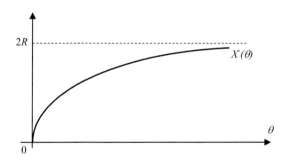

Figure 4.20 The curve of $X^-(\theta)$.

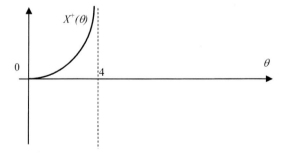

Figure 4.21 The curve of $X^+(\theta)$.

Institutional control corresponds to assigning certain quotas by the principal (restrictions on the maximal actions of agents); exceeding such quotas possibly applies appreciable penalties to agents.

Motivational control corresponds to variation of the parameters $\{\lambda_i\}$ (they can be treated as internal transfer prices). Several examples of motivational control are provided in [28, 113, 135] in the context of planning problems (when the principal fixes prices based on agents' messages – the efficiencies of their activity) and incentive problems (when agent's income is the incentive given by the principal).

Informational control corresponds to purposeful modification of information reported by the principal (the agents adopt this information in their decision-making). Indeed, the quantities (3)–(5) can be evaluated by the principal and agents using a priori available information. In this case, the estimate of the state of nature (the market price θ) enters them as a parameter. Hence, depending on the awareness of a system participant, the computed values of the parameters (3)–(5) may vary.

Imagine that the market price is unknown to the agents (alternatively, they know its inaccurate value). In the model under consideration, **informational regulation** lies in the following. The principal reports to the agents an estimate $\theta_0 \in \Theta$ of the state of nature (i.e., the principal employs a uniform strategy). By virtue of the trust principle (agents are confident in the principal and adopt the information reported by him/her in their decision-making), the agents choose the actions $\{x_i^*(\theta_0)\}$ defined by (4). Consequently, we obtain the total action $X(\theta_0)$ determined by (3).

Let $\Phi(x, \theta)$ be the goal function of the principal and θ_0 denote his/her information about the state of nature (by assumption, this information is adequate). Informational control problem lies in maximizing (by an appropriate message $\theta^* \in \Theta$) the guaranteed value of the principal's goal function on the set of agents' equilibrium states (under this message):

(6) $\Phi(x^*(\theta^*), \theta_0) \rightarrow \max_{\theta^* \in \Theta}.$

Active forecast consists in that the principal reports some information on future results of agents' activity to them (e.g., the total action of agents).

Suppose that $X^* \in \mathfrak{R}_+^1$ is the message of the principal. Using (5), agents can uniquely restore the state of nature adopted by the principal (if the latter expects the reported total action from agents).

The problem of active forecasting concerns maximization (by an appropriate message X^*) the goal function of the principal on the set of agents' equilibrium states (under this message):

(7) $\Phi(x^*(\Theta(X^*)), \theta_0) \rightarrow \max_{X^* \in \mathfrak{R}_+^1}.$

The relationships (3)–(5) being known, the problems (6) and (7) are the ones of standard optimization.

Note that, in the current Example, the efficiencies of active forecast and informational regulation do coincide. Indeed, the estimate of the state of nature is unambiguously restored based on the result of activity.

Consider the case when the costs of each agent increase with respect to the actions of other agents (the minus mark in the denominator of the cost function (2)).

Assume that the principal knows for sure that the external price $\theta_0 = 1$; the agents merely know that $\theta \in \Theta = [0; 3]$. We accept that the principal's goal function is defined by the total income $1 \cdot X$ (after deduction of the total costs of agents).

Consequently, the problem (6) acquires the form

(8) $X(\theta) - c_1(x^*(\theta), r_1) - c_2(x^*(\theta), r_2) \to \max\limits_{\theta \in [0;3]}$,

where $X(\theta)$ and $x^*(\theta)$ are described by the expressions (3) and (4), respectively. By supposing the homogeneity of agents ($r_1 = r_2 = 1$) and substituting formulas (3) and (4) into (8), one naturally arrives at the following. The principal's goal function depends on his/her message through

(9) $\Phi(x^*(\theta), \theta) = \dfrac{\theta(4 - \theta)}{4 + \theta}$.

On the segment $[0; 3]$, the function (9) attains its maximum at the point $\theta^* = 4(\sqrt{2} - 1)$. Hence, solution to the informational regulation problem lies in reporting the estimate θ^* to the agents. Actually, this estimate varies from the "true" one $\theta_0 = 1$ (the principal benefits by data misrepresentation).

Still, what would the agents' gains be in the case of $\theta_0 = 1$ and the principal's message of $\theta^* = 4(\sqrt{2} - 1)$. Evidently, each agent gains less than expected, but greater than in the case of principal's truth-telling. The conclusion seems somewhat paradoxical – agents benefit if the principal conceals the true value of the price! In the present situation, the Nash equilibrium does not appear Pareto-optimal; by his/her message, the principal "shifts" the informational equilibrium to the Pareto optimum.

Now, study the problem of active forecasting (7). Its solution lies in (a) evaluating $X(\theta^*)$ (based on the information on θ^* and formula (5)) and (b) reporting this value to agents as the forecast X_0 of their total action. Clearly, the principal's message $X_0 = 2(2 - \sqrt{2})$ stimulates agents to recover the estimate θ^* of the state of nature and to choose the actions required by the principal.

Imagine that the principal benefits nothing by data manipulation. Consequently, truth-telling regarding the state of nature would motivate the agents to reach the state mostly desired by the principal. In other words, that $\theta^* = \theta_0$ represents a special case ("an accident").

Under $\theta_0 = 1$, the principal's goal function becomes

$$\Phi(x, 1) = x_1 + x_2 - \frac{x_1^2}{2 - x_2} - \frac{x_2^2}{2 - x_1}.$$

Its maximum on the set $0 \le x_1 < 2, 0 \le x_2 < 2$ is observed at the points $x_1 = x_2 = 2 - \sqrt{2}$. These points result from the above informational impacts (informational regulation and active forecast). It suffices to maximize the principal's goal function with respect to all pairs (x_1, x_2). And so, this control (forming an awareness structure of depth 1) turns out optimal on the set of all awareness structures. Notably, the principal's endeavor to create more complicated awareness structures of agents (i.e., to perform reflexive control) would not enhance efficiency in comparison with informational regulation and active forecast.

4.9 MARKET COMPETITION

This section deals with a model, where the price of a product manufactured by agents depends on the total product output. As uncertain parameters, the model involves agents' characteristics (their efficiencies). We consider the problem of active forecasting; seeking to minimize the market price of the product, the principal reports the total product output to the agents (their total action). It appears that, under some assumptions, the message with accurate forecast minimizes the price.

Recall that the model in Section 4.8 has proceeded from the following. A market is not saturated, and the departments of a firm (agents) can sell any quantities of products at a fixed price (an uncertain parameter – the state of nature). Contrariwise, the present model presumes a preset demand in the form of $X(\lambda)$, where X means the total product output (the total action of agents) and λ indicates the market price. If $X(\cdot)$ is a monotonically decreasing function, there exists the inverse function $\lambda(X)$ reflecting the relation between the market price and the offer. The inverse function is also monotonically decreasing and continuous.

Suppose that agents do not know the objective efficiencies (parameters of cost functions) of each other. Yet, for each $i \in N$, agent i knows his/her type r_i, possesses certain beliefs about the opponents' types r_{ij} ($j \in N$), as well as considers these values of types as a common knowledge. In other words, the awareness structure of the game is defined by the expressions $r_{i\sigma j} = r_{ij}, i, j \in N, \sigma \in \Sigma$.

According to agent i, the goal function of agent j makes up

$$(1) \quad f_j(x, r_{ij}) = \lambda(X)\, x_j - \frac{x_j^2}{2r_{ij}}, \quad i, j \in N,$$

where $x = (x_1, x_2, \ldots, x_n) \in X'$ indicates the action vector of agents having the separable cost functions $c_j(x) = x^2/2r_j, j \in N$.

Each agent believes that he/she plays the game with a common knowledge. The equilibrium conditions yield the actions of agents:

$$(2) \quad x_{ij}^* = \frac{r_{ij}\lambda(X)}{1 - \lambda'(X)r_{ij}}, \quad i, j \in N.$$

Next, we analyze two ways of concretizing the function $\lambda(X)$.

Set $\lambda(X) = \lambda_0 - \gamma X, \lambda_0, \gamma > 0$. Then substituting $\lambda(X)$ into (2) gives

$$(3) \quad x_{ij}^* = \frac{r_{ij}\lambda_0}{(1 + \gamma r_{ij})(1 + \gamma\alpha_i)}, \quad i, j \in N,$$

where

$$(4) \quad \alpha_i = \sum_{j \in N} \frac{r_{ij}}{1 + \gamma r_{ij}}, \quad i \in N.$$

Study the informational equilibrium in the case of two agents ($n = 2$) and active forecasting applied by the principal. Let $r_1 = r_2 = \lambda_0 = \gamma = 1$.

Formula (3) implies that the total action equals

$$(5) \quad X(r_{12}, r_{21}) = \frac{2 + \alpha_1 + \alpha_2}{2(1 + \alpha_1)(1 + \alpha_2)},$$

with $\alpha_1 = \frac{1}{2} + r_{12}/(1 + r_{12})$, $\alpha_2 = \frac{1}{2} + r_{21}/(1 + r_{21})$. The expression (5) depends on the beliefs of agents about the partner's type.

The principal's problem is the guarantee the total action X. Which forecast X_0 solves this problem?

Agent 1 has been reported the forecast X_0. He thinks that the goal functions are defined by

$$f_1(x_1, x_2) = (1 - x_1 - x_2)x_1 - \frac{x_1^2}{2},$$

$$f_2(x_1, x_2) = (1 - x_1 - x_2)x_2 - \frac{x_2^2}{2r_2}.$$

The parameter r_2 appears unknown to agent 1. To evaluate a Nash equilibrium, agent 1 addresses the first-order necessary conditions of optimality and uses the forecast. This leads to the following system of equations:

$$\begin{cases} 1 - x_2 - 3x_1 = 0, \\ 1 - x_1 - \left(\frac{1}{r_2} + 2\right)y_2 = 0, \\ x_1 + x_2 = X_0. \end{cases}$$

The derived system admits the unique solution

$$x_1 = \frac{1 - X_0}{2}, \quad x_2 = \frac{3X_0 - 1}{2}, \quad r_2 = \frac{3X_0 - 1}{3 - 5X_0}.$$

Obviously, these values must be positive. Hence, the admissible messages of the principal belong to $\frac{1}{3} < X_0 < \frac{3}{5}$.

And the equilibrium strategy of agent 1 is $x_1^* = \frac{1 - X_0}{2}$. Similarly, we obtain that the equilibrium strategy of agent 1 constitutes $x_2^* = \frac{1 - X_0}{2}$.

Thus, by the reporting the forecast X_0, the principal ensures the total action $X = x_1^* + x_2^* = 1 - X_0$. Clearly, the unique accurate forecast of the principal in this game is $X_0 = \frac{1}{2}$.

Now, suppose that the relation between the price and the total action of two agents is defined by $\lambda(X) = 1/X$. Then their goal functions become

$$f_1(x_1, x_2) = \frac{x_1}{x_1 + x_2} - \frac{x_1^2}{2}, \quad f_2(x_1, x_2) = \frac{x_2}{x_1 + x_2} - \frac{x_2^2}{2r_2}.$$

Again, apply the first-order necessary conditions of optimality to find the equilibrium strategy of agents:

$$x_1^* = \frac{\sqrt[4]{r_1 r_2}}{\sqrt{r_1} + \sqrt{r_2}} \sqrt{r_1}, \quad x_2^* = \frac{\sqrt[4]{r_1 r_2}}{\sqrt{r_1} + \sqrt{r_2}} \sqrt{r_2}.$$

Assume that $r_1 = r_2 = 1$ and each agent knows his/her type (but not the type of another agent). That is, agent 1 (agent 2) is unaware of r_2 (r_1, respectively).

We believe that the principal strives for maximizing the price (by an appropriate price forecast). What forecast is optimal (as minimizing the actual price)?

If the principal reports the price forecast λ, agent 1 can calculate the type $\tilde{r}_2$ of agent 2 from the equation

$$\frac{1}{\lambda} = x_1^* + x_2^* = \sqrt[4]{r_1 \tilde{r}_2} = \sqrt[4]{\tilde{r}_2}.$$

We have $\tilde{r}_2 = \frac{1}{\lambda^4}$, leading to $x_1^* = \frac{\sqrt[4]{\tilde{r}_2}}{1 + \sqrt{\tilde{r}_2}} = \frac{1/\lambda}{1 + \frac{1}{\lambda^2}} = \frac{\lambda}{\lambda^2 + 1}$.

Repeating this line of reasoning for agent 2, one naturally obtains his/her equilibrium strategy: $x_2^* = \frac{\lambda}{\lambda^2 + 1}$.

Thus, under the principal's message of the forecast λ, the actual price Λ becomes

$$\Lambda(\lambda) = \frac{1}{x_1^* + x_2^*} = \frac{\lambda^2 + 1}{2\lambda}.$$

Evidently, the function $\Lambda(\lambda)$ reaches its minimum at $\lambda = 1$; this value is optimal. Moreover, $\Lambda(1) = 1$, i.e., the optimal forecast of the principal is an accurate forecast, and all agents gain their expected payoffs.

4.10   LUMP SUM PAYMENTS

In this section, we discuss an incentive model (see the surveys of game-theoretic models of motivation in [27, 134]). This model reflects a situation when the payment for a collective of agents takes the following form. Each agent obtains a fixed incentive, if the aggregated result of agents' activity (e.g., their total action) exceeds a preset norm; otherwise, agents receive nothing.

Agents have an hierarchy of beliefs about this norm. Besides a common knowledge, several variants are studied, namely,

– the agents' beliefs about the norm are pairwise different; then either nobody works, or one agent performs all work;
– the awareness structure possesses depth 2, and each agent subjectively believes that he/she plays a game with an asymmetric common knowledge; then the set of feasible equilibrium outcomes is maximal and coincides with the set of individually rational actions;
– the awareness structure possesses depth 2, and a symmetric common knowledge takes place at its lower level; then the set of feasible equilibrium outcomes is again maximal possible.

The derived results corroborate the following intuitively clear conclusion. In a collective of employees, joint activity is possible (represents an equilibrium) if there exists a common knowledge about the volume of work to-be-performed for gaining a reward. Furthermore, a slight modification in the awareness structure causes appreciable variations in the informational equilibrium.

Interestingly, the model below admits the following stable informational equilibrium. Each agent believes that the whole volume of works was performed exactly owing to his/her efforts and everybody knows this (or even this fact forms a common knowledge).

Consider an organizational system composed of a principal and n agents performing a joint activity.

The strategy of agent i lies in choosing an action $y_i \in X_i = \mathfrak{R}_+^1$, $i \in N$. On the other hand, the principal chooses an incentive scheme which determines the reward of each agent depending on the result of their joint activity. Imagine the following scheme of interaction among the agents. Attaining a necessary result requires that the sum of their actions is not smaller than a given threshold $\theta \in \Omega$. In this case, agent i yields a fixed reward σ_i, $i \in N$; if $\sum_{i \in N} y_i < \theta$, the reward vanishes for all agents.

Implementation of the action $y_i \geq 0$ makes it necessary that agent i incurs the costs $c_i(y, r_i)$, where $r_i > 0$ denotes his/her type (a parameter representing individual characteristics of the agent), $i \in N$.

Concerning the cost functions of the agents, assume that $c_i(y, r_i)$ is a continuous function, increasing with respect to y_i and decreasing with respect to r_i, and

$$\forall y_{-i} \in X_{-i} = \prod_{j \in N \setminus \{i\}} X_j, \quad \forall r_i > 0 : c_i(0, y_{-i}, r_i) = 0, \quad i \in N.$$

The stated model of interaction is known as the lump-sum payment.
Now, define the set of individual rational actions of the agents:

$$IR = \left\{ y \in X' = \prod_{i \in N} X_i \mid \forall i \in N \quad \sigma_i \geq c_i(r_i) \right\}.$$

If the agents' costs are separable (i.e., the costs $c_i(y_i, r_i)$ of each agent depend on his/her actions only), we obtain $IR = \prod_{i \in N} [0; y_i^+]$, where

$$y_i^+ = \max\{y_i \geq 0 \mid c_i(y_i, r_i) \leq \sigma_i\}, \quad i \in N.$$

Introduce the following notation:

$$Y(\theta) = \left\{ y \in X' \mid \sum_{i \in N} y_i = \theta \right\}, \quad Y^*(\theta) = \text{Arg} \min_{y \in Y(\theta)} \sum_{i \in N} c_i(y, r_i).$$

Next, we analyze different variants of agents' awareness about the parameter $\theta \in \Theta$. Obviously, even slight complexification of the awareness structure could lead to substantial changes in the set of informational equilibria within the reflexive game.

Variant I. Suppose that the value $\theta \in \Theta$ is a common knowledge. Then the equilibrium in the game of agents is provided by a parametric Nash equilibrium belonging to the set

(1) $E_N(\theta) = IR \cap Y(\theta)$.

We also define the set of Pareto-efficient actions of the agents:

(2) $Par(\theta) = IR \cap Y^*(\theta)$.

Since $\forall \theta \in \Theta$: $Y^*(\theta) \subseteq Y(\theta)$, the expressions (1)–(2) imply that the set of Pareto-efficient actions represents a Nash equilibrium. However, the set of Nash equilibria may be wider; in particular, under $\theta \geq \max\limits_{i \in N} y_i^+$, it always includes the vector of zero actions.

We give an example. Consider $n = 2$ agents with the Cobb-Douglas cost functions $c_i(y_i, r_i) = r_i \varphi(y_i / r_i)$, where $\varphi(\cdot)$ is a smooth strictly increasing function such that $\varphi(0) = 0$.

In this case, the point $y^*(\theta) = \{y_i^*(\theta)\}$, where $y_i^*(\theta) = \theta r_i / \sum_{j \in N} r_j$, $i \in N$, is a unique Pareto-efficient point (e.g., see [135]).

Let us evaluate $y_i^+ = r_i \varphi^{-1}(\sigma_i / r_i)$, $i \in N$. Then under the condition

(7) $\sigma_i \geq r_i \varphi \left(\theta / \sum\limits_{j \in N} r_j \right)$, $i \in N$,

the Pareto set is nonempty.

The sets of Nash equilibria in the game of $n = 2$ agents (with $\theta_2 > \theta_1$) are illustrated by Fig. 4.22; note that the point $(0; 0)$ is a Nash equilibrium in both cases.

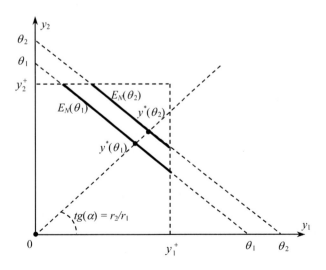

Figure 4.22 The parametric Nash equilibrium in agents' game.

Thus, we have studied the elementary awareness structure of agents (corresponding to a situation when the value of $\theta \in \Theta$ is a common knowledge). Now, consider the next variant with a greater level of complexity of agents' awareness; here a common knowledge is represented by the individual beliefs $\{\theta_i\}$ of the agents about the value $\theta \in \Theta$.

Variant II. Suppose that the beliefs of the agents about the uncertain parameter are pairwise different, yet make up a common knowledge. In other words, an asymmetric common knowledge takes place.

Without loss of generality, we renumber the agents such that their beliefs form an increasing sequence: $\theta_1 < \ldots < \theta_n$. In this situation, the structure of feasible equilibria is defined in the following statement.

Assertion 4.10.1. Consider a lump-sum payment game, where $\theta_i \neq \theta_j$ for $i \neq j$. Depending on the existing relation among the parameters, the following $(n + 1)$ outcomes may be an equilibrium: $\{y^* \mid y_i^* = 0, \ i \in N\}$; $\{y^* \mid y_k^* = \theta_k, \ y_i^* = 0, \ i \in N, \ i \neq k\}$, $k \in N$. In practice, this means that either nobody works, or merely agent k does (by choosing the action θ_k).

Proof.   Let the action vector $y^* = (y_1^*, \ldots, y_n^*)$ be an equilibrium, i.e., $y_i^* \leq y_i^+$ for any $i \in N$. Assume that there exists a quantity $k \in N$ such that $y_k^* > 0$. Below we demonstrate that, in this case, $\sum_{i \in N} y_i^* = \theta_k$.

Indeed, if $\sum_{i \in N} y_i^* < \theta_k$, then agent k does not expect any reward. Hence, he/she may increase the individual (subjectively expected) gain from a negative value to zero by choosing the zero action. On the other hand, with $\sum_{i \in N} y_i^* > \theta_k$, agent k reckons on a reward; still, he/she may increase the individual gain by the action $\max\{0, \theta_k - \sum_{i \in N \setminus \{k\}} y_i^*\} < y_k^*$ instead of y_k^*. Therefore, if $\sum_{i \in N} y_i^* \neq \theta_k$, agent k may increase his/her gain; this fact contradicts the equilibrium character of the vector y^*.

We have shown that, if $y_k^* > 0$, then $\sum_{i \in N} y_i^* = \theta_k$. Due to $\theta_i \neq \theta_j$, $i \neq j$, this equality holds for a single number $k \in N$. Consequently, $y_k^* > 0$ implies $y_i^* = 0$ for all $i \neq k$. And, clearly, $y_k^* = \theta_k$. •

Now, consider the relations among the parameters θ_i, $y_i^+ (i \in N)$ that ensure each equilibrium in Assertion 4.10.1.

The vector $(0, \ldots, 0)$ is an equilibrium when none of agents may perform sufficient work independently (according to his/her view) for receiving a proper reward (alternatively, his/her action equals y_i^+ and the gain of agent i remains zero). Formally, the discussed condition is expressed by $y_i^+ \leq \theta_i$ for any i.

The vector $\{y^* \mid y_k^* = \theta_k, \ y_i^* = 0, i \neq k\}$ is an equilibrium if $\theta_k \leq y_k^+$ and all agents with the numbers $i > k$ (actually, they believe no reward would be given) are not efficient enough for compensating the quantity $(\theta_i - \theta_k)$ themselves. Formally, we have $\theta_k + y_i^+ \leq \theta_i$ for any $i > k$.

Feasible equilibria in the game of two agents are presented in Fig. 4.23. It should be emphasized that (in contrast to Variant I) there exists an equilibrium-free domain.

To proceed, consider the general case when agents' beliefs may coincide: $\theta_1 \leq \cdots \leq \theta_n$. Similarly to Variant I, this could lead to a whole domain of equilibria. For instance, suppose that $\theta_m = \theta_{m+1} = \cdots = \theta_{m+p}$, $\theta_i \neq \theta_m$ for $i \notin \{m, \ldots, m+p\}$. Under the conditions $\sum_{k=m}^{m+p} y_k^* \geq \theta_m$ and $\theta_m + y_i^+ \leq \theta_i$, $i > m$, the equilibrium is then given by any vector y^* such that $\sum_{k=m}^{m+p} y_k^* = \theta_m$, $y_k^* \leq y_k^+$, $k \in \{m, \ldots, m+p\}$; $y_i^* = 0, i \notin \{m, \ldots, m+p\}\}$.

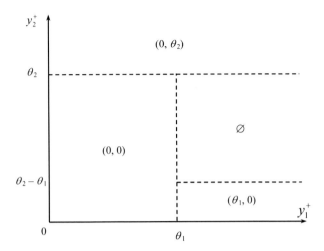

Figure 4.23 Equilibria in the game of two agents (the equilibrium-free domain is indicated by the symbol "∅").

The corresponding interpretation consists in the following. In an equilibrium, all work is performed by the agents with identical beliefs of the volume required for being rewarded.

Variant III. Let the awareness structure have depth 2 and each agent believe he/she participates in the game with an asymmetric common knowledge. In this case, the set of feasible equilibrium outcomes becomes maximal: $\prod_{i \in N} [0; y_i^+]$. And the following statement takes place.

Assertion 4.10.2. Consider a lump-sum payment game; for any action vector $y^* \in \prod_{i \in N} [0; y_i^+]$, there exists a certain awareness structure with depth 2 such that the vector y^* provides a unique equilibrium. Each agent then participates in the game with an asymmetric common knowledge.

Proof. For any $i \in N$, it suffices to select

$$\theta_i = \begin{cases} y_i^*, & y_i^* > 0; \\ y_i^+ + \varepsilon, & y_{i=0}^* \end{cases}$$

(here ε stands for an arbitrary positive number) and to choose any values $\theta_{ij} > \sum_{i \in N} y_i^+$, $j \in N \setminus \{i\}$. Then agent i expects zero actions from the opponents, while his/her own subjectively equilibrium action is given by y_i^*. •

Remark 1. We have constructed the equilibrium which is (objectively) Pareto-efficient if the sum $\sum_{i \in N} y_i^*$ equals the actual value of the uncertain parameter θ.

Remark 2. The action $y_i^* = y_i^+$ is an equilibrium provided that $\theta_i = y_i^+$. However, under this condition, the action $y_i^* = 0$ also forms an equilibrium. In both cases, the subjectively expected gain of agent i makes zero.

Variant IV. Now, imagine that the awareness structure of the game possesses depth 2 and the symmetrical common knowledge is at the lower level. In other words, each phantom agent believes that the uncertain parameter equals θ and this fact represents a common knowledge.

It turns out that (even in this case) the set of equilibrium outcomes is maximal: $\prod_{i\in N}[0; y_i^+]$. Furthermore, the following fact is true.

Assertion 4.10.3. Consider a lump-sum payment game with a symmetric common knowledge at the lower level; for any action vector $y^* \in \prod_{i\in N}[0; y_i^+]$, there exists a certain awareness structure of depth 2 such that the vector y^* provides a unique equilibrium.

Proof. Take any value $\theta > \sum_{i\in N} y_i^+$ and suppose this is a common knowledge among the phantom agents. Then a unique equilibrium in the game of phantom agents is, in fact, the zero action chosen by each agent. Next, for each number $i \in N$, we select

$$
\theta_i = \begin{cases} y_i^*, & y_i^* > 0, \\ y_i^+ + \varepsilon, & y_i^* = 0, \end{cases}
$$

with ε being an arbitrary positive number. Apparently, the best response of agent i to the zero actions of the opponents (he/she expects such actions) lies in choosing the action y_i^*. •

Remarks 1–2 *expressis verbis* remain in force for variant IV.

Therefore, we have studied the structure of informational equilibria in a lump-sum payment game under different variants of agents' awareness. The derived results confirm the following (intuitively verisimilar) observation. Within a collective of employees, joint work is possible (forms an equilibrium) if there is a common knowledge regarding the volume of works to-be-performed for receiving a reward.

Now, address the issue of informational equilibrium stability. We will analyze variant II under an asymmetric common knowledge. Assume that, as the result of the game, a common knowledge of the agents is the fact of paid or unpaid reward.

Clearly, the equilibrium $(0, \ldots, 0)$ is always stable; i.e., nobody works, nobody expects a reward and nobody receives a reward.

It has been demonstrated above that the equilibrium $\{y^* \mid y_k^* = \theta_k, y_i^* = 0, i \in N, i \neq k\}$, $k \in N$, where $\theta_1 < \cdots < \theta_n$, is feasible if $\theta_k \leq y_k^+$, $\theta_k + y_i^+ \leq \theta_i$ for any $i > k$. Then i-agents $(i \leq k)$ expect a reward, and this is not the case for their colleagues with the numbers $i > k$. Hence, stability is guaranteed only by the condition $k = n$. Therefore, we obtain the following stability condition:

$$
(8) \quad \theta_n \leq y_n^+.
$$

Similarly, for $\theta_1 \leq \cdots \leq \theta_{m-1} < \theta_m = \cdots = \theta_n$ any set of the form $\{y^* \mid \sum_{k=m}^{n} y_k^* = \theta_m, y_k^* \leq y_k^+, k \in \{m, \ldots, n\}; y_i^* = 0, i \notin \{m, \ldots, m + p\}\}$ appears stable.

According to Assertion 4.10.2, the principal may apply informational control for making the agents choose any set of actions $y^* \in \prod_{i\in N}[0; y_i^+)$ (e.g., this is achieved by forming a certain structure such that each agent subjectively plays a game with an asymmetric common knowledge). It turns out that there exists stable informational control ensuring the stated result. We demonstrate this in the case of $y_i^* > 0$.

Take a given set $y^* \in \prod_{i \in N} (0; y_i^+)$, $\sum_{i \in N} y_i^* \geq \theta$. For each $i \in N$, set $\theta_i = y_i^*$; on the other hand, for each $j \in N \setminus \{i\}$, choose any θ_{ij} such that $\theta_{ij} < \theta_i$. Then i-agent subjectively satisfies the stability condition (8) and y_i^* represents his/her unique equilibrium action. Note that:

1) the work will be completed, and the agents will receive rewards;
2) receiving the rewards forms an expected outcome for all real and phantom agents.

The corresponding interpretation is the following. Each agent believes that exactly he/she has performed all work (and that this is a common knowledge).

4.11   SELLERS AND BUYERS

In this section, we consider a model, where a buyer and a seller have a hierarchy of mutual beliefs about the value of a certain product. They negotiate the price for the deal.

A necessary condition of such contract lies in the following. According to both participants, the subjective prices of all (real and phantom) sellers do not exceed the subjective prices of each (real and phantom) buyer.

The deal is concluded with the price required by the principal under the following elementary awareness structure formed for the buyer and seller. Both participants must be sure that, for all their phantom agents (the buyer in seller' beliefs, the seller in buyer's belief, and so on), the value of the product coincides with the price required by the principal.

Suppose that the seller and buyer (agents s and b, respectively) must make a compromise over the price of a product or a service, works, etc.

Introduce the following notation: θ_b is buyer's belief about his/her value of the product (the maximal price he/she is willing to pay); θ_s stands for seller's belief about his/her value of the product (the minimal price he/she is willing to sell); θ_{bs} indicates the buyer's belief about the seller's beliefs, θ_{sb} corresponds to the seller's belief about the buyer's belief; θ_{sbs} means the seller's belief about the buyer's belief about the seller's belief, and so on. Assume that $\theta_\tau \in \Re^1_+$, where τ is an arbitrary finite sequence of indexes (probably, empty) from the set $\{s; b\}$ of participants. Recall that the set of various finite sequences of indexes is designated by Σ_+, while the union of Σ_+ and the empty sequence is denoted by Σ.

What properties of the mutual beliefs of the buyer and seller make this deal possible? The deal takes place only if the value of the product for the buyer is not smaller than that for the seller. Consequently, we have

(1) $\forall \tau \in \Sigma: \theta_{\tau b} \geq \theta_\tau, \quad \theta_{\tau s} \leq \theta_\tau.$

Formula (1) implies that the subjective size of the domain of compromise can be rewritten as

(2) $\Delta_\tau = \theta_{\tau b} - \theta_{\tau s}, \quad \tau \in \Sigma.$

Furthermore, $\forall \tau \in \Sigma$: $\Delta_\tau \geq 0$.

The above domain of compromise turns out nonempty due to (1)–(2). Now, discuss admissible mechanisms of compromise. Under the above subjective beliefs (and the given domain of compromise), different allocation procedures (options) for the "profit" Δ appear feasible.

Option 1 lies in defining a mapping $\pi = (\pi_s, \pi_b)$: $\Re_+^2 \to \Re_+^2$, which meets the following properties for all $\theta_b \geq \theta_s$:

$$\pi_s(\theta_s, \theta_b) + \pi_b(\theta_s, \theta_b) = \Delta,$$

$$\frac{\partial \pi_s(\theta_s, \theta_b)}{\partial \theta_s} \leq 0, \quad \frac{\partial \pi_s(\theta_s, \theta_b)}{\partial \theta_b} \geq 0,$$

$$\frac{\partial \pi_b(\theta_s, \theta_b)}{\partial \theta_s} \leq 0, \quad \frac{\partial \pi_b(\theta_s, \theta_b)}{\partial \theta_b} \geq 0.$$

Their practical interpretations seem obvious. Take the compromise mechanism (being invariant with respect to additive shifts of beliefs):

$$\pi_s = \alpha(\theta_b - \theta_s), \quad \pi_b = (1 - \alpha)(\theta_b - \theta_s), \quad \alpha \in [0; 1].$$

We have ealier emphasized that axiomatic characterization of different compromise mechanisms represents a challenging problem (yet, it goes beyond this book).

Option 2 (direct evaluation of a compromise point) is defining a mapping π: $\Re_+^2 \to \Re_+^1$ with the following properties for all $\theta_b \geq \theta_s$:

$$\theta_s \leq \pi(\theta_s, \theta_b) \leq \theta_b, \quad \frac{\partial \pi(\theta_s, \theta_b)}{\partial \theta_s} \geq 0, \quad \frac{\partial \pi(\theta_s, \theta_b)}{\partial \theta_b} \geq 0.$$

Here practical interpretations are clear, as well. An example concerns the the compromise mechanism $\pi = \alpha\theta_b + (1 - \alpha)\theta_s$ ($\alpha \in [0; 1]$); again, it enjoys invariance with respect to additive shifts of beliefs.

Apparently, these options (variants of compromise mechanisms) appear equivalent. In the sequel, keep in mind option 2 for definiteness.

The reflexive game between the buyer and seller will be formalized as follows. The admissible action of each player – the seller or buyer – is reporting "his/her own" price – x_s or x_b, respectively. This is done simultaneously with the opponent and independently of the latter. On the ground of players' messages, the deal takes place under the price $\pi(x_s, x_b)$ (if $x_s \leq x_b$) and does not (if $x_s > x_b$). The payoff functions have the form

$$f_s(\theta_s, x_s, x_b) = \begin{cases} \pi(x_s, x_b) - \theta_s, & x_s \leq x_b, \\ -\varepsilon_1, & x_x > x_b, \end{cases}$$

$$f_b(\theta_b, x_s, x_b) = \begin{cases} \theta_b - \pi(x_s, x_b), & x_s \leq x_b, \\ -\varepsilon_2, & x_s > x_b, \end{cases}$$

where ε_1 and ε_2 are arbitrary positive numbers (actually, these are the costs of a message, if the deal fails). Moreover, suppose that each agent can abandon the negotiations; the deal takes no place and the agent without a message gains nothing.

To proceed, describe the awareness of players. Assume that the feasible actions and goal functions form a common knowledge (the accuracy is within the quantities θ_s and θ_b). Next, the seller and buyer possess point-type awareness structures of a finite complexity, namely, $I_s = (\theta_s, \theta_{sb}, \theta_{sbs}, \ldots)$ and $I_b = (\theta_b, \theta_{bs}, \theta_{bsb}, \ldots)$. By virtue of the axiom of self-awareness, indexes s and b alternate.

Which informational equilibria are admissible in this reflexive game? For definiteness, first we follow the reasoning of the seller.

To evaluate the equilibrium action x_s^* of the seller, we should find the equilibrium actions of all phantom agents existing in his/her beliefs. Thus, evaluation of x_s^* requires finding all $x_{s\tau}^*$, $\tau \in \Sigma$.

Lemma 4.11.1. According to the seller's view, a set of actions $x_{s\tau}^*$, $\tau \in \Sigma$, is an informational equilibrium (and the seller would not abandon negotiations) if $x_{s\tau}^* \equiv x_s^*$ for any $\tau \in \Sigma$ and $\theta_s' \leq x_s^* \leq \theta_s''$, where $\theta_s' = \max\limits_{\tau \in \Sigma} \theta_{s\tau s}$, $\theta_s'' = \max\limits_{\tau \in \Sigma} \theta_{s\tau b}$.

Proof. ($\Rightarrow$) Let $x_{s\tau}^*$, $\tau \in \Sigma$, be an informational equilibrium. Consider an arbitrary nonempty τ and an equilibrium action $x_{s\tau s b}^*$ By definition of an informational equilibrium, the action $x_{s\tau s}^*$ maximizes the function $f_s(\theta_{s\tau s}, x_{s\tau s}, x_{s\tau s b}^*)$ with respect to $x_{s\tau s}$. In other words, the $s\tau s$-agent (the seller) expects the action $x_{s\tau s b}^*$ from the $s\tau s b$-agent (the buyer). Next, the expression $\theta_{s\tau s} > x_{s\tau s b}^*$ means that the $s\tau s$-agent (the seller) expects from the opponent a smaller message than his/her subjective price. Hence, the subjectively optimal choice of this agent is abandoning the negotiations; thus, the deal fails, and this contradicts the assumption. Thus, $\theta_{x\tau s} \leq x_{s\tau s b}^*$ (the subjective price of the seller does not exceed the price reported by the buyer). However, in this case, we have $x_{s\tau s}^* = x_{s\tau s b}^*$ – the optimal choice of the seller coincides with the price of the buyer.

By analogy, show that, if $x_{s\tau b s}^*$ is an equilibrium action, then $\theta_{s\tau b} \geq x_{s\tau b s}^*$ and $x_{s\tau b}^* = x_{s\tau b s}^*$.

Therefore, for an arbitrary τ we have the relations $x_{s\tau}^* = x_s^*$, $\theta_{s\tau s} \leq x_s^* \leq \theta_{s\tau b}$. The awareness structure possesses a finite complexity; and so, there is a finite number of pairwise different elements $\theta_{s\tau}$. Thus, the last inequality implies $\theta_s' \leq x_s^* \leq \theta_s''$.

($\Leftarrow$) Let x_s^* be such that $\theta_s' \leq x_s^* \leq \theta_s''$. Then for any $\tau \in \Sigma$ we obtain $\theta_{s\tau s} \leq x_s^* \leq \theta_{s\tau b}$, $\theta_{s\tau s} \leq x_s^* \leq \theta_{s\tau}$. The set of actions $x_{s\tau}^* = x_s^*$, $\tau \in \Sigma$, provides informational equilibria (for the seller); and the latter would not abandon negotiations (note that the expressions (1) hold true). •

Clearly, a similar fact applies to the buyer. By combining them, we arrive at

Assertion 4.11.1. The set of actions x_σ^*, $\sigma \in \Sigma_+$, forms informational equilibria (and the deal is concluded) if for any $\tau \in \Sigma$:

$$x_{s\tau}^* = x_s^*, \quad x_{b\tau}^* = x_b^* \quad \text{and} \quad \theta_s' \leq x_s^* \leq \theta_s'', \quad \theta_b' \leq x_b^* \leq \theta_b'',$$

where $\theta_s' = \max\limits_{\tau \in \Sigma} \theta_{s\tau s}$, $\theta_s'' = \min\limits_{\tau \in \Sigma} \theta_{s\tau b}$, $\theta_b' = \max\limits_{\tau \in \Sigma} \theta_{b\tau s}$, $\theta_b'' = \min\limits_{\tau \in \Sigma} \theta_{b\tau b}$.

In particular, Assertion 4.11.1 shows how one should form the awareness structure of the game when the principal (1) can form any awareness structure and (2) strivies for the simplest structure.

For instance, assume that the principal is interested in concluding the deal with a price such that $\theta_s \leq \theta^* \leq \theta_b$. In other words, the principal wants to make θ^* the unique equilibrium price. It suffices to form the following awareness structures for the agents: $I_s = (\theta_s, \theta^*, \theta^*, \theta^*, \ldots), I_b = (\theta_b, \theta^*, \theta^*, \theta^*, \ldots)$. Obviously, then $\theta'_s = \theta''_s = \theta'_b = \theta''_b = \theta^*$. By the assertion, the only informational equilibrium is such that $x^*_s = x^*_b = \theta^*$. This informational equilibrium appears stable – the deal is made under the price requested by the agents (in their messages).

Another relevant remark relates to the following (see the Introduction, as well). We proceed from that the principal is able to form *any* awareness structure for the agents. And we are concerned with the question, "*What* is the desired awareness structure?" The issue *how* should the principal form this structure goes beyond the present book. Actually, it requires the multidisciplinary approach (psychology, sociology, etc.).

Consider an interesting example. Let the subjective price of the seller (buyer) be 20 (50, respectively). Imagine that the principal aims at making the deal under the price of 40. Then he/she should report the following messages: "The buyer believes that the subjective prices of the buyer and seller equal 40, and this is a common knowledge" (to the seller), and "The seller believes that the subjective prices of the seller and buyer equal 40, and this is a common knowledge" (to the buyer). Thus, the principal forms the following awareness structures: $I_s = (20, 40, 40, 40, \ldots)$, $I_b = (50, 40, 40, 40, \ldots)$. Both agents report 40, and the deal occurs.

4.12   CUSTOMERS AND EXECUTORS

This section focuses on a model, where the efficiency of executor's work (known to the latter!) is an uncertain parameter for a customer. In this case, by appropriate variations of the customer's beliefs about the efficiency of executor's work (within the awareness structure of depth 2), any amount of a contract from a specific interval can be made a stable informational equilibrium.

This model bases on analyzing the optimal action of the executor, i.e., the action maximizing the difference between the customer's income and the executor's costs. Suppose that the customer has the goal function

$$\Phi(y, \sigma) = H(y) - \sigma(y).$$

This is the difference between his/her income $H(y)$ gained by the activity (action) $y \in A = \Re^1_+$ of the executor and the incentive paid to the latter.

Assume that the executor's goal function takes the form

$$f(y, \sigma, \theta) = \sigma(y) - c(y, \theta),$$

i.e., represents the difference between the amount of a contract and the costs $c(y, \theta)$ that depend on the executor's action $y \in A$ and the scalar parameter $\theta \in \Theta \subseteq \Re^1_+$. Furthermore, $\forall \theta \in \Theta$: $c(0, \theta) = 0$ and $\forall y \in A$ the function $c(y, \theta)$ is nonincreasing with respect to $\theta \in \Theta$. In other words, the parameter θ can be treated as the executor's level of proficiency (efficiency of actions).

Therefore, the model under consideration incorporates one uncertain parameter – the executor's efficiency of activity $\theta \in \Theta$. This parameter is known to the executor (but not to the customer).

Just imagine that the value of θ represents a common knowledge. Then the following action of the executor would be optimal:

(1) $y^*(\theta) = \arg \max_{y \in A}[H(y) - c(y, \theta)].$

For instance, if $H(y) = y$, $c(y, \theta) = y^2/2\theta$, then $y^*(\theta) = \theta$.

However, the customer does not know the executor's efficiency for sure. And he/she has to adopt some procedure of uncertainty elimination. Below we outline the major alternatives.

First, the customer can use the principle of the maximin action:

$$y^g = \arg \max_{y \in A}[H(y) - \max_{\theta \in \Theta} c(y, \theta)],$$

which gives the guaranteed profit $\max_{y \in A}[H(y) - \max_{\theta \in \Theta} c(y, \theta)]$.

Second, the customer can apply certain mechanisms with information revelation by the executor (about his/her efficiency of activity) [135] and/or suggest a corresponding menu of contracts to the executor.

And third, the customer can make concrete assumptions regarding the properties of the executor's cost function. Subsequently, the customer employs them in the expression (1) or performs informational reflexion with respect to the values of the parameter $\theta \in \Theta$. We discuss the last alternative in a greater detail.

The awareness structure of this reflexive game acquires the form $I_s = (\theta_s, \theta_{sb}, \theta_{sbs}, \ldots)$, $I_b = (\theta_b, \theta_{bs}, \theta_{bsb}, \ldots)$. However, some components are dependent. Indeed, the actual value of θ is known to the executor $(\theta_s = \theta)$, and this forms a common knowledge. Therefore, for any $\tau \in \Sigma$: $\theta_{\tau s} = \theta_\tau$.

Recall that the model with a common knowledge has been studied earlier (see formula (1)). In this case, the graph of reflexive game becomes $B \leftrightarrow S$. Thus, let us analyze a slightly complicated model, where the graph of reflexive game is $S \leftarrow B \leftrightarrow BS$. Suppose that the customer "moves first"; he/she suggests a contract with the amount $c(y^*(\theta_b), \theta_b)$ to the executor. According to (1), the customer agrees if

(2) $\theta_b \leq \theta.$

The customer gains the profit $u_b^0 = H(y^*(\theta_b)) - c(y^*(\theta_b), \theta_b)$. On the other hand, the executor's profit makes up

(3) $u_s^0 = c(y^*(\theta_b), \theta_b) - c(y^*(\theta_b), \theta),$

where $y^*(\cdot)$ follows from (1).

Yet, under $\theta_b > \theta$, the interaction between such customer and executor seems impossible. Due to individual rationality, the latter never signs a contract whose amount does not compensate the costs.

Therefore, this model allows making any point θ_b an informational equilibrium by a proper variation of $\theta_b \leq \theta$. Similarly to the model of seller-buyer interaction, an informational equilibrium turns out stable – the customer expects contract acceptance by the executor (and this is the case).

Considering complex awareness structures in this model seems unjustified. As a matter of fact, this initiative yields nothing new (in comparison with the expressions (1)–(3)). The executor is a passive participant of this situation. He/she can just accept or reject the contract "dictated" by the customer (who makes the first move). And the values of $\theta_{sb}, \theta_{sbs}, \ldots$ are of no importance.

Meanwhile, the customer also knows about such "passivity" of the executor. Thus, defining a contract, he/she accounts only θ_b (and disregards the values associated with higher reflexion ranks – $\theta_{bs}, \theta_{bsb}, \ldots$).

4.13   CORRUPTION

The current section focuses on the model of corruption. Each governmental official possesses subjective beliefs about penalties for the fact of a bribery (these penalties depend on the "total level of corruption"). If the officials observe the total level of corruption, in a stable outcome this level is independent of the mutual beliefs of the corruptionists about their types. Whether such beliefs are true or false is actually not important.

Consequently, it seems impossible to influence the level of corruption only by modifying the mutual beliefs of the officials. Therefore, any stable informational control leads to the same level of corruption.

Consider the following game-theoretic model of *corruption*. There are n agents (government officials), and the additional income of each official is proportional to the total bribe $x_i \geq 0$ taken by him/her, $i \in N = \{1, \ldots, n\}$. We will assume that a bribe offer is unbounded. Let each agent be characterized by the type $r_i > 0$ $(i \in N)$, which is known to him/her (and not to the rest agents). The type may be interpreted as the agent's subjective perception of the penalty "strength."

Irrespective of the scale of corruption activity $(x_i \geq 0)$, agent i may be penalized by the function $\chi_i(x, r_i)$ which depends on the actions $x = (x_1, x_2, \ldots, x_n) \in \mathfrak{R}^n_+$ of all agents and on the agent's type.

Consequently, the goal function of agent i is defined by

(1) $f_i(x, r_i) = x_i - \chi_i(x, r_i), \quad i \in N.$

Suppose that the penalty function has the form

(2) $\chi_i(x, r_i) = \varphi_i(x_i, Q_i(x_{-i}), r_i).$

Formula (2) means that the penalty of agent i depends on his/her action and on the aggregated opponents' action profile $Q_i(x_{-i})$ (in the view of agent i, this is the "total level of corruption of the rest officials").

Assume that the number of agents and the general form of the goal functions are a common knowledge. Moreover, each agent has an hierarchy of beliefs about the parameter $r = (r_1, r_2, \ldots, r_n) \in \mathfrak{R}_+^n$. Denote by r_{ij} the belief of agent i about the type of agent j, by r_{ijk} the belief of agent i about the beliefs of agent j about the type of agent k, and so on $(i, j, k \in N)$.

In addition, assume that the agents observe the total level of corruption. Thus, informational equilibrium stability takes place under any beliefs about the types of real or phantom agents such that the corresponding informational equilibrium leads to the same value of the aggregate $Q_i(\cdot)$ for any $i \in N$.

In this case, the goal functions (1)–(2) of the agents obviously satisfy the conditions of Assertion 2.8.1. Hence, for any number of the agents and any awareness structure, all stable equilibria of the game are true. In other words, the following statement holds.

Assertion 4.13.1. Let the set of actions $x_\tau^*, \tau \in \Sigma_+$, be a stable informational equilibrium in the game (1)–(2). Then this equilibrium is true.

Corollary. The level of corruption in a stable opponents' action profile does not depend on the mutual beliefs of corrupt officials about their types. Whether these beliefs are adequate or not appears not important.

It is then impossible to influence the level of corruption only by modifying the mutual beliefs. Therefore, any stable informational equilibrium results in the same level of corruption.

Suppose that

$$\varphi_i(x_i, Q_i(x_{-i}), r_i) = x_i(Q_i(x_{-i}) + x_i)/r_i, Q_i(x_{-i}) = \sum_{j \neq i} x_j, \quad i \in N$$

and all types are identical: $r_1 = \cdots = r_n = r$.

Evidently, the equilibrium actions of the agents are $x_i = \frac{r}{n+1}$, $i \in N$, while the total level of corruption constitutes $\sum_{i \in N} x_i = \frac{nr}{n+1}$. The latter quantity may be changed only by a direct impact on the agents' types.

4.14 BIPOLAR CHOICE

This section introduces a model, where agents choose between two alternatives (known as positive and negative poles for generality, see Section 4.3.4). For instance, take a candidacy for an election (voting "for" or "against"), a product or service (purchasing it or not), ethical choice (acting in a "good" or a "bad" manner), and so on.

There are agents of three types. The first definitely choose a positive pole, the second choose a positive or negative pole (depending on the behavior of the rest agents), and the third unconditionally choose a negative pole.

Now, suppose that the principal can influence the situation. He/she strives for increasing the probability of positive choice (the positive pole) in the whole "population". For this, the principal is able to exert an impact on agents from the second and third group (agents belonging to the first group choose the positive pole anyway). On the one hand, the principal can influence the third group by spending a certain resource (e.g., finances); thus, some share of agents is moved to the second group. On the other hand, the principal can influence the second group by modifying the beliefs of its members about the share of agents belonging to the third group. Notably, such impact aims at forming the following belief for the second group: "a certain share of agents moved from the third group to the second one." This procedure requires specific costs.

In other words, the principal is able to change the real or "phantom" share of agents in the third group. The total resources available to the principal (a budget) are fixed. The principal's problem lies in the following: allocate the fixed resources among informational impacts such that the share of agents making positive choice attains maximum.

If the principal has "not very much resources," then optimal control is spending all available resources to change only the real or only the "phantom" share of agents in the third group (depending on model's parameters). Let us describe the corresponding model.

Consider a situation when agents belonging to an infinitely large community choose between two alternatives.

To solve the control problem for the whole community, suppose that the choice of a specific agent is not important (due to the infinite number of the agents). Actually, the share of agents choosing the positive pole appears relevant. This statement could be reformulated as follows. The action of the "aggregated" agent is the probability x related to his/her choice of the positive pole.

Introduce the following assumptions:

1) there exist n different types of the agents;
2) the share of agents having type i makes α_i, $0 \leq \alpha_i \leq 1$;
3) the action of an i-type agent (an agent from group i) is defined by the *response function on expectation*:

$$\pi(p), \pi: [0, \ 1] \to [0, 1].$$

Here p means the probability (expected by the agents) of the event that an arbitrary agent from the community would choose the positive pole. In other words, if an agent expects that the share of agents choosing the positive pole equals p, then his/her action x_i is determined by

$$x_i = \pi_i(p).$$

4) assumptions 1–3 are a common knowledge among the agents.
   Denote by $x_i \in [0, 1]$ the action of an i-type agent. Consequently, the share of such agents choosing the positive pole constitutes $p = \sum_{j=1}^{n} \alpha_j x_j$.

Define a *bipolar choice equilibrium* as a set of actions x_i such that

$$(1) \quad x_i = \pi_i \left(\sum_{j=1}^{n} \alpha_j x_j \right), \quad i = 1, \ldots, n.$$

Making a digression, let us underline that formulas (1) represent a possible way to describe the bipolar choice. Another approach is developed by V. Lefebvre [96], T. Taran [169] and others. It proceeds from that a decision-maker performs *the first-kind reflexion* [127], i.e., observes his/her behavior, ideas and feelings. In other words, several interrelated levels exist within the agent; his/her final decision depends on the impact of an external environment and the states of the above-mentioned levels. In the present book, we consider an agent as an individual, i.e., an "indivisible person" performing *the second-kind reflexion* (regarding the decisions made by his/her opponents).

We get down to the discussion of bipolar choice equilibria. Formulas (1) specify a self-mapping of the unit hypercube $[0, 1]^n$:

$$(2) \quad (x_1, \ldots, x_n) \rightarrow \left(\pi_1 \left(\sum_{j=1}^{n} \alpha_j x_j \right), \ldots, \pi_n \left(\sum_{j=1}^{n} \alpha_j x_j \right) \right).$$

The functions $\pi_i(\cdot)$ being continuous (this seems natural), the mapping (2) is also continuous. Then, according to the fixed point theorem, the system (1) admits (at least) one solution.

Consider the following example. There are agents of three types ($n = 3$), and their actions are given by the functions

$$\pi_1(p) \equiv 1, \quad \pi_2(p) = p, \quad \pi_3(p) \equiv 0.$$

In practice, this means that first-type agents choose the positive pole (regardless of anything), while third-type ones prefer the negative pole. The agents of the second type hesitate in their choice, and their actions coincide with the expected action of the whole community.

In this case, the system (1) is reduced to $x_1 = 1, x_2 = \alpha_1 x_1 + \alpha_2 x_2 + \alpha_3 x_3, x_3 = 0$.

This yields $x_1 = 1, x_2 = \frac{\alpha}{1-\alpha_2}, x_3 = 0$ (provided that $0 < \alpha_i < 1$, $i = 1, 2, 3$). At the same time, we have

$$(3) \quad p = \alpha_1 x_1 + \alpha_2 x_2 + \alpha_3 x_3 = \alpha_1 + \alpha_2 \frac{\alpha}{1 - \alpha_2}.$$

Now, imagine a principal is able to influence the situation; he/she strives for increasing the probability of positive choice made by the whole community (in fact, this is the quantity p). To succeed, the principal may exert a certain impact on the agents belonging to the second or the third group (i.e., the first-type agents choose $x_1 = 1$ themselves). Suppose the principal may have an impact on the third group by moving a certain share y of its members to the second group. This procedure requires some resources (e.g., costs) in the amount of $C_2 y$. The principal may also influence the second group, changing the beliefs of its members about the parameter α_3 (irrespectively of the actual value). In particular, the impact consists in the following belief formed for the second group: "the share x of the members belonging to the third-group has joined the second group." The corresponding costs to form such belief are $C_1 x$.

In other words, the principal modifies either the real or a "phantom" share of agents having the third type. Note that the total quantity of the resources available to the principal (a budget) makes C.

Principal's problem is distributing the resource C (i.e., choosing the shares x and y) to maximize the probability p. Formally, the optimization problem of the principal is posed as follows (see (3)):

$$(4) \quad p(x, y) = \alpha_1 + (\alpha_2 + y\alpha_3) \frac{\alpha_1}{1 - (\alpha_2 + x\alpha_3)} \to \max$$

under the constraints

$$(5) \quad C_1 x + C_2 y \leq C, \quad 0 \leq x \leq 1, \quad 0 \leq y \leq 1.$$

Evidently, the problem (4) is reduced to maximization of the function $\varphi(x, y) = \frac{\alpha_2 + y\alpha_3}{1 - (\alpha_2 + x\alpha_3)}$. The latter increases with respect to both arguments, x and y; thus, the first constraint in (5) becomes an equality. Hence, we have obtained the maximization problem for the function

$$\psi(x) = \frac{\alpha_2 + \alpha_3(C - C_1 x)/C_2}{1 - (\alpha_2 + x\alpha_3)} = \frac{C_1}{C_2} \frac{\alpha_2 C_2/C_1 + \alpha_3 C/C_1 - x\alpha_3}{1 - \alpha_2 - x\alpha_3}.$$

Apparently, the function $\psi(x)$ is monotonically increasing (decreasing or constant) if the quantity

$$(6) \quad \frac{\alpha_2 C_2}{C_1} + \frac{\alpha_3 C}{C_1} - (1 - \alpha_2)$$

is positive (negative or zero, respectively).

Denote $k_1 = \frac{C_1}{C}$ and $k_2 = \frac{C_2}{C}$. Then the positivity condition for (6) takes the form

$$(7) \quad \alpha_3 > k_1 - \alpha_2(k_1 + k_2).$$

In the sequel, we assume that $C_1 > C$ and $C_2 > C$. This means that the principal is unable to "turn" all agents of the third type into the second-type ones. At the same time, an optimal choice of the principal is when all available resources are invested to increase the real or phantom share of the second-type agents (under the condition (7)).

The relationship between the optimal choice of the principal and the parameters (α_2, α_3) is illustrated by Fig. 4.24.

The shaded domain in Fig. 4.24. corresponds to the true condition (7); i.e., the optimal strategy of the principal lies in consuming all available resources to modify the beliefs:

$$(8) \quad x = \frac{C}{C_1}, y = 0.$$

The solution (8) describes a situation when the share α_2 of the second-type agents is sufficiently large. Fig. 4.24 makes it clear that, under $\alpha_2 > \frac{k_1}{k_1 + k_2}$, the solution (8) is always optimal. However, if

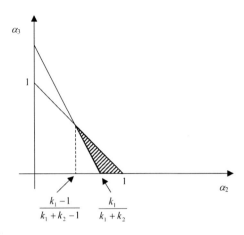

Figure 4.24 The optimal choice of the principal.

(9) $\quad \dfrac{k_1 - 1}{k_1 + k_2 - 1} < \alpha_2 < \dfrac{k_1}{k_1 + k_2},$

the solution (8) is optimal for a sufficiently large α_3. In practice, this case possesses the following interpretation. For a certain range of α_2 (*viz*, the inequality (9) being true), it is optimal to impact the beliefs if they are too pessimistic. In particular, if α_3 is sufficiently large, the probability p of the negative pole choice is large, as well.

In conclusion, we emphasize that the elementary case of informational control under the conditions of bipolar choice has been studied here. Promising directions of future investigations include (a) further extension of the model (increasing the number of agents' types, complexifying the awareness structure and the response function on expectation) and (b) verifying the model using the results of actions made by economic agents (customers) and political agents (voters).

4.15 ACTIVE EXPERTISE: INFORMATIONAL REFLEXION

In the current section, we treat a model of collective decision-making [170] by a group of experts. Each of them possesses specific beliefs about the result of an expertise. And so, an expert demonstrates a strategic behavior – endeavors to influence the result of the expertise. The problem of agents' strategic behavior is traditional in collective choice theory. However, researchers almost do not study the problem of data manipulation in an expertise (by a principal, i.e., the organizer of an expertise) [39, 92, 101, 113].

Assume that the principal can form beliefs about opponents' opinions for experts. In this case, a sufficiently wide range of collective decisions is implementable as an informational equilibrium (sometimes, as a unanimous decision) in the reflexive game of experts.

Consider an example of reflexive control of agents by the principal in the model of an active expertise. First, we describe the model and well-known analysis results

[25, 28, 113, 118, 135] for *expertise mechanisms*. An expertise mechanism serves for acquiring and processing of some information from experts (having a great authority in problem domains).

Suppose there are n experts (agents) assessing a certain object (e.g., a candidate for a position, possible directions of financing, and so on). Expert $i \in N$ reports an estimate $s_i \in [d; D]$, where d and D are the minimal and maximal feasible estimates, respectively. The final assessment (the collective decision) $x = \pi(s)$ to-be-used in decision-making is a function of the experts' messages, $s = (s_1, s_2, \ldots, s_n)$. Denote by r_i the subjective opinion of expert i (this is his/her true belief about the object assessed). Assume that the expertise procedure $\pi(s)$ is a strictly increasing continuous function meeting the *unanimity condition*: $\forall a \in [d, D]$: $\pi(a, a, \ldots, a) = a$.

Generally, experts are supposed to report their actual opinions $\{r_i\}_{i \in N}$. Imagine that each expert makes a small mistake (unconsciously – depending on his/her skill level). Then the average estimate $\frac{1}{n}\sum_{i=1}^{n} r_i$ provides an objective and accurate assessment of the object. The experts being concerned with specific expertise results would not (surely) report the actual opinions. Notably, the mechanism $\pi(\cdot)$ can be subjected to manipulation ($s_i \neq r_i$).

Let us formalize the interests of an agent. Assume that each agent (due to his/her professional or/and timeserving interests) strives for making the expertise result x as close to his/her opinion r_i as possible.

Consider an example of manipulation. Set $n = 3, d = 0, D = 1, r_1 = 0.4, r_2 = 0.5$, $r_3 = 0.6$ (the agents are sorted in the ascending order of their peak points) and fix the following estimation procedure (expertise procedure): $x = \pi(s) = \frac{1}{3}\sum_{i=1}^{n} s_i$. If $s_i \equiv r_i, i = \overline{1, 3}$ (all agents adhere to truth-telling), then $x = 0.5$. Moreover, the final result coincides with the actual opinion of expert 2 (the latter is fully satisfied). In contrast, the rest experts (1 and 3) are not satisfied, as far as $r_1 < 0.5$ and $r_3 > 0.5$. And so, $s^* = (0; 0.5; 1)$ is a Nash equilibrium.

Introduce the following values: $w_1 = \pi(d, D, D) = \pi(0, 1, 1) = 2/3; w_2 = \pi(d, d, D) = \pi(0, 0, 1) = 1/3$ (note that $\pi(0, 0, 0) = 0$ and $\pi(1, 1, 1) = 1$). Moreover, $w_2 \leq r_2 \leq w_1 (1/3 \leq 1/2 \leq 2/3)$. In other words, within the segment $[w_2; w_1]$, expert 2 is "a *dictator* with restricted authorities" (the restrictions are the boundaries of this segment). Now, construct a mechanism, where all experts benefit from truth-telling, and the final assessment coincides with that of the mechanism $\pi(\cdot)$.

The principal may ask the experts for the actual opinions $r = \{r_i\}_{i \in N}$ and use them as follows (an equivalent direct mechanism). First, sort the experts in the ascending order of the reported peak points. Second, if there exists $q \in \overline{2, n}$ such that $w_{q-1} \geq r_{q-1}, w_q \leq r_q$, then $x^* = \min(w_{q-1}; r_q)$. The uniqueness of q is easily shown. In the present example, $q = 2$ and $1/2 = \min(2/3; 1/2)$.

Evidently, $s_i^* = d, i < q, s_i^* = D, i > q$. Therefore, the principal finds a Nash equilibrium s^* using the message r and the values w_1, w_2.

Obviously, the constructed direct mechanism makes truth-telling a Nash equilibrium for the agents. The final assessment coincides with that in the original mechanism.

Now, following [135], consider the general case with an arbitrary number of experts. Suppose that all r_i differ and are sorted in the ascending order (i.e., $r_1 < r_2 < \cdots < r_n$), while x^* forms a Nash equilibrium ($x^* = \pi(s^*)$).

Readers would easily make certain of the following. If $x^* > r_i$, then $s_i^* = d$; if $x^* < r_i$, then $s_i^* = D$. In the case of $d < s_i^* < D$, one obtains $x^* = r_i$. Note that $x^* = r_q$ yields $\forall j < q: s_j^* = d, \forall j > q: s_j^* = D$. The quantity s_q^* is defined by the condition

$$
\pi \left(\underbrace{d, d, \ldots, d}_{q-1}, s_q^*, \underbrace{D, D, \ldots, D}_{n-q} \right) = r_q.
$$

Thus, equilibrium evaluation requires just finding a positive integer q. To succeed, let us define $(n+1)$ numbers:

$$
w_i = \pi \left(\underbrace{d, d, \ldots, d}_{i}, \underbrace{D, D, \ldots, D}_{n-i} \right), \quad i = \overline{0, n}.
$$

Here $w_0 = D > w_1 > w_2 > \cdots > w_n = d$, and if $w_i \leq r_i \leq w_{i-1}$, then $x^* = r_i$ (expert i is a dictator on the segment $[w_i; w_{i-1}]$). Evidently, there exists a unique expert q such that $w_{q-1} \geq r_{q-1}, w_q \leq r_q$.

By defining q, one can calculate the final equilibrium assessment: $x^* = \min(w_{q-1}, r_q)$. Truth-telling ($\tilde{r}_i \equiv r_i)_{i \in N}$ is a dominant strategy [135].

Now, we reject the hypothesis that the agents' type vector is a common knowledge. Consequently, the stable informational equilibrium results from the following beliefs of (real and phantom) agents:

$$
r_{\sigma q(r)} \in [\min\{w_{q(r)-1}; r_{q(r)}\}; r_{q(r)}], \quad \sigma \in \Sigma,
$$

$$
r_{\sigma i} \leq \min\{w_{q(r)-1}; r_{q(r)}\}, \quad \sigma \in \Sigma, \ i < q(r),
$$

$$
r_{\sigma i} \geq \min\{w_{q(r)-1}; r_{q(r)}\}, \quad \sigma \in \Sigma, \ i > q(r).
$$

Consider an example. Set $n = 3, r_1 = 0.4, r_2 = 0.5$, and $r_3 = 0.6$. The principal adopts the estimation mechanism defined by $x = \pi(s) = \frac{1}{3}\sum_{i=1}^{n} s_i$. If $s_i \equiv r_i, i = \overline{1, 3}$ (all experts choose truth-telling), then $x = 0.5$. The final assessment coincides with the actual belief of expert 2 (he/she is completely satisfied with the result). The rest ones (experts 1 and 3) are not happy, so long as $r_1 < 0.5$ and $r_3 > 0.5$. Hence, they will report other estimates s_1 and s_3. Assume they report $s_1^* = 0, s_2^* = 0.5, s_3^* = 1$. In this case, $x^* = \pi(s_1^*, s_2^*, s_3^*) = 0.5$. The final assessment remains the same, but the "new" vector of messages does represent a Nash equilibrium. Notably, in our example $w_0 = 1, w_1 = 2/3, w_2 = 1/3$, and $w_3 = 0$; hence, $q = 2$ and

$$
r_2 = 1/2 = \min(2/3; 1/2).
$$

Thus, the stable informational equilibrium results from the following beliefs of (real and phantom) agents: $r_{\sigma 2} = 1/2, r_{\sigma 1} \leq 1/2, r_{\sigma 3} \geq 1/2, \sigma \in \Sigma$.

We have established an important fact. For any expertise mechanism $\pi(\cdot)$, it is possible to build an equivalent direct mechanism, where truth-telling makes a Nash equilibrium. And so, if the principal is interested in agents' truth-telling, he/she can guarantee this by a strategy-proof direct mechanism. However, the principal may pursue other goals.

For instance, the principal is concerned with making the expertise result as close to a value $x_0 \in [d; D]$ as possible. Assume that the principal knows the agents' opinions $\{r_i \in [d; D]\}_{i \in N}$, but each agent knows nothing about the actual opinions of the others. For each agent, reflexive (informational) control by the principal consists in forming specific beliefs about the opponents' opinions such that the information reported by the agents (as a subjective informational equilibrium) leads to the most beneficial result for the principal (i.e., as close to x_0 as possible).

Denote by $x_{0i}(a_i, r_i)$ a solution to the equation

$$(1) \quad \pi(a_i, \ldots, a_i, x_0, a_i, \ldots, a_i) = r_i,$$

where x_0 holds position i $(i \in N)$.

The condition (1) means the best response of agent i to the message a_i unanimously reported by the rest agents.

The monotonicity and continuity of the mechanism $\pi(\cdot)$ implies that $x_{0i}(a_i, r_i)$ is a continuous decreasing function of a_i (under a fixed type r_i of agent i). Suppose that $x_0 \in [d; D]$, then $\forall a_i \in \Re^1, \forall r_i \in [d; D]$

$$(2) \quad x_0 \in [d_i(r_i); D_i(r_i)], \, i \in N,$$

$$(3) \quad d_i(r_i) = \max\{d; x_{0i}(D, r_i)\}, D_i(r_i) = \min\{D; x_{0i}(d, r_i)\}, \quad i \in N.$$

Assertion 4.15.1. Let the principal know the types of all experts (but they do not know the opponents' types). Any results x_0 such that

$$(4) \quad x_0 \in [\max_{i \in N} d_i(r_i); \min_{i \in N} D_i(r_i)]$$

can be implemented as a unanimous collective decision by an appropriate reflexive control.

Proof. According to the above structure of a Nash equilibrium in the mechanism of active expertise, the set of informational equilibria is $[d; D]^n$.

Consider the following awareness structure of agent i: $r_{ij} = a_i, j \neq i, r_{ijk} = a_i, k \in N$. In other words, (in the eyes of agent i) all opponents have identical peak points a_i (see (1)), believe that he/she has the same peak point and that this fact is a common knowledge.

Therefore, agent i expects from all opponents the message a_i as an informational equilibrium of their game (note that the principal should not construct complex or deep awareness structures and evaluate informational equilibria for them). His/her best response (by the definition (1) of a_i) lies in reporting $x_{0i}(a_i, r_i)$, whose range is described by (2)–(3). We have obtained that $X_i(r_i) = [d_i(r_i); D_i(r_i)], i \in N$.

Recall that a unanimous decision is necessary; thus, find the intersection of the sets (2)–(3) with respect to all agents. This procedure yields formula (4).

Thus, all agents report x_0, and (by the unanimity condition) this decision is made (a detached onlooker would hardly cavil at the "democratic principles" of the decision mechanism and its results). •

Let us apply Assertion 4.15.1 to the linear anonymous expertise mechanism $\pi(s) = \frac{1}{n}\sum_{i\in N} s_i$, s_i, $r_i \in [0; 1]$, $i \in N$ [135]. (Recall that any mechanism being symmetrical with respect to agents' permutation is said to be *anonymous* [118]). Evaluate $a_i = \frac{nr_i - x_0}{n-1}$, $i \in N$. Since $a_i \in [0; 1]$ (see also (2)–(4)), one obtains the range of unanimously implemented collective decisions:

(5) $\max\{0; n(\max\limits_{i\in N} r_i - 1) + 1\} \le x_0 \le \min\{1; n \min\limits_{i\in N} r_i\}$.

Note that the condition (5) implies the constraint

$$\max\limits_{i\in N} r_i - \min\limits_{i\in N} r_i \le 1 - \frac{1}{n}$$

on the feasible variations of the experts' opinions for ensuring the existence of (at least, one) result x_0 implemented by reflexive control as a unanimous collective decision.

On the other hand, (5) means that $x_0 \in [0; 1]$ if

$$\max\limits_{i\in N} r_i \le 1 - \frac{1}{n}, \quad \min\limits_{i\in N} r_i \ge \frac{1}{n}.$$

This indicates that, in a linear anonymous expertise mechanism, the following represents a sufficient condition of unanimous implementation of any collective decision by reflexive control: there exist no experts with extremely low or extremely high opinions.

Now, reject the requirement of collective decisions unanimous implementation. Consider two vectors:

$$d(r) = (d_1(r_1), d_2(r_2), \ldots, d_n(r_n)), \quad D(r) = (D_1(r_1), D_2(r_2), \ldots, D_n(r_n)).$$

Assertion 4.15.2. Suppose that the principal knows the type of each expert (but they do not know the opponents' types). Any result x_0 such that

(6) $x_0 \in [\pi(d(r)); \pi(D(r))]$

can be implemented as a collective decision by an appropriate reflexive control.

Proof. Assertion 4.15.2 differs from Assertion 4.15.1 in the following. On the one hand, the former does not presume identical equilibrium messages of the agents. On the other hand, it extends the constraint imposed on the collective decision implementable as an informational equilibrium (the condition (4) is replaced by the condition (6)).

Fix the vector $r \in [d; D]^n$ of agents' peak points. The described structure of an equilibrium implies that each agent reports the minimal request (0) or the maximal request (1), or else his/her actual type (if this agent is a dictator). Recall that it is possible to form arbitrary beliefs about the types of rest agents, their beliefs, etc. (for a given agent). Thus, each agent can be convinced that the set of feasible action profiles constitutes $[d; D]^{n-1}$.

For instance, it suffices to form the following awareness structure of depth 3: agent ij must be a dictator and this fact must be a common knowledge for ijk-agents.

In the course of proof of Assertion 4.15.1, we have established that $X_i(r_i) = [d_i(r_i); D_i(r_i)], i \in N$. The informational structures of agents are formed independently. Hence, we obtain that the vector of minimal (maximal) requests is $d(r)$ ($D(r)$, respectively). The monotonicity and continuity of the decision procedure $\pi(\cdot)$ lead to (6). •

Again, apply Assertion 4.15.2 to the linear anonymous expertise mechanism $\pi(s) = \frac{1}{n}\sum_{i \in N} s_i, s_i, r_i \in [0; 1], i \in N$. Evaluate the message s_i of expert i being subjectively optimal for him/her under the opponents' action profile s_{-i}:

(7) $\quad s_i(r_i, S_{-i}) = nr_i - S_{-i}, \quad i \in N,$

with $S_{-i} = \sum_{j \neq i} s_j \in [0; n - 1])$.

Thus, $X_i(r_i) = [\max\{0; 1 - n(1 - r_i)\}; \min\{1; nr_i\}], i \in N$. Under (7), substitute the left and right boundaries of the sets $X_i(r_i)$ into the linear anonymous planning mechanism to obtain

(8) $\quad x_0 \in \left[\sum_{i \in N} \frac{1}{n} \max\{0; 1 - n(1 - r_i)\}; \sum_{i \in N} \frac{1}{n} \min\{1; nr_i\} \right].$

Assertions 4.15.1 and 4.15.2 bring us to the following conclusion (see their proofs describing the minimal awareness structure which implements a given collective decision).

Corollary. Solving reflexive control problems in the mechanisms of active expertise, one can be confined with rank 2 of experts' reflexion.

Study a numerical example with three agents whose peak points are $r_1 = 0.4$, $r_2 = 0.5$ and $r_3 = 0.6$. Set $x_0 = 0.8$. Under the conditions of truth-telling, the indirect mechanism yields the result $x = 0.5$ (the same result is provided by the corresponding direct (strategy-proof) mechanism). Yet, the principal wants each agent to report a greater estimate, making the final assessment closer to 0.8.

The condition (5) is satisfied. Find the quantities

$0.8 + 2a_1 = 3 \times 0.4 \rightarrow a_1 = 0.2;$

$0.8 + 2a_2 = 3 \times 0.5 \rightarrow a_2 = 0.35;$

$0.8 + 2a_3 = 3 \times 0.6 \rightarrow a_3 = 0.5.$

For agent 1, the principal forms the belief that the types (peak points, actual opinions) of the rest agents constitute 0.2, while they believe the type of agent 1 is 0.2 (and this is a common knowledge). Similar "beliefs" are formed by the principal for agents 2 and 3 (0.35 and 0.5, respectively).

The best response of agent 1 (leading to the collective decision coinciding with his/her peak point) to the message 0.2 reported by the rest agents equals 0.8. Moreover, this is the best response of agents 2–3, as well. Thus, all agents unanimously report 0.8.

In the studied numerical example, the condition (8) holds for any $x_0 \in [0; 1]$, i.e., $n(\max_{i \in N} r_i - 1) + 1 \le 0, n \min_{i \in N} r_i \ge 1$.

Let us consider another example: $n = 2, r_1 = 0.2, r_2 = 0.7$. Then (5) implies there exists a unique x_0 (actually, 0.4) being implemented as a unanimous collective decision. At the same time, according to Assertion 4.15.2, the set of implementable collective decisions makes up the segment $[0.2; 0.7]$.

The boundaries of this segment have matched the agents' types incidentally. For instance, for $r_1 = 0.1$ and $r_2 = 0.5$, the unanimous collective decisions lie within the segment $[0; 0.2]$. Within the framework of Assertion 4.15.2, they belong to the segment $[0; 0.6]$.

To complete the discussion of reflexive control in the mechanisms of active expertise, we emphasize the following aspect. The derived results (Assertions 4.15.1–4.15.2) proceed from the assumption that the coordinator of the expertise (the principal) knows the type of each agent (but the agents know nothing about the types of each other!). A somewhat more realistic assumption is that each participant (the principal and experts) possesses specific beliefs about the ranges of opponents' types, i.e., the managerial capabilities of the principal are limited. In this case, a promising direction of further research consists in analyzing the set of collective decisions implemented as informational equilibria.

4.16 THE COURNOT OLIGOPOLY: INFORMATIONAL REFLEXION

This section deals with the Cournot oligopoly model, where agents make decisions regarding the amount of products manufactured by them. The market price of products decreases as the total output grows; moreover, it depends on demand. Imagine that the uncertain parameter is demand and each agent has a specific hierarchy of beliefs about it; then an informational equilibrium essentially depends on the mutual beliefs of agents. If the uncertain parameter lies in agents' costs, then agents may attain a true informational equilibrium in dynamics by observing chosen actions.

Consider an organizational system which engages n agents with the following goal functions:

$$(1) \quad f_i(r_i, x) = \left(Q - \alpha \sum_{j \in N} x_j \right) x_i - \frac{x_i^2}{2r_i},$$

where $x_i \ge 0, i \in N, \alpha > 0$.

In practical interpretation, x_i means the output of agent i, and Q corresponds to demand for the product. Thus, the first term in the goal function makes up sales proceeds (the price multiplied by sales volume), and the second term forms manufacturing costs. The parameter r_i (the type of agent i) reflects the efficiency (skill level) of his/her activity.

The best response of agent i takes the form

(2) $BR_i(x_{-i}, r_i) = (Q - \alpha \sum\limits_{j \neq i} x_j)/(2\alpha + 1/r_i), \quad i \in N.$

Assume that each agent observes the price $(Q - \alpha \sum_{j \neq i} x_j)$. The conditions of Assertion 2.8.1 are satisfied; therefore, this model admits no false equilibria (only true stable informational equilibrium is possible). We provide an illustrative numerical example. Choose $n = 2$, $\alpha = 1$, $Q = 5$, and $r_1 = r_2 = 1$. Evaluate the parametric Nash equilibrium:

(3) $x_1(r_1, r_2) = \dfrac{\left[1 - \frac{2r_2+1}{r_2}\right]Q}{1 - \frac{(2r_1+1)(2r_2+1)}{r_1 r_2}},$

(4) $x_2(r_1, r_2) = \dfrac{\left[1 - \frac{2r_1+1}{r_1}\right]Q}{1 - \frac{(2r_1+1)(2r_2+1)}{r_1 r_2}}.$

Formulas (3) and (4) lead to agents' types making the observed action vector (x_1, x_2) a stable informational equilibrium:

(5) $r_1(x_1, x_2) = 1/[(Q - x_2)/x_1 - 2],$

(6) $r_2(x_1, x_2) = 1/[(Q - x_1)/x_2 - 2].$

Now, consider the dynamic model for the mutual beliefs of agents. Suppose that each agent independently chooses an action by substituting his/her type and beliefs about the opponent's type into (3) or (4). Having observed the opponent's choice, each agent evaluates a new estimate of opponent's type (using (5) or (6)). According to the hypothesis of indicator behavior, the agent corrects his/her beliefs. And the procedure repeats.

The dynamics of agent's beliegs are shown in Fig. 4.25, Fig. 4.26 and Fig. 4.27.

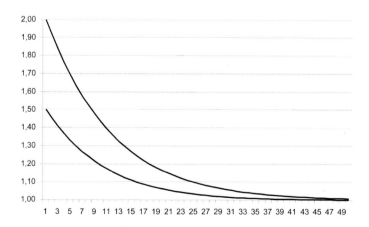

Figure 4.25 Both agents initially overestimate each other $(r_{12} = 2, r_{21} = 1.5)$.

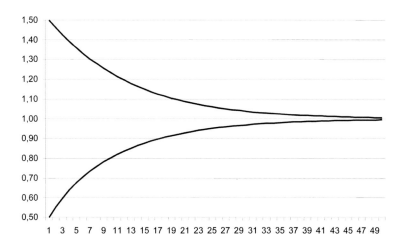

Figure 4.26 Agent 1 initially overestimates agent 2, while agent 2 underestimates agent 1 ($r_{12} = 1.5, r_{21} = 0.5$).

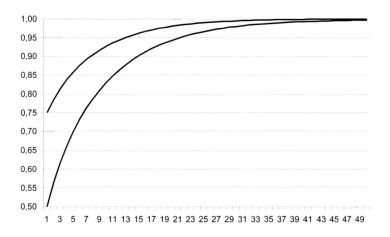

Figure 4.27 Both agents initially underestimate each other ($r_{12} = 0.75, r_{21} = 0.5$).

Clearly, the beliefs of each agent about the opponent's type monotonically converge to the true value (1; 1). Moreover, the agents' actions tend to the true informational equilibrium – the parametrized Nash equilibrium, where parameters equal the true types of agents.

4.17 RESOURCE ALLOCATION

This section is dedicated to a model of the following situation. A principal allocates a limited resource among agents based on their requests. Due to their active property, agents can distort information (demonstrate a strategic behavior). In an equilibrium, some agents (called *dictators* [28]) receive an optimal quantity of the resource, whereas the rest one obtain less.

Assume that agents possess a certain hierarchy of beliefs about necessary quantities of the resource for each of them. In many resource allocation mechanisms, stable inadequate beliefs may exist only about agents not being dictators. However, the vector of allocated resources appears the same as in the case of complete knowledge. Let us describe the model rigorously.

The above models have analyzed stability of an informational equilibrium representing a generalization of a Nash equilibrium. Several models admit the existence of a stronger equilibrium, namely, a *dominant strategies' equilibrium* (DSE). It is defined as a vector of absolutely optmal actions of agents (being independent of opponent's action profile and the types vector of the rest agents) [75, 121].

Assertion 4.17.1. A DSE is a stable informational equilibrium.

This result follows from the notions of a DSE and an informational equilibrium. Assertion 4.17.1 implies that the analysis problem for an informational equilibrium gets degenerated in systems with a DSE. Yet, the matter by no means applies to the problem of equilibrium stability analysis. The models of resource allocation mechanisms and the mechanisms of active expertise indicate the following. Agents can choose dominant strategies within wide ranges of their mutual beliefs.

Consider the set of agents $N = \{1, 2, \ldots, n\}$ having the goal functions $f_i(x_i, r_i)$; here $x_i \geq 0$ is the resource quantity allocated to agent i and $r_i \geq 0$ designates his/her type (the optimal quantity of the resource). In other words, $f_i(x_i, \cdot)$ is a single-peaked function with the *peak point* $r_i, i \in N$.

Each agent knows for sure his/her peak point (type), but the principal has no information on these points.

The principal's problem consists in allocating a given quantity R of the resource based on agents' requests $s_i \in [0; R]$, $i \in N$. Notably, in this model, the agents' actions are their messages reported to the principal. A decision rule $x_i = \pi_i(s)$ ($i \in N$, $s = (s_1, s_2, \ldots, s_n)$) adopted by the principal is called a *planning mechanism (procedure)*.

According to [135], suppose that

1) $\pi_i(s)$ is a continuous and strictly increasing function of $s_i, i \in N$;
2) $\pi_i(0, s_{-i}) = 0 \ \forall s_{-i} \in [0; R]^{n-1}, i \in N$;
3) the resource allocation mechanism turns out anonymous, i.e., an arbitrary permutation of agents causes an appropriate permutation of their quantities of resources.

The following facts are well-known in the case when the agents' types are a common knowledge (e.g., see [135]). First, the situation of the Nash equilibrium $s^*(r)$ in the game of agents possessesthe following structure (for mechanisms meeting properties 1–2): $\forall i \in N, \forall r \in \Re^n$

(1) $\pi_i(s^*(r)) < r_i \Rightarrow s_i^*(r) = R,$
(2) $s_i^*(r) < R \Rightarrow \pi_i(s^*(r)) < r_i.$

Second, all resource allocation mechanisms enjoying properties 1–3 are equivalent to (lead to the same equilibria as) the mechanism of proportional allocation: agents obtain the resource in the quantities

$$(3) \quad \pi_i(s) = \begin{cases} s_i, & \text{if } \sum_{j=1}^{n} s_j \leq R \\ \min\left\{ s_i, R\, s_i / \sum_{j=1}^{n} s_j \right\}, & \text{if } \sum_{j=1}^{n} s_j > R \end{cases},$$

And so, we can focus on the mechanism (3).

There exists a famous search algorithm for a Nash equilibrium in the game of agents (e.g., see [135]); this algorithm involves the *procedure of serial resource allocation*[15]. Its idea consists in the following:

0. Agents are sorted in the ascending order of their peak points. The set of dictators[16] is empty.
1. All resource quantity is divided among agents in equal shares.
2. If $r_1 < R/n$, agent 1 enters the set of dictators, all agents receive the resource in the quantity of r_1 (under $r_1 < R/n$, the set of *dictators* turns out empty, all agents report identical requests in an equilibrium and obtain the same quantity of the resource; and the algorithm terminates).
3. By setting $r_i := r_i - r_1, i := i - 1, R := R - nr_1$, repeat Step 2 (clearly, the number of repetitions does not exceed the number of agents).

The procedure of serial resource allocation yields the set of dictators $D(r) \subseteq N$, who gain the optimal quantities of the resource (defined by their peak points). Due to the anonymous property of this mechanism, the rest agents (non dictators) receive the same quantity:

$$x_0(r) = \left(R - \sum_{j \in D(r)} r_j\right) / (n - |D(r)|).$$

Now, reject the assumption that the agents' types form a common knowledge; analyze informational equilibria and their stability. Suppose that the observation function of each agent is the opponents' action vector (see Section 2.9). According to the results established in Sections 2.7–2.9, only true informational equilibria are possible. However, the beliefs about opponents' types can be adequate or inadequate.

---

[15] First, the procedure of serial resource allocation is a direct mechanism (uses information on the peak points of agents). Second, this procedure appears strategy-proof, i.e., each agent benefits from truth-telling about his/her peak point (provided that the principal undertakes to apply the procedure of serial resource allocation).

[16] Recall that a dictator is an agent obtaining the absolutely optimal plan ($x_i = r_i$).

Consider the following cases:

Case 1. The vector of agent's actual types is such that $D(r) = N$. This appears possible if $\sum_{j \in N} r_j \leq R$. Then the actual types of agents are a common knowledge.

Case 2. The vector of agent's actual types is such that $D(r) = \emptyset$. This appears possible if

(4) $\quad \min_{i \in N}\{r_i\} > R/n.$

Then the best response of each agent is independent of his/her subjective beliefs meeting (4); any combination of such beliefs forms a true equilibrium.

A particular interest of researchers belongs to the intermediate case (there are dictators and non dictators among agents).

Case 3. The vector of agent's actual types is such that $D(r) \neq \emptyset$, $D(r) \neq N$. By virtue of observability of chosen actions, in an equilibrium a dictator receives the quantity coinciding with his/her type (by definition). Hence, none of the agents have stable inadequate beliefs about their types (see Case 1):

(5) $\quad r_{\sigma i} = r_i, \quad \sigma \in \Sigma, \quad i \in D(r).$

Meanwhile, stable inadequate belifes about the types of agents from the set $N \setminus D(r)$ may exist:

(6) $\quad r_{\sigma i} \geq \min_{j \in N \setminus D(r)} r_j, \quad \sigma \in \Sigma, \quad i \in N \setminus D(r).$

Consider an example. Select $n = 3, R = 1, r_1 = 0.2, r_2 = 0.3,$ and $r_3 = 0.6$. Then $s_1^* = 0.4, s_2^* = 0.6, s_3^* = 1, x_1^* = r_1 = 0.2, x_2^* = r_2 = 0.6,$ and $x_3^* = 0.5$. Formulas (5)–(6) imply that

$$\forall \sigma \in \Sigma : r_{\sigma 1} = r_1, \quad r_{\sigma 2} = r_2, \quad r_{\sigma 3} \geq 0.5.$$

Therefore, within monotonous anonymous mechanisms of resource allocation, stable inadequate beliefs may exist only regarding the types of agents not entering the set of dictators. However, the vector of allocated resources is the same as in the case of a complete knowledge.

4.18 INSURANCE

This section addresses a model, where insurants have a hierarchy of mutual beliefs about the probabilities of insured accidents and report to an insurer the desired insurance premiums.

Interestingly, here an informational equilibrium is any set of messages with the following property. According to any (real or phantom) agent, the total message of real insurants equals the expected total damages from insured accidents, whereas each message does not exceed the expected insurance indemnity. Furthermore, all equilibrium actions of real insurants are attained within the framework of their subjective awareness of each other. Let us pass to model description.

Formal models of risk management (including insurance models studied in actuarial mathematics) generally disregard the active property of insurants and insurers; this property manifests itself in their reflexion, desire to distort information, and so on. Probably, the exceptions are the paper [29] and the book [135]. The cited works consider the models of mutual and endowment insurance, where am insurer adopts the information reported by insurants to establish the parameters of insurance contracts. In addition, the so-called "discount mechanism" making truth-telling beneficial to an insurant is proposed. In the present section, we analyze models of mutual insurance, where agents (participants of a mutual insurance system) possess a certain hierarchy of beliefs about the probabilities of insured accidents for each agent.

Consider a union of n insurants (regarded as an insurer in the model of mutual insurance). Agents have the goal functions defined through the expected utilities

(1) $Ef_i = g_i - r_i + p_i[h_i - Q_i], \quad i \in N.$

We use the following notation: g_i represents the income gained by the economic activity of insurant i; r_i is the insurance premium; h_i stands for the insurance indemnity; p_i gives the probability of an insured accident (suppose that the insured accidents for different agents are independent events); Q_i designates the losses inflicted by an insured accident; and finally, $N = \{1, 2, \ldots, n\}$ is the set of insurants. For simplicity, let us describe the interaction of insurants over a time interval, where the premiums are collected and indemnification is paid once. We assume here that the remaining funds, if positive, are used as the oddments of funds for the next time interval.

According to (1), assume that all insurants regard risk identically, but differ in their estimates of the probabilities of insured accidents and corresponding losses. As is well-known, risk reallocation appears profitable only for agents differing in their attitude toward risks. Therefore, one can suppose that, on the one hand, all insurants are neutral to risks and, on the other hand, the main phenomenon to-be-studied in mutual insurance is reflexion by insurants and their incomplete awareness. Since all insurants have an identical attitude toward risk, it is admissible to reallocate it arbitrarily, (provided that all insurants have are complete aware of each other). Under an incomplete awareness or no common knowledge takes place, the requirement of balanced premiums and expected payments can be violated.

Under complete awareness, the total insurance premium makes up $R = \sum_{i \in N} r_i$, while the expected insurance indemnity equals $H = \sum_{i \in N} p_i h_i$. Recall we focus on the mutual (noncommercial) insurance. By virtue of the equivalence principle [29], there must be $R = H$, i.e.,

(2) $\sum_{i \in N} r_i = \sum_{i \in N} p_i h_i.$

Note that (2) reflects the equality between the total insurance premium and the expected insurance indemnity. In other words, the problems of mutual insurance bankruptcy are not discussed.

Imagine that, in the case of an insured accident ($h_i = Q_i, i \in N, H = \sum_{i \in N} p_i Q_i$), the damages are refunded completely (the assumption of incomplete reparation of damages – an a priori fixed level of insurance indemnity – will not affect qualitatively the main results of analyzing the mechanisms of mutual insurance). Then, under the complete awareness, the following mechanism of mutual insurance could be used:

(3) $r_i = p_i Q_i, \quad i \in N.$

Within its framework, the insurance premium of each insurant is equal precisely to its expected loss (the amount at risk coincides with the losses, and the insurance rate coinciding with the net premium is defined by the corresponding probability of an insured accident).

However, when each insurant knows only his/her individual parameters (not observed by the rest insurants), the mechanism (3) turns out inapplicable. Therefore, we consider two options as follows. First, insurants report the information on the probabilities of insured accidents. Second, the mechanism of mutual insurance, which meets the system of mutual beliefs of agents about essential parameters, is analyzed.

Mechanisms with revelation of information. Assume that the estimates $\{s_i\}$ of the probabilities of insured accidents can be reported by insurants to each other. Generally, all insurants will strive for reducing the probabilities of insured accidents. Consequently, revelation of the minimal estimates forms an equilibrium. Therefore, we consider some alternative mechanisms of mutual insurance.

Denote by $s = (s_1, s_2, \ldots, s_n)$ the vector of agents' messages. Let an insurance contract makes a clear provision that the insurance premium of each insurant is defined by the reported estimates of the probabilities of insured accident, i.e., $r_i(s_i) = s_i Q_i$. Moreover, the insurance contract states that, upon occurrence of the insured accidents, the damages are refunded proportionally to the collected insurance fund $R(s) = \sum_{i \in N} r_i(s)$:

(4) $h_i(s) = \alpha(s) Q_i, \quad i \in N.$

Here $\alpha(s)$ is the unique share of the insurance indemnity (the ratio of the insurance indemnity $h_i(s)$ to the amount at risk Q_i) resulting from the relationship between the insurance fund $R(s)$ and the necessary amount of insurance indemnity H. Control strategy consists in choosing $\alpha(\cdot)$.

By substituting (4) into (1), we obtain that for insurant i the condition of beneficial participation in the mutual insurance is determined by

(5) $s_i \leq \alpha(s) p_i, \quad i \in N.$

If the control strategy

(6) $\alpha(s) = \min\{R(s)/H; 1\}$

is involved, the balance condition (2) definitely holds true. It follows from (5) that insurant's message does not exceed the actual value of the probability of insured accident: $s_i \leq \alpha(s) p_i, i \in N.$

By substituting (4) and (6) into (1) and differentiating with respect to s_i, we obtain that mechanism (6) is not strategy-proof. Truth-telling becomes unbeneficial to insurants. Each insurant endeavors to reduce the probability of insured accident (in this case, reduction in the insurance premium is greater than in the share of the insurance indemnity).

The following mechanism of mutual insurance is an alternative to (5). Let insurants conclude a contract stating that (a) at the beginning of the period under consideration, they must report their estimates of the probabilities of insured accidents (at the beginning of the period, premiums are not collected!), (b) at the end of the period (when the insured accidents are realized), they compensate to the "aggrieved persons" their losses and (c) the contribution of each insurant is defined by the estimates reported at the beginning of the period. In this case, the expected reparation of damages is $H = \sum_{i \in N} p_i Q_i$. Consequently, the sum of premiums must be

$$(7) \quad \sum_{i \in N} r_i(s) = H,$$

where the dependencies $r_i(\cdot)$ represent the control mechanism. The expected value of the insurants goal function acquires the form

$$(8) \quad Ef_i = g_i - r_i(s), \quad i \in N.$$

Accordingly, the condition of beneficial participation in mutual insurance becomes:

$$(9) \quad r_i(s) \leq p_i Q_i, \quad i \in N.$$

Choose the following control mechanism:

$$(10) \quad r_i(s) = \frac{s_i Q_i}{\sum_{i \in N} s_i Q_i} H, \quad i \in N.$$

Here the premium of each insurant is proportional to the reported expected losses. Then (8) reaches its maximum for the minimal messages, *viz.*, the mechanism (10) is not strategy-proof.

Analysis of the conditions (9)–(10) suggests an idea. To reduce distortion of information, one must select a control mechanism, where the insurance premium goes down with increasing the message of an insurant. We provide a corresponding example. Consider the mechanism

$$(11) \quad r_i(s) = \frac{1/s_i}{\sum_{i \in N} (1/s_i)} H, \quad i \in N.$$

By substituting (11) into (8), we get that the mechanism (11) does not stimulate an insurant to reduce his/her message. However, this mechanism does not ensure truth-telling.

Therefore, each of the mechanisms (10) and (11) possesses advantages. The mechanism (10) is balanced and agrees with the condition (7); still, insurants reduce their messages. On the other part, the mechanism (11) stimulates insurants to increase their messages, but fails to provide the "balance" in the sense of (7). To construct a mechanism enjoying all these advantages, one should probably get a reasonable trade-off between the increase and decrease of the insurant's goal function with respect to his/her message. For mutual insurance, however, this balance appears impossible. Due to its noncommercial orientation, mutual insurance represents a "zero-sum game" (from the insurant's point of view). The condition (2) implies that the total premiums must be equal to the total reparation of damages. Therefore, reducing the insurance premium by an insurant leads to the following situation. This reduction is compensated by all insurants (including those who distorted their information but to a smaller extent – see formula (10) or (11)). Therefore, additional resources must be used to "suppress" distortion of information. The dependence of their volume on the messages of insurants must stimulate their truth-telling. For instance, mobilize the financial resources of the third party (in contrast to the above participants of an insurance contract). This approach brings to endowment insurance.

Reflexive model. Consider a situation, where the principal possesses some information on the losses from insured accidents $\{\tilde{Q}_i\}$ and the probabilities of insured accidents $\{\tilde{p}_i\}$. Assume that the quantities $\{\tilde{Q}_i\}$ and $\{\tilde{p}_i\}$ form a common knowledge (generally, $\tilde{Q}_i \neq Q_i$ and $\tilde{p}_i \neq p_i$). Each insurant reports his/her premium s_i to the principal (alternatively, an insurant refuses insurance). If all insurants report their premiums, the principal verifies the condition

$$(12) \quad \sum_{i \in N} s_i \geq H = \sum_{i \in N} \tilde{p}_i \tilde{Q}_i.$$

Inequality (12) being valid, a mutual insurance contract is signed. Otherwise (or if, at least, one insurant refuses insurance), the contract is not signed.

In the described situation, the goal function of insurant i has the form

$$f_i(p_i, s_i, \ldots, s_n) = \begin{cases} p_i \tilde{Q}_i - s_i, & \sum_{i \in N} s_i \geq H, \\ -\varepsilon_i, & \sum_{i \in N} s_i < H, \end{cases}$$

where ε_i means an arbitrary positive constant (organizational costs in the case of unsigned contract). Moreover, suppose that an agent refusing from mutual insurance gains nothing.

The information available to players is described by their beliefs about the parameters p_i (the probabilities of insured accidents). Denote by p_{ij} the belief of agent i (an insurant) about the value of p_j, by p_{ijk} the belief of agent i about the beliefs of agent j about the value of p_k, and so on $(i, j, k \in N)$. In the aggregate, these beliefs form the corresponding awareness structure.

Informational equilibria in this reflexive game of insurants are characterized by the following assertion (recall that Σ indicates the set of all possible finite sequences of indexes from N, including the empty sequence).

Assertion 4.18.1. Let insurants possess an awareness structure of a finite complexity. The set of actions $s^*_{\sigma i}, \sigma \in \Sigma, i \in N$, is an informational equilibrium (a mutual insurance contract is signed) if the conditions

$$\sum_{i \in N} s^*_j = H, \quad \forall i \in N \quad s^*_i \leq p_i \tilde{Q}_i$$

form a common knowledge. The latter implies that for any $\sigma \in \Sigma$ we have

(13) $\forall i \in N : \sum_{i \in N} s^*_{\sigma ij} = H, \quad s^*_{\sigma i} \leq p_{\sigma i} \tilde{Q}_i.$

Proof. Assume that the set of actions $s^*_{\sigma i}, \sigma \in \Sigma, i \in N$, represents an informational equilibrium (leading to signing of a mutual insurance contract). Fix arbitrary values of $\sigma \in \Sigma$ and $i \in N$. The action $s^*_{\sigma i}$ maximizes the goal function $f_i(p_{\sigma i}, s^*_{\sigma i1}, \ldots, s^*_{\sigma i,i-1}, s_{\sigma i}, s^*_{\sigma i,i+1}, \ldots, s^*_{\sigma i1})$ of σi-agent with respect to $s_{\sigma i}$. Thus, he/she chooses the minimal action meeting $\sum_{j \in N} s^*_{\sigma ij} \geq H$; evidently, the inequality becomes the equality. The goal function must possess a nonnegative value (otherwise, σi-agent benefits from refusing mutual insurance). And so, we derive $p_{\sigma I} \tilde{Q}_i - s^*_{\sigma i} \geq 0$.

Next, suppose that for any $\sigma \in \Sigma$ the condition (13) holds true. Then each action $s^*_{\sigma i}$ maximizes the goal function $f_i(p_{\sigma i}, s^*_{\sigma i1}, \ldots, s^*_{\sigma i,i-1}, s_{\sigma i}, s^*_{\sigma i,i+1}, \ldots, s^*_{\sigma i1})$ of σi-agent. Therefore, the choice of actions $s^*_{\sigma i}, \sigma \in \Sigma, i \in N$, is an informational equilibrium. •

Finally, note that a mutual insurance contract fails if, at least, one real or phantom agent refuses insurance. Formally, the refusal of all agents defines an equilibrium. See the reservation in parentheses in Assertion 4.18.1.

Reflexion ranks of insurants and equilibria. Consider the complexity of the subjective beliefs of an insurant, guaranteeing the reachability of all possible informational equilibria. In other words, study the problem of maximal rational reflexion rank. The following assertion demonstrates that this rank equals 1.

Assertion 4.18.2. In the reflexive game of insurants, all possible actions of real agent i ($i \in N$) are reachable within his/her subjective common knowledge about the set $(p_1, \ldots, p_n)$, i.e., within the awareness structure such that

$\forall \sigma \in \Sigma \quad \forall j \in N : p_{i\sigma j} = p_{ij}.$

Proof. If an agent refuses to participate, Assertion 4.18.2 is trivial (it suffices to announce $p_{ij} = 0$ for all $j \in N$ as a common knowledge).

Consider the situation from the viewpoint of agent i ($i \in N$). Assume that his/her action s^*_i is subjectively equilibrium in a certain equilibrium $s^*_{i\sigma j}, \sigma \in \Sigma, j \in N$. By virtue of Assertion 4.18.1, for all $j \in N$ we have the formulas

$$\sum_{j \in N} s^*_{ij} = H, \quad 0 \leq s^*_{ij} \leq \tilde{Q}_j.$$

Set $p_{i\sigma j} = 1$ for all $\sigma \in \Sigma, j \in N$ (i.e., form an awareness structure, where a common knowledge takes place according to agent i). Obviously, the set of actions $w^*_{i\sigma j} = s^*_{ij}$

subjectively satisfies the condition (13) (in the view of agent i). Consequently, the set $w^*_{i\sigma j}$, $\sigma \in \Sigma, i, j \in N$, is subjectively an informational equilibrium. Moreover, in this equilibrium the action of agent i coincides with his/her action in the original equilibrium $s^*_{i\sigma j}$. •

Let us summarize the outcomes. In this section, we have studied the model of mutual insurance with informational reflexion of insurants. We have characterized the set of informational equilibria leading to signining of an insurance contract. Finally, all equilibrium actions of an insurant have been shown reachable under a subjective common knowledge of insurants about each other.

4.19 PRODUCT ADVERTIZING

Consider a model, where an agent purchases a certain product based on his/her preferences and on the information about the share of other agents planning such purchase (see Section 4.2). Interestingly, most advertizing campaigns can be described by the model of informational control with agents' reflexion of rank 1 or 2.

Suppose that there are agents of two types. The ones having type 1 incline to purchase products regardless of advertising, while agents of type 2 would not. Denote by $\theta \in [0; 1]$ the share of type 1 agents.

Type 2 agents (their share makes up $(1 - \theta)$) are subject to advertising effect; however, they do not comprehend this. Let us reflect social impact [77, 153, 178] as follows. Assume that type 2 agents choose the action a with the probability $p(\theta)$ and the action the action r with the probability $1 - p(\theta)$. The relationship $p(\cdot)$ (the choice probability as a function of the share of type 1 agents) characterizes their reluctance of being contrarians.

Imagine that the actual share θ of type 1 agents forms a common knowledge. Then agents expect that θ agents purchase the product. Nevertheless, in reality they observe that the product is purchased by

(1) $x(\theta) = \theta + (1 - \theta)p(\theta)$

agents (by supposition, agents do not realize the advertising effect). Since $\forall \theta \in [0; 1]$: $\theta \leq x(\theta)$, the indirect social impact becomes assuring: "Look, more people incline to purchase the product than we expected!"

To proceed, we analyze an asymmetrical awareness. Type 1 agents choose their actions independently. Thus, they can be treated as having an adequate awareness about the parameter θ and the beliefs of type 2 agents.

Consider the model of informational regulation, where a principal (an organizer of an advertizing campaign) forms the beliefs θ_2 about the parameter θ for type 2 agents.

Making a small digression, discuss the properties of the function $p(\theta)$. Suppose that $p(\cdot)$ is a nondecreasing function on $[0; 1]$ such that $p(0) = \varepsilon, p(1) = 1 - \gamma$. Here ε and δ indicate constants belonging to unit segment, $\varepsilon \leq 1 - \delta$. In practical interpretation, ε corresponds to "mistakes" of some type 2 agents (they purchase the product, even believing that the rest agents have type 2). The constant δ characterizes (in a certain sense) agents' susceptibility to influence. Indeed, a type 2 agent has a chance to be independent (by refusing to purchase, even if he/she believes that the rest agents will purchase the product). The special case of $\varepsilon = 0, \delta = 1$ describes independent agents of type 2 (who deny the purchase).

Agents have no idea of manipulation by the principal (see the trust principle discussed earlier and in [136]). Thus, they expect that θ_2 agents will purchase the product. Actually, they observe the share of purchasers

(2) $x(\theta, \theta_2) = \theta + (1 - \theta)p(\theta_2).$

Suppose that the principal's income is proportional to the share of agents purchasing the product and the advertizing costs $c(\theta, \theta_2)$ represent a nondecreasing function of θ_2. Consequently, the principal's goal function (the difference between the income and costs) without advertizing is defined by (1). Under the advertizing campaign, it takes the form

(3) $\Phi(\theta, \theta_2) = x(\theta, \theta_2) - c(\theta, \theta_2).$

Hence, the efficiency of informational regulation can be defined as the difference between (3) and (1). Accordingly, the problem of informational regulation is rewritten as

(4) $\Phi(\theta, \theta_2) - x(\theta) \to \max\limits_{\theta_2}.$

Now, let us address existing constraints of the problem (4). The first constraint lies in $\theta_2 \in [0; 1]$ (more specifically, $\theta_2 \geq \theta$).

Consider an example. Set $p(\theta) = \sqrt{\theta}, c(\theta, \theta_2) = (\theta_2 - \theta)/2\sqrt{r}$, where $r > 0$ specifies a constant. Then the problem (4) is reduced to

(5) $(1 - \theta)(\sqrt{\theta_2} - \sqrt{\theta}) - (\theta_2 - \theta)/2\sqrt{r} \to \max\limits_{\theta_2 \in [\theta; 1]}.$

Solution to the problem (5) takes the form $\theta_2(\theta) = \max\{\theta; r(1 - \theta)^2\}$, i.e., informational regulation becomes pointless for the principal if $\theta \geq \frac{(2r+1) - \sqrt{4r+1}}{2r}$ (the advertizing costs are not compensated, since sufficiently many agents purchase the product without advertizing).

Now, besides $\theta_2 \in [\theta; 1]$, require stability of an informational equilibrium. Under observability of the share of product purchasers, assume that type 2 agents observe the actual share of product purchasers (if it is smaller than the share reported by the principal). Notably, the stability condition acquires the form $x(\theta, \theta_2) \geq \theta_2$. Using (2), we obtain:

(6) $\theta + (1 - \theta)p(\theta_2) \geq \theta_2.$

Hence, the optimal stable solution of the informational regulation problem follows from maximizing the function (4) subject to the constraint (6).

Concluding this section, we point to the following. In our example, any informational regulation turns out stable in the sense of (6). Alternatively, stability can be understood as the equivalence of the expected and observed results (formula (6) becomes an equality). Then the unique stable informational regulation is the principal's message that all agents have type 1: $\theta_2 = 1$. Interestingly, this situation appears common for most advertizing campaigns.

4.20   THE HUSTINGS

This section concentrates on a model with the following informational control. Voters supporting certain candidacy are persuaded that the latter would not be elected and they should support other candidacies. It appears that the resulting informational control can be stable and even true.

Let us consider the example of reflexive control in the hustings (an election campaign). There exist three candidacies (a, b, and c). The election procedure is based on the simple majority rule (to win, a candidacy has to poll more than 50% percent of votes). It may happen that none of the candidacies receives the majority of votes; in this case, an additional round is organized with other candidacies (denote them by d). Suppose there are three groups of voters whose shares make up α_1, α_2 and α_3, respectively ($\alpha_1 + \alpha_2 + \alpha_3 = 1$). Table 4.1 below shows strict preferences of the groups of voters (this information is a common knowledge). The higher position of a candidacy in the table means he/she is more preferential to the corresponding group of voters.

Compare each pair of the candidacies by evaluating the number (share) of voters believing that one candidacy is better than the other: $S_{ab} = \alpha_1 + \alpha_3, S_{ac} = \alpha_1, S_{ba} = \alpha_2, S_{bc} = \alpha_1 + \alpha_2, S_{ca} = \alpha_2 + \alpha_3, S_{cb} = \alpha_3$.

Consider the game of voters, where the strategy set of each agent is $A = \{a, b, c\}$. By presuming that the vector $(\alpha_1, \alpha_2, \alpha_3) = (1/3, 1/3, 1/3)$ is a common knowledge, we arrive at the following result. The set of Nash equilibria includes six vectors:

$$(a, a, a) \rightarrow a, (b, b, b) \rightarrow b, (c, c, c) \rightarrow c,$$

$$(a, b, a) \rightarrow a, (a, c, c) \rightarrow c, (b, b, c) \rightarrow b.$$

Now, study a reflexive game, where agent 1 foists an awareness structure on agents 2–3. Agent 1 aims at "electing" candidacy a. Suppose that the awareness structure corresponds to the graph of reflexive game in Fig. 4.28.

First, group 1 aims at convincing group 3 that (i) the most preferential candidacy c (according to their viewpoint) will not be elected, (ii) this fact is a common knowledge and (iii) they should support candidacy a. It suffices that

$$\alpha_{32} + \alpha_3 < 1/2, \quad \alpha_{31} + \alpha_3 > 1/2, \quad \alpha_{31} + \alpha_3 + \alpha_{32} = 1.$$

Table 4.1   The preferences of the groups of voters.

α_1	α_2	α_3
a	b	c
b	c	a
c	a	b
d	d	d

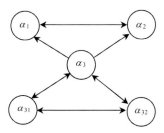

Figure 4.28 The graph of reflexive game (the hustings).

Second, group 1 has to persuade group 2 that (i) candidacy a will be elected and (ii) the actions of group 2 actually affect nothing (group 2 supports candidacy a in the last place). It suffices that group 2 is adequately informed about the beliefs of group 3 (see Fig. 4.28).

Election returns are independent of group 2. Thus, one may believe this group will support the most preferential candidacy (in fact, b); i.e., an informational equilibrium will be defined by the vector (a, b, a). This vector is a stable informational equilibrium. Moreover, since (a, b, a) is a Nash equilibrium under the common knowledge conditions (see the discussion above), it forms a true equilibrium (although the beliefs of group 3 can be false).

4.21   RANK-ORDER TOURNAMENTS

Analyze the following model. A principal (the organizer of a rank-order tournament, an auction) influences the results by some impacts on the beliefs of auction attendees (agents) about opponents' parameters. However, such informational control appears stable only if the actions of agents coincide with their actions in the common knowledge conditions.

Following [135], consider a rank-order tournament (an auction). A principal possesses R_0 units of a resource. The size of admissible request of each agent is fixed, equaling x_0. For simplicity, suppose that $k = R_0/x_0$ is an integer smaller than n (the number of agents attending the auction) – the so-called *scarcity hypothesis*. Agents report their prices $\{y_i\}$ of the resource to the principal; subsequently, the principal sorts them in the descending order of the reported prices and sells the resource (first, to the agent with the maximal price, then to the second agent in this sequence, and so on – until all available quantity is sold out).

Denote by x_i the quantity received by agent i ($i \in N$). Let $\varphi_i(x_i, r_i)$ be the income of agent i from resource utilization. This is a smooth concave function increasing in x_i such that $\varphi_i(0, r_i) = 0$; here r_i means the type of agent i characterizing the efficiency of resource utilization by him/her (i.e., $\varphi_i(\cdot)$ increases with respect to r_i, $i \in N$). The condition of individual rationality (nonnegativity of the goal function[17]

---

[17]By refusing from participation in the auction, an agent guarantees his/her zero gain.

$f_i(y, x_i, r_i) = \varphi_i(x_i, r_i) - y_i x_0)$ leads to the maximal price $p_i(r_i) = \varphi_i(x_0, r_i)/x_0$ paid by agent i for the quantity x_0 of the resource ($i \in N$).

Arrange agents in the descending order of their types[18]: $r_1 \geq r_2 \geq \cdots \geq r_n$. By virtue of the introduced assumptions, the order of agents by maximal prices will be totally the same: $p_1 \geq p_2 \geq \cdots \geq p_n$.

The following messages form an equilibrium under complete awareness (the so-called auction solution):

$$y_i^*(r) = p_{k+1} + \delta, \quad i = \overline{1, k}, \qquad y_i^*(r) = 0, \quad i = \overline{k+1, n}.$$

Here δ indicates an arbitrarily small strictly positive constant. In other words, first k agents (auction winners) purchase the resource almost at the price of first loser; all losers refuse to participate in the auction.

According to agents, the efficiency of this auction is defined by the ratio of their total income (effect) to the allocated quantity of the resource:

$$(1) \quad K(r, R_0) = \sum_{i=1}^{k} \varphi_i(x_0, r_i)/R_0.$$

According to the principal, the efficiency is the ratio of his/her income to the allocated quantity of the resource:

$$(2) \quad K_0(r, R_0) = p_{k+1} + \delta.$$

The efficiency does not increase as the number of agents goes up.

Making a digression, compare these formulas with the efficiency of *transfer pricing mechanisms* in the case of $\varphi_i(x_i, r_i) = 2\sqrt{r_i x_i}$:

$$(3) \quad \lambda(s) = \sqrt{\dfrac{\sum\limits_{i \in N} s_i}{R_0}},$$

$$(4) \quad x_i(s) = \dfrac{s_i}{\sum\limits_{j \in N} s_j} R_0, \quad j \in N.$$

The principal fixes the transfer price (3) for one unit of the resource. The goal functions of agents have the form

$$(5) \quad f_i(x_i, r_i) = \varphi_i(x_i, r_i) - \lambda x_i, \quad i \in N.$$

Consequently, the efficiencies (1) and (2) become

$$(6) \quad K(r, R_0) = 2 \sum_{i=1}^{k} \sqrt{r_i} \bigg/ \sqrt{k R_0},$$

---

[18]By supposition, there is a certain rule to sort two agents with identical types.

(7) $\quad K_0(r, R_0) = 2\sqrt{\dfrac{(k+1)r_{k+1}}{R_0}},$

where $R = \sum\limits_{i \in N} r_i$.

By analogy with (6)–(7), evaluate the efficiencies of the transfer pricing mechanism:

(8) $\quad K(r, R_0) = \sqrt{\dfrac{R}{R_0}},$

(9) $\quad K_0(r, R_0) = \sqrt{\dfrac{R}{R_0}}.$

Compare formulas (6)–(7) with (8)–(9) to obtain the following result. From agents' viewpoint, the efficiency of the rank-order tournament exceeds that of the transfer pricing mechanism if

(10) $\quad \sum\limits_{i \in N} r_i \le 4(k+1)r_{k+1}.$

From principal's viewpoint, this is the case provided that

(11) $\quad \sum\limits_{i=1}^{k} \sqrt{r_i/k} \ge \sqrt{\sum\limits_{i \in N} r_i / 2}.$

Revert to analysis of the auction. The constructed solution is implemented only if the actual types of all agents form a common knowledge. Imagine that agents possess inadequate information on the types of each other.

A detailed study of informational equilibria in this auction model can be found in [136]. Let us focus on stability of informational equilibria. Under given R_0 and k, the auction yields a set of winners and the price p_{k+1} (adopted by the principal for resource sales). Hence, for a given set of winners $Q \subseteq N$ and a price p, a stable result is any set of beliefs with the following properties. First, agents from the set Q appear first in the sequence of beliefs about types (sorted in the ascending order). Second, the beliefs of all (real and phantom) agents about the type of agent $(k+1)$ in this sequence make up p.

Clearly, all informational equilibria turn out true. Regardless of the mutual beliefs of agents, winners fix the price p, and the rest refuse to attend the auction.

4.22 EXPLICIT AND IMPLICIT COALITIONS IN REFLEXIVE GAMES

This section deals with the model of a market oligopoly, where market participants can form (explicit or implicit) coalitions [68]. All coalitionists strive for maximizing the total payoff, whereas the rest market participants maximize their own payoffs. In the case of implicit coalition, interaction modeling involves the concept of informational equilibrium in a reflexive game.

An oligopoly is a market structure with a few dominating sellers; high barriers complicate market entry by new manufacturers. An oligopoly arises in the following situation. The number of firms is so small that each has to consider competitors' response (while choosing its economic policy). Similarly to a chess player analyzing possible moves of an opponent, an oligopolist should be prepared for different (often alternative) scenarios of market development from various behaviors of competitors.

Everybody's interdependence gets revealed in the conditions of exacerbation of competitive struggle, leading to arrangement with other oligopolists and a tendency towards a monopoly.

For a firm, there exist two basic forms of behavior within oligopolies, *viz.*, noncooperative and cooperative behavior. In the former case, each seller establishes product price and output independently. In the latter case, all cooperating firms negotiate product price and output.

Under a high degree of uncertainty, oligoplists act differently. Some endeavor ignoring competitors and acting as if perfect competition takes place. On the contrary, others try to predict opponents' behavior and keep vigilant watch on them. The third prefer backstage collusion with opponent firms.

In this section, we consider an oligopoly, where several manufacturers form a coalition. Two situations will be studied: (a) coalitionists explicitly announce their coalition to the rest members of an oligopoly and (b) some manufacturers make backstage collusion (the rest guess whether coalition exists or does not).

The model without coalitions. Consider a system composed of n homogeneous players representing oligopolists. Each player can choose a certain action x_i. The goal functions of the players are defined by

$$(1) \quad f_i = (1 - x_1 - \cdots - x_n)x_i - \frac{x_i^2}{2r}.$$

The first term in (1) means the income of player i, the second term corresponds to his/her costs to manufacture x_i units of a product. A standard way of Nash equilibrium evaluation lies in maximization of the goal functions with respect to actions (and solution of the resulting system of n equations). In our example, in an equilibrium all players choose identical actions coinciding with

$$(2) \quad \frac{1}{n + 1 + \frac{1}{r}},$$

and their payoffs constitute $\dfrac{1}{2} \dfrac{r(2r + 1)}{(nr + r + 1)^2}$.

Explicit coalitions. Suppose that first m players have formed a coalition. In other words, choosing actions, these players would maximize not their own goal functions, but the aggregated function

$$(3) \quad F = \sum_{k=1}^{m} f_k.$$

Consider the case when all players know about the coalition. Then the goal functions become

(4)
$$\begin{cases} F_i = (1 - x_1 - \cdots - x_n)x_1 - \dfrac{x_1^2}{2r} + \ldots + (1 - x_1 - \cdots - x_n)x_m - \dfrac{x_m^2}{2r}, & i \leq m, \\[2ex] F_i = (1 - x_1 - \cdots - x_n)x_i - \dfrac{x_i^2}{2r}, & i > m. \end{cases}$$

Standard evaluation of Nash equilibria brings to the system of equations

(5)
$$\begin{cases} 1 - \displaystyle\sum_{k=1}^{n} x_k - \sum_{k=1}^{m} x_k - \dfrac{x_i}{r} = 0, & i \leq m, \\[3ex] 1 - \displaystyle\sum_{k=1}^{n} x_k - x_i - \dfrac{x_i}{r} = 0, & i > m. \end{cases}$$

The first group of equations describes actions of the coalitionists, whereas the second one corresponds to actions of the rest players.

By sequentially deducing equations within each group, transform the system to the following equivalent representation:

(6)
$$\begin{cases} 1 - \displaystyle\sum_{k=1}^{n} x_k - \sum_{k=1}^{m} x_k - \dfrac{x_1}{r} = 0, \\[3ex] \dfrac{1}{r}x_i - \dfrac{1}{r}x_{i+1} = 0, & i \leq m - 1, \\[2ex] 1 - \displaystyle\sum_{k=1}^{n} x_k - x_{m+1} - \dfrac{x_{m+1}}{r} = 0, \\[3ex] \left(1 + \dfrac{1}{r}\right)x_i - \left(1 + \dfrac{1}{r}\right)x_{i+1} = 0, & m + 1 \leq i \leq n - 1. \end{cases}$$

Therefore (evidently), all players in the coalition have identical actions. This also applies to the players lying outside the coalition. Set $y = x_i, i = 1, \ldots, m$ and $z = x_i$, $i = m + 1, \ldots, n$. Then the system can be rewritten as

(7)
$$\begin{cases} 1 - 2my - (n - m)z - \dfrac{y}{r} = 0, \\[2ex] 1 - my - (n - m)z - z - \dfrac{z}{r} = 0. \end{cases}$$

Its solution acquires the form

(8)
$$\begin{cases} y = \dfrac{(r + 1)r}{mr^2 n - m^2 r^2 + 2mr^2 + mr + nr + r + 1}, \\[3ex] x = \dfrac{(mr + 1)r}{mr^2 n - m^2 r^2 + 2mr^2 + mr + nr + r + 1}. \end{cases}$$

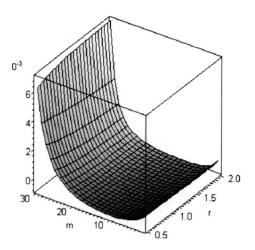

Figure 4.29 The payoffs variation for the first m players (coalition vs. independent activity) under $n = 30$.

What are players' payoffs in the equilibrium? Actually, each coalitionist gains

$$\frac{1}{2} \frac{(r+1)r(2mr^2 + 2mr + r + 1)}{(mr^2n - m^2r^2 + 2mr^2 + mr + nr + r + 1)^2}.$$

(By default, the total payoff is equally shared by all coalitionists).
The rest players (outside the coalition) have the payoffs

$$\frac{1}{2} \frac{(mr+1)r(2mr^2 + 2mr + 2r + 1)}{(mr^2n - m^2r^2 + 2mr^2 + mr + nr + r + 1)^2}.$$

For the first m players, find the variation of their payoffs (coalition vs. independent activity), which is defined by

$$-\frac{1}{2} \frac{r^3(-1 + 8m^2r^3n + 2nmr - 4mr^3n - 2r - 2mr^3n^2 + m)}{(mr^2n - m^2r^2 + 2mr^2 + mr + nr + 1)^2(nr + r + 1)^2}$$

$$-\frac{1}{2} \frac{r^3(8m^2r^2 + 4mr - 2nr - 2m^3r^2n + m^2r^2n^2 + 4m^2r^2n)}{(mr^2n - m^2r^2 + 2mr^2 + mr + nr + 1)^2(nr + r + 1)^2}$$

$$-\frac{1}{2} \frac{r^3(2m^2r^3n^2 - 2mr^3 - n^2r^2 - 2nr^2 - 8m^3r^3 + 8m^2r^3)}{(mr^2n - m^2r^2 + 2mr^2 + mr + nr + 1)^2(nr + r + 1)^2}$$

$$-\frac{1}{2} \frac{r^3(8m^3r^2 - 2m^3r + 2m^4r^3 - r^2 - 4m^3r^3n + m^4r^2)}{(mr^2n - m^2r^2 + 2mr^2 + mr + nr + 1)^2(nr + r + 1)^2}.$$

Select the special case of $n = 30$. Draw the curve of this function in the parameter space $<r, m>$ (see Fig. 4.29).

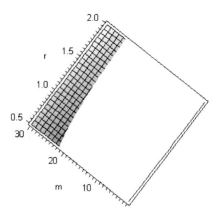

Figure 4.30 The domain of profitable coalitions under $n = 30$.

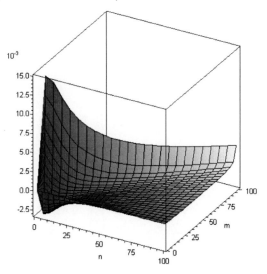

Figure 4.31 The payoffs variation for coalitionists under $r = 1$.

Fig. 4.29 demonstrates that the variation is positive only under sufficiently large numbers of coalitionists. Moreover, players have the maximal payoff if they all form the coalition (i.e., all market competitors act jointly).

Thus, coalitions are sometimes unbeneficial (far from always players increase their payoff by forming coalitions).

Interestingly, the "beneficial" number of coalitionists depends on the efficiency of all players. Fig. 4.30 highlights the domain, where players gain more by forming a coalition (as against independent activity).

Consider another special case. Let the efficiency of all players be $r = 1$. Analyze the payoffs variation for the players (again, coalition vs. independent activity) – see Fig. 4.31. Obviously, coalitions turn out beneficial under a sufficiently large number

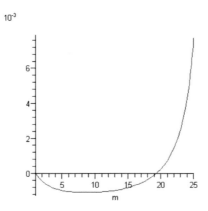

Figure 4.32 The payoffs variation for coalitionists under $r = 1$ and $n = 30$.

of players (m being comparable with n). If the number of players is fixed, coalitions are initially disadvantageous (coalitions start earning profits as their sizes grow). For instance, take the case of $n = 25$ (see Fig. 4.32).

Find the minimum point for the payoffs variation under $r = 1$. This function is described by the expression

$$(9) \quad \Delta = -\frac{1}{2} \frac{-4 - 4n - n^2 - 2nm + 3m^4 - 18m^3 + 4m + 15m^2}{(nm - m^2 + 3m + n + 2)^2(n + 2)^2}$$
$$+ \frac{1}{2} \frac{12m^2n + 3m^2n^2 - 6m^3n - 2mn^2}{(nm - m^2 + 3m + n + 2)^2(n + 2)^2}.$$

Consider the payoffs variation Δ as a function with argument m and parameter n. Apply the first-order necessary optimality conditions to obtain the following result. The function (9) attains its extremum (here – minimum) at the point

$$(10) \quad m = \frac{1}{6}n + \frac{1}{6} + \frac{1}{6}\sqrt{n^2 + 2n + 13}.$$

The corresponding curve can be found in Fig. 4.33. Note that, under $n = 25$, we have $m_{\min} \approx 8.7$ (this agrees with Fig. 4.32).

Implicit coalitions. We have studied the case of explicit coalitions (the existence of a coalition and its members are a common knowledge). However, the following situation may occur. Only coalitionists are aware of their coalition (the rest players suspect nothing).

Assume that the first m players have formed a coalition and act jointly; the rest $(n - m)$ players do not know this. An adequate description of such situation is a reflexive game, where n real players interact with m phantom ones. In the present case, phantom players are the first m players in the beliefs of the others. Indeed, the first m players actually form a coalition (but act independently in the beliefs of the rest $(n - m)$ players).

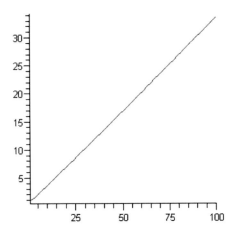

Figure 4.33 The least profitable size of a coalition.

Goal functions (the actual ones and the ones in the beliefs of players) have the following form:

$$(11) \quad \begin{cases} f_i = (1 - x_1 - \cdots - x_n)x_1 - \dfrac{x_1^2}{2r} + \cdots + (1 - x_1 - \cdots - x_n)x_m - \dfrac{x_m^2}{2r}, & i \leq m, \\[2mm] f_i = (1 - x_{i1} - \cdots - x_{in})x_i - \dfrac{x_i^2}{2r}, & i > m, \\[2mm] f_{ji} = (1 - x_{j1} - \cdots - x_{jn})x_{ji} - \dfrac{x_{ji}^2}{2r}, & j > m, \ i \leq n, \\[2mm] x_{ii} = x_i, & i \leq n. \end{cases}$$

Here x_{ij} indicates the action of player j in the belief of player i. By supposition, x_{ii} and x_i coincide (each player possesses an adequate awareness about his/her actions).

Computing an informational equilibrium in this game, we arrive at the system of equations

$$(12) \quad \begin{cases} 1 - \displaystyle\sum_{k=1}^{m} x_k - \sum_{k=1}^{n} x_k - \dfrac{x_i}{r} = 0, & i \leq m, \\[4mm] 1 - \displaystyle\sum_{k=1}^{n} x_{ik} - x_i - \dfrac{x_i}{r} = 0, & i > m, \\[4mm] 1 - \displaystyle\sum_{k=1}^{n} x_{jk} - x_{ji} - \dfrac{x_{ji}}{r} = 0, & j > m, \ i \leq n, \\[4mm] x_{ii} = x_i, & i \leq n. \end{cases}$$

Similarly to the case of explicit coalitions, deduce sequentially the equations within each group. This yields the system

(13)
$$
\begin{cases}
1 - \sum_{k=1}^{n} x_k - \sum_{k=1}^{m} x_k - \dfrac{x_1}{r} = 0, \\[2mm]
\dfrac{1}{r}x_i - \dfrac{1}{r}x_{i+1} = 0, \quad i \le m - 1, \\[2mm]
1 - \sum_{k=1}^{n} x_{ik} - x_{m+1} - \dfrac{x_{m+1}}{r} = 0, \\[2mm]
\left(1 + \dfrac{1}{r}\right)x_i - \left(1 + \dfrac{1}{r}\right)x_{i+1} = 0, \quad m + 1 \le i \le n - 1, \\[2mm]
1 - \sum_{k=1}^{n} x_{jk} - x_{j1} - \dfrac{x_{j1}}{r} = 0, \quad j > m, \\[2mm]
\left(\left(2 + \dfrac{1}{r}\right)x_{j1} + \sum_{k=2}^{n} x_{jk}\right) - \left(\left(2 + \dfrac{1}{r}\right)x_{j+1,1} + \sum_{k=2}^{n} x_{j+1,k}\right) = 0, \quad j > m, \\[2mm]
\left(1 + \dfrac{1}{r}\right)x_{ji} - \left(1 + \dfrac{1}{r}\right)x_{j,i+1} = 0, \quad j > m, \ i < n, \\[2mm]
x_{ii} = x_i, \quad i \le n.
\end{cases}
$$

By analogy, the actions of players in each group do coincide. Introduce the following notation for actions of players in groups:

- denote by x the real actions of coalitionists;
- denote by y the real actions of players outside a coalition;
- denote by z the actions of phantom players, *viz.*, coalitionists, in the beliefs of players outside a coalition.

The system (13) can be rewritten as

(14)
$$
\begin{cases}
\left(2m + \dfrac{1}{r}\right)x + (n - m)y = 1, \\[2mm]
\dfrac{y}{x} + (n + 1)z = 1, \\[2mm]
\left(n + 1 + \dfrac{1}{r}\right)z = 1, \\[2mm]
z = y.
\end{cases}
$$

Consequently, the actions of players acquire the form

(15)
$$
\begin{cases}
x = \dfrac{r}{2mr + 1}\dfrac{mr + r + 1}{nr + r + 1}, \\[2mm]
y = \dfrac{r}{nr + r + 1}.
\end{cases}
$$

Note an important aspect. If $m = 1$ (no coalition), the actions of players coincide with the actions in the case of a coalition-free game.

The coalitionists' payoff makes up

$$(16) \quad f = \frac{1}{2} \frac{r(mr + r + 1)^2}{(nr + r + 1)^2 (2mr + 1)}.$$

The payoffs variation (coalition vs. Independent activity) is given by

$$(17) \quad \Delta f = \frac{1}{2} \frac{r^3(m^2 - 2m + 1)}{(nr + r + 1)^2 (2mr + 1)}.$$

Again, "freeze" the efficiency of all players at $r = 1$; draw the curve of this function under different number of players (see Fig. 4.34).

Clearly, "backstage collusion" ensures the maximal payoff to coalitionists under the maximal number of coalitionists (similarly to the case of an explicit coalition!). But an obvious distinction here consists in the following. Backstage collusion is always beneficial (increases the payoffs of coalitionists).

Consider the payoffs and their variation for non-coalitionists (as against the case of no coalition). The corresponding formulas are

$$(18) \quad g = -\frac{1}{2} \frac{2mn - 8m + 4n - 3 - 4m^2}{(2m + 1)(n + 2)^2}.$$

and

$$(19) \quad \Delta g = -\frac{mn - m + 2n - 2m^2}{(2m + 1)(n + 2)^2}.$$

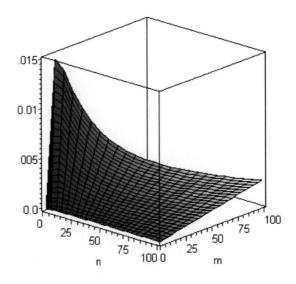

Figure 4.34 The payoffs variation for coalitionists under $r = 1$.

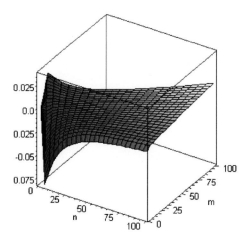

Figure 4.35 The payoffs variation for non-coalitionists under $r = 1$.

Analyze the relationship between the payoffs variation (under different number of players) and the number of coalitionists (see Fig. 4.35).

Generally, players outside a coalition lose payoffs. However, non-coalitionists benefit from sufficiently many players in a coalition.

What is the size of an implicit coalition such that non-coalitionists still gain profits? Naturally, the expression (19) implies the following. Non-coalitionists have profits (against the case of no coalition) under sufficiently large m – starting from

$$(20) \quad m = \frac{1}{4}n - \frac{1}{4} + \frac{1}{4}\sqrt{n^2 + 14n + 1}.$$

For infinite n, this formula yields $m \approx n/2$.

Concluding analysis of implicit coalitions, note that the resulting informational equilibrium appears unstable (see Section 2.7). In fact, the outcome of the game is unexpected for non-coalitionists. To characterize the dynamics of agents' beliefs in reflexive games, one should adopt more sophisticated methods (see Section 2.14). This issue goes beyond the scope of the book.

Therefore, the current section has treated the model of an oligopoly with explicit or implicit coalitions. Let us summarize the results:

– an explicit coalition does not necessarily give additional profits to coalitionists;
– an explicit coalition affecting all (or almost all) market participants (a monopoly) ensures the maximal acceleration of profits to all coalitionists;
– an implicit coalition always guarantees additional profits to coalitionists;
– an implicit coalition can be fruitful to no-coalitionists.

A promising direction of future investigations concerns (a) formation of explicit and implicit coalitions (including several coalitions) in other types of games and (b) identification of general laws of interaction under incomplete information on existing coalitions.

4.23   ACTIVE FORECAST

General notion of an active forecast. At all times, people strive for reducing the impact of uncontrolled factors on the results of their activity (by acquiring additional information on what is unknown or incompletely known). Perhaps, this aspect explains the popularity of various forecasts (weather forecasts, marketing forecasts, economic forecasts, scientific and technical advances' forecast, etc.).

Information a subject obtains by forecasting (including a collective subject) can be simply noted or even change subject's behavior (in comparison with the absence of such information). In the former case, the matter concerns a *passive forecast*, whereas the latter case corresponds to an *active forecast* [136]. This discrimination seems rather relative; in principle, any forecast can be treated as an active one. Generally, an active forecast is a forecast leading to a purposeful modification of subject's behavior.

Consider basic approaches to the definition of passive and active forecasts. According to Merriam Webster Dictionary, a forecast is a prophecy, estimate, or prediction of a future happening or condition.

There exists the positive (constructive) approach to forecasting as an active impact on future by planning, programming, designing and managing of phenomena and processes. Within such framework, forecasting is not an end in itself; rather, forecasting underlies decision-making. A forecast must incorporate control variables as parameters (i.e., forecasting serves to analyze different scenarios)[19] [113].

In other words, defining a decision depending on current state reflects the policy of a decision maker (i.e., a decision mechanism). Thus, we have a game engaging an active player and a passive one (referred to as "nature" in operations research). The influence of his/her forecast on the behavior of other subjects is considered implicitly (in a passive way – through forecasts of uncontrolled variables). In this case, a forecast represents not a control method, but the one of uncertainty elimination.

Next, researchers distinguish between the descriptive approach and the normative approach. The *descriptive approach* defines possible states of a forecast object in future. A problem of *normative forecasts* consists in choosing the ways and periods of reaching desired states of an explored object in future. A normative forecast represents prophecies attracting interest and stimulating some actions. Therefore, a normative forecast can be viewed as explicit control.

The book [107] underlines the following. An expert making decisions based on forecasts tries to prevent implementation of unfavorable forecasts (and to increase the probability of a favorable forecast). There are two "extremes" in the impact of a forecast on control. A self-implementing forecast is a forecast which becomes reliable only by having been made. A self-canceling forecast is a forecast which becomes unreliable (or avoidable) only by having been made.

It is possible to differentiate between *active and passive forecasts*. Here one faces the relevant problem of prior/posterior assessment of forecast quality [136]. "A passive forecast is a forecast whose result does not affect (and cannot affect) a forecast object. If the impact of a forecast on a forecasted object might not be neglected (an active

---

[19]Forecasting *in abstracto* is an integral stage of control. This book focuses on situations when control lies in purposeful revelation of forecasted (or some other) information.

forecast), the logic of forecasting abruptly changes and gets complicated. Indeed, a forecast must then consider the effect of forecasting results." Hence, any normative forecast is active; similarly, descriptive forecasts used in managerial decision-making are active.

Passive and active forecasts *inter alia* differ in the following aspect. First, consider the accuracy of a passive forecast. There exists a system whose future states are forecasted. Such forecast bases on a certain model of the system (a model of the forecast object). The posterior difference ||the forecast − the fact||, where ||·|| reflects the "distance" between system states, can be treated as the *accuracy of a passive forecast* and as the adequacy criterion (between the model and the system).

Imagine that the system is passive (contains no active elements being able to modify their behavior under new information) or system participants do not know the forecast. Then any forecast turns out passive.

In the case of an active forecast, the above difference seems inapplicable for defining its accuracy. Indeed, it is impossible to assess the "actual" state of the system (the "pure" state − without informational impact) and to compare it with the forecasted state.

The following occasion is described by the famous social psychologist D. Myers in [120, p. 162]. "On the evening of January 6, 1981, Joseph Granville, a popular Florida investment adviser, wired his clients: "Stock prices will nosedive; sell tomorrow." Word of Granville's advice soon spread, and January 7 became the heaviest day of trading in the previous history of the New York Stock Exchange. All told, stock values lost $40 billion."

We provide another example from the same field. In the early 1970s, intensive research on mathematical models of stock exchange yielded the so-called Black-Scholes option pricing formula. With the course of time, this formula was included in most textbooks on economics. Everybody adopts it to compute real prices of options (without a moment's hesitation in its adequacy). In fact, the Black-Scholes model forms the "reality" [128].

The problem of active forecasting has been discussed at the qualitative level. Now, we proceed with game-theoretic analysis.

Active forecast as a method of equilibria choice. We have earlier dodged the following important question. What action would an agent choose in case of several equilibria? This issue is a subject of debates in game theory from the very beginning (formulation of the equilibrium concept by J. Nash). Voluminous literature is dedicated to equilibria choice (refinement); however, such analysis exceeds the limits of the book (see the monograph [79] written by two Nobel Prize winners).

In the context of active forecasting, consider just one concept, *viz.*, *the focal point effect* (this term was pioneered in [154]). The effect lies in the following. In a game with several equilibria, players choose an equilibrium being remarkable in a certain sense (the so-called *focal equilibrium*).

We give an example [98]. An inmate is in a prison cell (Fig. 4.36). His partner is outside, trying to release him. Each of them cannot independently punch a hole in the wall. But if both punch a hole simultaneously in the same place, the hole will be made. The wall has a uniform thickness.

Suppose that the hole can be made only in corners (nodes *A*, *B*, *C*, *D*, *E*, *F*, and *G*). Partners have no contact before and during work (nobody knows for sure the partner's decision). What is the behavior of intelligent and rational partners?

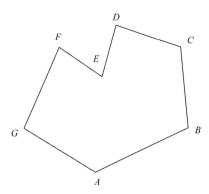

Figure 4.36 The focal point effect: node E is remarkable against others.

Clearly, each game participant chooses among 7 strategies – punching a hole in point A, point B,..., point G. And we have 7 admissible equilibria: (A, A), (B, B),...,(H, H). However, a glance at Fig. 4.36 indicates that the focal equilibrium is (E, E): intelligent and rational partners would surely punch a hole in this corner.

Now, get back to the general case with multiple equilibria. Imagine that someone (e.g., a principal) has claimed that the game would result in some equilibrium. The fact of such forecast separates this equilibrium. Thus, the "forecasted" equilibrium becomes focal (the focal point) and, most probably, it will be implemented.

Active forecast as informational impact. Consider a situation with an uncertain parameter in the goal functions of agents (at least, one agent). Performing informational regulation and reflexive control, a principal reports a certain "fact" (information on the uncertain parameter or the beliefs of opponents) to an agent. In the problem of active forecasting, a principal reports a forecast, i.e., a certain quantity $z \in Z$ with known dependence on the uncertain parameter and the actions of agents (e.g., the total action of agents). The principal informs agents, "If you act rationally (choose equilibrium actions), the result will coincide with my forecast." In this sense, a forecast reported to an agent can be the latter's action.

Next, each agent "recovers" information on the state of nature based on the forecast. Similarly to informational regulation, this information serves for equilibrium actions' evaluation (including the agent's action).

We express the aforesaid formally. A reflexive game is defined by a tuple $\{N, (X_i)_{i \in N}, f_i(\cdot)_{i \in N}, \Theta, I\}$, where $N = \{1, 2, ..., n\}$ means the set of game participants (players, agents), X_i indicates the set of feasible actions of agent i, $f_i(\cdot)$: $\Theta \times X' \to \Re^1$ gives his/her goal function ($i \in N$), and I stands for an awareness structure. Moreover, there is a function $z(\cdot)$: $\Theta \times X' \to Z$, $i \in N$, mapping a vector (θ, x) into an element z of a certain *forecasts' set* Z. An element $z \in Z$ is the forecast reported by the principal to agents[20].

---

[20] Assume that the principal adopts a homogeneous strategy (reports the same forecast to all agents).

The forecast z being received, agents solve a system of $(n + 1)$ equations $x_i \in BR_i(\theta, x_{-i})$, $i \in N$, $z(\theta, x) = z$. Thus, they define the values of the uncertain parameter θ and compute an informational equilibrium[21] $(x_1, \ldots, x_n)$. This forms an awareness structure of depth 1 (composed of one element).

Several examples of active forecast have been discussed above.

4.24 SOCIAL NETWORKS

In recent years, much attention of investigators is drawn by social networks (e.g., see the monographs [74, 89]). At qualitative level, a *social network* is understood as a social structure consisting of a set of network elements or agents (independent or collective subjects, e.g., individuals, families, groups, organizations) and a set of relations defined on it (a totality of relations among the agents, e.g., acquaintance, friendship, partnership, communication). Formally, a social network represents a certain graph $G(N, E)$, where $N = \{1, 2, \ldots, n\}$ stands for a finite set of nodes and E specifies the set of arcs reflecting the interaction among agents.

When modeling social networks, one generally proceeds from the assumption that the primary characteristic of an element (an opinion regarding some issue, "infection rate" in epidemic spread models, etc.) varies according to a specific law with respect to characteristics of "neighbor" elements. In such models, an element is *per se* passive; we will call them first-type models for convenience. Numerous models of informational control of opinions, reputation and confidence levels of passive agents in social networks are described in [14, 60, 72–74].

Somewhat less common are models, where a network element chooses a characteristic (e.g., an action or inaction) himself/herself, having in mind individual abilities and interests. Within such models (referred to as the second-type models), a network element appears active; in other words, he/she possesses individual interests (that can be formalized in terms of a goal function) and freedom of choice.

This section considers an example of the complex model [60] which accounts for both passive and active aspects of elements belonging to a social network. Each of the stated elements is described by a specific parameter varying under the impact of other agents and a control subject (the principal). At the same time, each agent has an individual goal function and chooses an action from a set of feasible actions.

Description of the game-theoretic model. Suppose there exists a finite set of agents, $N = \{1, 2, \ldots, n\}$; each agent is characterized by the type r_i [133] $(0 < r_i < 1)$ and by the goal function f_i. The agent chooses the action x_i (a nonnegative real value). The goal function of agent i has the following form:

$$(1) \quad f_i(x_1, \ldots, x_n) = x_i(x_1 + \cdots + x_n - 1) - \frac{x_i^2}{r_i}.$$

The corresponding practical interpretation states that agents apply the effort x_i to some joint action. It appears successful (provides positive contribution to their goal functions) when the total effort exceeds a specific threshold (equal to 1). The action

---

[21] By supposition, this system has a unique solution.

being successful, the agent's gain (the first term in the goal function) increases as the agent's effort grows. On the other hand, the agent's effort itself results in negative contribution to the goal function (see the second term) which depends on the type r_i. The larger is the type variable, the "easier" the agent applies the effort (for instance, psychologically the agent shows a greater loyalty or liking for the joint action).

We introduce the following system of equations to evaluate a Nash equilibrium:

$$(2) \qquad \frac{\partial f_i(x_1, \ldots, x_n)}{\partial x_i} = x_i - \frac{2}{r_i} x_i + (x_1 + \cdots + x_n - 1) = 0, \quad i \in N.$$

Note all actions are positive (i.e., for each agent the goal function attains its maximum within the domain $x_i > 0$).

Actually, the set of actions satisfying the above system yields a Nash equilibrium, since the expression

$$(3) \qquad \frac{\partial^2 f_i(x_1, \ldots, x_n)}{\partial x_i^2} = 2 - \frac{2}{r_i} < 0$$

takes place.

To solve the system, reexpress the action x_i through the sum of actions chosen by all agents:

$$(4) \qquad x_i = \frac{r_i}{2 - r_i} \left(\sum_j x_j - 1 \right).$$

Here and in the sequel, we are summing up from 1 to n.

To proceed, perform summation over all agents:

$$(5) \qquad \sum_{j=1}^{n} x_i = \sum_{i=1}^{n} \frac{r_i}{2 - r_i} \left(\sum_{j=1}^{n} x_j - 1 \right).$$

and denote

$$(6) \qquad k_i = \frac{r_i}{2 - r_i},$$

$$(7) \qquad \sigma = \sum_i k_i.$$

Using the notation, one may rewrite

$$(8) \qquad \sum_j x_j = \frac{\sigma}{\sigma - 1}.$$

This formula shows that the system of equations is feasible if

$$(9) \qquad \sigma > 1.$$

Under the stated condition, we derive the following expression defining the equilibrium actions of agents:

$$(10) \quad x_i = k_i \left(\frac{\sigma}{\sigma - 1} - 1 \right),$$

$$(11) \quad x_i = \frac{k_i}{\sigma - 1}.$$

The described argumentation proves the following statement.

Assertion 4.24.1. A nonzero equilibrium exists if $\sigma > 1$.

Note that the established equilibrium is non-unique; in some cases, the agent's goal function achieves its maximum on the boundary of the admissible actions' set (i.e., under $x_i = 0$). Let us formulate two additional statements defining the structure of game equilibria.

Assertion 4.24.2. The set of actions $(0, 0, \ldots, 0)$ is a Nash equilibrium.

Proof. Evidently,

$$(12) \quad f_i(0, \ldots, x_i, \ldots, 0) = x_i (x_i - 1) - \frac{x_i^2}{r_i}, \quad i \in N.$$

The first and second derivatives of this function with respect to x_i possess negative values at the zero point; thus, $x_i = 0$ is the maximum point. This completes the proof.

Assertion 4.24.3. Suppose that the action of (at least) one agent equals zero and the system is in a Nash equilibrium. Then the equilibrium constitutes $(0, 0, \ldots, 0)$.

Proof. Without loosing generality, assume that agent 1 has zero action. As far as zero point represents the maximum with respect to x_1 (this is the equilibrium action), the corresponding partial derivative appears negative at that point:

$$(13) \quad \frac{\partial f_1(x_1, \ldots, x_n)}{\partial x_1} = \left(2 - \frac{2}{r_1} \right) x_1 + \sum_{j \neq 1} x_j - 1 < 0.$$

Consequently,

$$(14) \quad \left(2 - \frac{2}{r_1} \right) x_1 = 0 < 1 - \sum_{j \neq 1} x_j.$$

The last formula gives a necessary condition for the existence of a Nash equilibrium under zero action of agent 1:

$$(15) \quad \sum_j x_j < 1.$$

Now, involve the derivative of the goal function of agent k:

$$(16) \quad \frac{\partial f_k(x_1,\ldots,x_n)}{\partial x_k} = \left(2 - \frac{2}{r_k}\right)x_k + \sum_{j \neq k} x_j - 1.$$

The actions are bounded from below and the function is continuously differentiable; therefore, the necessary condition of a nonzero local minimum consists in zero derivative. It could be rewritten as

$$(17) \quad 2\left(\frac{r_k - 1}{r_k}\right)x_1 = 1 - \sum_{j \neq k} x_j.$$

The obtained equality makes a contradiction; indeed, the left-hand side is negative, while its right-hand counterpart is positive. This concludes the proof of Assertion 4.24.3.

Informational control problem: evaluating the desirable types of agents. Let us study the problem of informational control. Consider the principal which strives for controlling the agents' types fro bringing them to a (most desirable) Nash equilibrium. Suppose that the principal's problem is to find the set of network states (the agents' types) ensuring the maximum of his/her goal function.

Several possible goal functions of the principal are discussed in detail in the monograph [74]. Here, we select the following goal function:

$$(18) \quad F = -\sum_j x_j.$$

In practice, it means that the agents' actions are desirable to the principal; thus, his/her problem lies in minimizing the total action of the agents. An example is provided by protection systems responding to actions of intruders.

Obviously, the principal benefits from zero actions of the agents. Hence, he/she should seek for eliminating a nonzero equilibrium (making the latter impossible). According to Assertion 4.23.1, a nonzero equilibrium does not exist if

$$(19) \quad \sigma \leq 1.$$

Reexpressing this inequality in terms of the agents' types, we finally obtain the following condition which guarantees infeasibility of a zero equilibrium:

$$(20) \quad \sum_i \frac{r_i}{2 - r_i} \leq 1.$$

Under identical types of agents, slight modifications transform it to

$$(21) \quad r \leq \frac{2}{n + 1}.$$

We have thereby derived the condition (to-be-imposed on the agents' types) when the principal attains his/her goal. The stated condition implies there is no nonzero

Nash equilibrium in the system; thus, zero equilibrium appears the only (unique) one, and it provides the maximal value to the principal's goal function.

Informational control problem: choosing optimal impact on the agents' types. In this subsection, we analyze control capabilities of the principal related to modifying the initial opinions (types) of the agent. Clearly, in this case (see (6), (10), and (17)), the principal is interested in reducing the agents' opinions. Hence, we focus only on the impacts decreasing the types. The relationship between the agent's type and impact exerted by the principal on the agent must possess the following properties:

- being continuous;
- approaching the zero point asymptotically as the principal's impact grows infinitely;
- being equal to the initial agent's type (before the principal's impact) in the zero point;
- being monotonically decreasing.

These requirements are satisfied, for example, by the function

$$(22) \quad r_j^u = r_j^0 e^{-\frac{u_j}{\alpha_j}},$$

where j indicates the agent's number, u_j is the effort applied by the principal (a nonnegative real value), r_j^0 means the initial type of the agent (before the principal's impact), r_j^u represents the final type of the agent (after the principal's impact), and α_j represents a constant (a nonnegative real value) describing the level of the principal's impact on the given agent. In what follows, we introduce an upper bound for u_j.

In a social network, agents communicate and exchange opinions. This changes the agent's type according to the types of agents he/she trusts. When interaction among agents has sufficient duration, their opinions get stabilized, i.e., converge to a common resulting type.

The resulting type is rewritten as the sum of agents' initial types with positive weights w_j characterizing the influence level of the agents:

$$(23) \quad r^\infty = w_1 r_1^u + \cdots + w_n r_n^u.$$

That the total effort of the principal is bounded from the above seems a natural assumption. Denote by U the maximal total effort of the principal (to-be-called a resource). Let k be the number of agents that can be influenced by the principal. Without loss of generality, we believe these are the first k agents. Then the optimization problem with respect to u_j is defined by

$$(24) \quad \sum_{i=1}^n w_i r_i^0 e^{-\frac{u_i}{\alpha_i}} \rightarrow \min,$$

$$(25) \quad u_j \geq 0; \quad \forall j \leq k,$$

$$(26) \quad \sum_{j=1}^k u_j = U; \quad u_{k+p} = 0; \quad p = 1, \ldots, n-k.$$

To find the solution, employ Lagrange's method of multipliers. The corresponding Lagrange function takes the form

(27) $\quad L = \sum_{i=1}^{n} w_i r_i^0 e^{-\frac{u_i}{\alpha_i}} + \lambda \left(\sum_{j=1}^{k} u_j - U \right).$

Hence, one may write down the system of equations

(28) $\quad \begin{cases} \dfrac{\partial}{\partial u_j} L = -\dfrac{1}{\alpha_i} w_i r_i^0 e^{-\frac{u_j}{\alpha_j}} + \lambda = 0, \\[3mm] \dfrac{\partial}{\partial \lambda} L = \sum_{j=1}^{k} u_j - U = 0. \end{cases}$

Solving the system, we arrive at the following formula:

(29) $\quad u_i = \alpha_i \ln \dfrac{\alpha_i w_i r_i^0}{\lambda}.$

Now, reexpress λ using the second equation and the derived formula for u_i to obtain

(30) $\quad \sum_{i=1}^{k} \alpha_i \ln \dfrac{\alpha_i w_i r_i^0}{\lambda} - U = 0.$

Certain transformations lead to the equality

(31) $\quad \ln \lambda = \left(\sum_{j=1}^{k} \alpha_j \right)^{-1} \left(\sum \alpha_i \ln \alpha_i w_i r_i^0 - U \right).$

Therefore, λ is given by

(32) $\quad \lambda = e^{-\left(\sum\limits_{j=1}^{k} \alpha_j \right)^{-1} \left(\sum\limits_{j=1}^{k} \alpha_i \ln \alpha_i w_i r_i^0 + U \right)}.$

This allows deriving optimal control actions that ensure the minimal resulting type of the agents (under the existing constraint):

(33) $\quad u_i = \alpha_i \left(\ln \alpha_i w_i r_i^0 - \ln \lambda \right),$

(34) $\quad u_i = \left(\sum_{j=1}^{k} \alpha_j \right)^{-1} \alpha_i U + \left(\sum_{j=1}^{k} \alpha_j \right)^{-1} \sum_{p=1}^{k} \alpha_p \ln \alpha_p w_p r_p^0 - \alpha_i \ln \alpha_i w_i r_i^0.$

The systems lead to a certain requirement to-be-imposed on the resource quantity, such that optimal control of the principal guarantees the absence of nonzero Nash equilibria:

$$(35) \quad \sum_{i=1}^{k} e^{-\frac{U}{\alpha_i} \left(\sum_{j=1}^{k} \alpha_j \right)^{-1}} w_i r_i^0 e^S \leq \frac{2}{n+1} - \sum_{i=k+1}^{n} w_j r_j^0,$$

where

$$(36) \quad S = \ln \alpha_i w_i r_i^0 - \frac{1}{\alpha_i} \left(\sum_{j=1}^{k} \alpha_j \right)^{-1} \sum_{i=1}^{k} \alpha_i \ln \alpha_i w_i r_i^0.$$

For identical α_j (identical capabilities of the principal to influence agents), the last expression can be simplified:

$$(37) \quad U \geq \alpha \left(\sum_{j=1}^{k} \ln w_j r_j^0 + \left(\sum_{j=1}^{k} \alpha_j \right) \ln \left(\sum_{j=1}^{k} \alpha_j \right) + \left(\sum_{j=1}^{k} \alpha_j \right) \ln \left(\frac{2}{n+1} - \sum_{j=k+1}^{n} w_j r_j^0 \right) \right).$$

Thus, we have established condition (for the total resource quantity) ensuring that the principal is able to prevent activity of the agents (i.e., ensuring that zero Nash equilibrium is unique). In other words, if the principal has the available resources U meeting the condition (35) (taking into account (36)), he/she guarantees inactivity (zero actions) of the agents via the control (34).

At the same time, under the shortage of the resources (the condition (35) being violated), there exists a nonzero Nash equilibrium, where the agents perform a collective action being undesirable to the principal.

In this section, we have analyzed the model of informational control in a social network, describing the principal's impact exerted on the opinions (types) of the agents. The principal seeks to ensure a beneficial result of interaction among the agents. In general case, of particular interest is describing the relationship between the principal's capabilities regarding the impact on the opinions of social network members (on the one part) and the results of their interaction (on the other part). This problem seems to have no explicit solution. Therefore, studying special cases (with practical interpretations) would be a promising direction of further investigations.

4.25   MOB CONTROL

Model of a mob. According to [23], consider the following *model of a mob*. There is a set $N = \{1, 2, \ldots, n\}$ of *agents* choosing between two *decisions*, "1" (being active, e.g., participating in mass riots) or "0" (being passive). Agent $i \in N$ is characterized by

- the *influence* on agent j, denoted by $t_{ji} \geq 0$ (a certain "weight" of his/her opinion for agent j); for each agent j, we have the normalization conditions $\sum_{i \neq j} t_{ji} = 1, t_{ii} = 0$;
- the decision $x_i \in \{0; 1\}$;
- the *threshold* $\theta_i \in [0; 1]$, defining whether agent i acts under a certain *opponents' action profile* (the vector x_{-i} comprising the decisions of the rest agents). Formally,

define the action x_i of agent i as the best response to the existing opponents' action profile:

$$
(1) \quad x_i = BR_i(x_{-i}) = \begin{cases} 1, & \text{if } \sum_{j \neq i} t_{ij} x_j \geq \theta_i, \\ 0, & \text{if } \sum_{j \neq i} t_{ij} x_j < \theta_i. \end{cases}
$$

The behavior described by (1) is called *threshold behavior* [22]. A *Nash equilibrium* is an agents' action vector x_N such that $x_N = BR(x_N)$.

Take the *following dynamic model of collective behavior*. At an initial instant, all agents are passive. At each subsequent instant, agents act simultaneously and independently according to the procedure (1). Introduce the notation

$$
(2) \quad Q_0 = \{i \in N | \theta_i = 0\}, \quad Q_k = Q_{k-1} \cup \left\{ i \in N \middle| \sum_{j \in Q_{k-1}, j \neq i} t_{ij} \geq \theta_i \right\}, \quad k = 1, 2, \ldots, n-1.
$$

Clearly, $Q_0 \subseteq Q_1 \subseteq \cdots \subseteq Q_{n-1} \subseteq Q_n = N$. Let $T = \{t_{ij}\}$ be the influence matrix of agents and $\theta = (\theta_1, \theta_2, \ldots, \theta_n)$ correspond to the vector of their thresholds. Evaluate the following rate:

$$
(3) \quad q(T, \theta) = \min\{k = \overline{0, n-1} \mid Q_{k+1} = Q_k\}.
$$

Define the *collective behavior equilibrium* x^* (CBE) by

$$
(4) \quad x_i^*(T, \theta) = \begin{cases} 1, & \text{if } i \in Q_{q(T, \theta)} \\ 0, & \text{if } i \in N \backslash Q_{q(T, \theta)} \end{cases}, \quad i \in N.
$$

The following result was established in [22]. For any influence matrix T and agents' thresholds θ, there exists a unique CBE (4) representing a Nash equilibrium in the game with the best response (1).

We underline that the above definition of a CBE possesses constructive character. Indeed, its evaluation based on (2)–(4) seems easy. Moreover, a reader should notice an important fact. Without agents having zero thresholds, passivity of all agents makes up a CBE. In the sense of control, this circumstance means the following. Most attention should be paid to the so-called *"ringleaders,"* i.e., agents deciding "to be active" even when the rest remain passive.

Control problem. The aggregated rate of mob state is the number of active agents:

$$
(5) \quad K(T, \theta) = |Q_{q(T, \theta)}|.
$$

In the anonymous case, we have $K(m) = p(m)$.

Let T^0 and θ^0 be the vectors of initial values of influence matrices and agents' thresholds, respectively. Suppose that the following parameters are given: *the admissible sets* of the influences and thresholds of agents (T and Θ, respectively), the *principal's gain* $H(K)$ from achieved mob state K and his/her *costs* $C(T, \theta, T^0, \theta^0)$ to modify the reputations and thresholds of agents.

As control efficiency criterion, select the principal's goal function representing the difference between the gain $H(\cdot)$ and the costs $C(\cdot)$. As a result, *control problem* takes the form:

(6) $\quad H(K(T,\theta)) - C(T,\theta,T^0,\theta^0) \rightarrow \max_{T\in T, \theta\in\Theta}$.

In the anonymous case, the control problem (6) becomes:

(7) $\quad H(p(m)) - C(m,m^0) \rightarrow \max_{m\in M}$,

where M is the admissible set of threshold vectors in the anonymous case, while m and m^0 designate the terminal and initial threshold vectors, respectively. Solution to the threshold control problem is presented in [23]. Consider the problem of informational control.

Informational control. Let us analyze the capabilities of *reflexive control* – the principal influences the beliefs of agents about their parameters, the beliefs about beliefs, etc. Select agents' thresholds as the subject of control. *Informational control* concerns formation of the following awareness structures of agents: θ_{ij} – the beliefs of agent i about the threshold of agent j (an awareness structure of rank 2 or depth 2); θ_{ijk} – the beliefs of agent i about the beliefs of agent j about the threshold of agent k (an awareness structure of rank 3 or depth 3), and so on. Possessing a certain awareness structure, agents choose actions as an *informational equilibrium* (see Section 2.3). Notably, each agent chooses a specific action as the best response to the actions expected by him/her from the opponents (according to his/her awareness structure).

Recall the expression (4) which characterizes the thresholds leading to a desired CBE. For convenience, we believe that any result (being attainable via a real variation of thresholds) can be implemented by an informational control (a modification of agents' beliefs about their thresholds). And so, informational control of thresholds turns out equivalent to threshold control (in a common sense) [23]. Apparently, the former type of control is "softer" than the latter.

Nevertheless, implementation of informational control in mob control problems faces an obstacle. One property of "good" informational control concerns its *stability* (see Section 2.7) when all agents observe the results they actually expected.

Assume that, in the suggested mob model, each agent a posteriori observes the number of agents decided to "be active." In fact, this is a rather weak assumption in comparison with the mutual observability of individual actions. Then in a stable informational control each agent observes the number of active agents he/she actually expected. Stability requirement is essential under long-term interaction between the principal and agents. Imagine that (under an unstable informational control) agents just once doubt the truth of information reported by the principal. Clearly, they would have good reason to doubt it in the future.

Assertion 4.25.1. In the anonymous case, there exists no stable informational equilibrium such that the number of active agents is strictly smaller than in a CBE [23].

Proof of Assertion 4.25.1. Denote by Q_Σ the set of agents that act in a stable informational equilibrium. Suppose that their number does not exceed the number of agents acting in a CBE: $|Q_\Sigma| \le |Q_{p(\theta)}|$. Due to stability of the informational equilibrium, each agent $i \in Q_\Sigma$ meets the condition $|Q_\Sigma| - 1 = \sum_{j \neq i} x_j \ge (n-1)\theta_i$. Hence, $|Q_{p(\theta)}| - 1 \ge (n-1)\theta_i$. This implies $i \in Q_{p(\theta)}$. Therefore, $Q_\Sigma \subseteq Q_{p(\theta)}$. If passive agents exist in $Q_{p(\theta)} \backslash Q_\Sigma$, this equilibrium appears unstable for them. And so, $Q_\Sigma = Q_{p(\theta)}$ for the stable informational equilibrium. •

The "negative" result of Assertion 4.26.1 witnesses to the complexity of implementing long-term informational control of threshold behavior of a mob.

4.26 THE REFLEXIVE PARTITIONS METHOD

Nowadays, there are several applied models illustrating the effects of strategic reflexion in models of collective behavior [91, 92, 132, 133]. Let us outline the basic results testifying that the presence of reflexing agents can appreciably change the collective behavior.

In **the diffuse bomb model** (Section 4.26.1), introducing reflexing agents ("intelligent" and adaptive ones that forecast the opponents' behavior) essentially enhances the efficiency of collective behavior.

The **Colonel Blotto game model** (Section 4.26.2) witnesses that the reflexive "superstructure" over the classic game-theoretic model (the Colonel Blotto game) enlarges the set of "equilibrium" outcomes. However, simulation results show that this possibly requires high reflexion ranks of game participants.

The **Cournot oligopoly model** is treated in Section 4.26.3 (as the basic model, we involve the classic oligopoly [6, 108, 147]). For a specific range of initial actions of the agents, it is possible to realize Pareto-efficient or Nash equilibrium levels of production (by introducing agents with reflexion ranks 1 and 2).

The **consensus problem** (Section 4.26.4) possesses the following interpretation. Actions of the agents represent certain positions on a straight line (e.g., coordinates in a space, opinions, and so on – see the overviews in [157]), while the aggregated outcome means the average value of the agents' coordinates. The goal function of an agent is defined as his/her "deviation" from the aggregated outcome. The efficiency criterion lies in the "variance" of the agents' positions (the principal's goal function depends on the aggregated outcome of the game and the whole agents' action vector). It appears that introducing reflexing agents enlarges the set of action vectors chosen by the agents, as well as increases the value of the efficiency criterion.

In the model of **active expertise** (Section 4.26.5), the presence of reflexing agents even with rank 1 appreciably extends the range of feasible expertise results [92] manipulatable by an expertise organizer. The presence of reflexing agents can cause negative consequences (according to the viewpoint of the whole group; see models of team building in [135]). In addition, reflexion rank 2 makes it possible to realize a Nash equilibrium.

In the **model of transport flows and evacuation** (Section 4.26.6), the presence of reflexing agents with rank 1 ensures the minimal (optimal from the "centralized" viewpoint) evacuation time

The **model of a stock exchange** (Section 4.26.7) demonstrates that only a certain "critical mass" of reflexing agents can change the outcome (in comparison with nonreflexive decision-making). Interestingly and probably, stock exchange represents an object of research, where investigators often adopt "reflexive" reasoning – e.g., see [161].

4.26.1 Diffuse bomb

This subsection considers the so-called *diffuse bomb problem* (the problem of collective penetration through a defense system [91]). We perform simulation analysis by comparing six patterns with different "intelligence" levels in the behavior of agents – moving objects – their adaptivity, capacity for reflexion, forecasting, etc. Finally, we demonstrate that endowing the agents with the ability of considering the parameters of a defense system and forecasting the behavior of other agents improves the efficiency of collective penetration through a defense system.

In many applications, one faces control problems for a group of moving objects (MOs) executing a certain task jointly. For instance, a possible task concerns searching for mobile or immobile objects in a given area, penetration in a given area, hitting *targets*, and so on. Generally, a group of MOs operates in a conflict environment, i.e., in *counteraction conditions* (detection, informational counteraction, annihilation) by objects of search, a security system or a defense system (whose elements are referred to as "*sensors*") protecting the area, targets, etc.

Introduce the following *system of classifications* for collective control problems in counteraction conditions:

1) by the task of a group of MOs: search or penetration in a given area;
2) by the target of search or hitting (the target object (TO)): one target or several targets; a mobile target or an immobile target;
3) by type of motion: in a two-dimensional or three-dimensional space;
4) by the hitting time of a TO: fixed hitting time or minimized hitting time; bounded hitting time or arbitrary hitting time;
5) by the number of sensors: one sensor, two sensors or several sensors;
6) by the type of a sensor network: a network a priori known or unknown to MOs; mobile or immobile sensors;
7) by the number of MOs: one or several objects;
8) by the speed of MOs: a fixed speed or a variable speed;
9) by the constraints imposed on the speed and acceleration of MOs: such constraints do or do not exist;
10) by the distribution of TOs (and/or tasks, functions, etc.) among MOs: centralized distribution or autonomous (decentralized) distribution; programmed distribution or real-time distribution;
11) by path planning for MOs (including collision avoidance): centralized path planning or autonomous path planning; programmed path planning or real-time path planning;
12) by the interaction of MOs (forecasting and coordination depending on detection probability): the interaction is considered or not;
13) by the detection risk functional of MOs (a function of MOs speed or a function of distances to sensors; summation of signals at sensors, and so on – these aspects are discussed below).

Each combination of values taken by the classification attributes of MOs characterizes a corresponding class of problems.

The present subsection does not aim at analyzing all problems generated by the above system of classifications. Instead, we dwell on one possible statement. Notably, the target of a *group* of several MOs advancing in a two-dimensional space consists in "searching for" (hitting) an immovable TO; the hitting time of the TO appears not fixed; there are several immovable sensors; MOs advance with a given fixed (by the absolute value) speed, i.e., moving direction can be changed; their path planning takes place in real-time in the decentralized (autonomous) way. The *awareness* of MOs (the information on defense system parameters and parameters of other MOs, being available to MOs at the moment of path planning), as well as the type of the detection risk functional (see (2) and (5)) will be specified below. For a group of MOs, the efficiency criterion represents the number of objects K that have reached the TO. This class of problems can be called *the diffuse bomb problem*.

A distinctive feature of the model studied in this subsection relates to "cooperative" decentralized decision-making by MOs (their choice of motion paths) when detection/annihilation probability for each object depends on the relative arrangement of all objects in a group. The methods and results of solving similar problems in the case of one TO are overviewed in [91].

We begin with formulating the "noncooperative" problem and some common methods of its solution. Subsequently, we address the case when the probability of individual detection and/or annihilation depends on the relative arrangement of all MOs.

Path planning in counteraction conditions. Consider the following problem. Suppose that the initial positions $(x_j(0), y_j(0)), j = \overline{1, K_0}$ of K_0 moving objects are given on the plane. Their target is to reach a point with the coordinates (x^*, y^*). Denote by $(x_j(t), y_j(t))$ the position of MO j at instant $t \geq 0$, by $v_j(t) = \sqrt{(\dot{x}_j)^2 + (\dot{y}_j)^2}$ its speed and by T_j the time of hitting the point (x^*, y^*).

There exist N immovable sensors with the coordinates $(a_i, b_i), i = \overline{1, N}$, being able to sum up the incoming signals (the signals are supplied simultaneously). The distance between MO j and sensor i will be defined by $\rho_{ij}(t) = \sqrt{(x_j(t) - a_i)^2 + (y_j(t) - b_i)^2}$.

In the general case, the *detection risk* for MO j by the system of sensors is described by the following functional:

$$(1) \quad R_j = \int_0^{T_j} \sum_{i=1}^{N} \frac{(v_j(t))^m}{(\rho_{ij}(t))^k} \, dt.$$

Here the "signal" at a sensor (the summand in (1)) depends on the speed of the MO and the distance to the sensor. Formula (1) implies that detection risk for MOs represents a function of "signal values" at different sensors. The exponent k characterizes the physical field of detection, whereas the exponent m specifies the relationship between the signal intensity and the object's speed (e.g., signals of a primary sonar field).

"Noncooperative" model. Suppose that all MOs advance with a fixed (by the absolute value) speed v_0. Being aware of sensors location and their nonnegative sensitivity levels $\{c_i\}$, $i = \overline{1, N}$, we proceed by analogy to formula (1). Notably, for each point (x, y) in a two-dimensional space, define the detection risk (detection probability)

$$(2) \quad r(x, y) = \min \left\{ \sum_{i=1}^{N} \frac{c_i}{\left(\sqrt{(x - a_i)^2 + (y - b_i)^2} \right)^k} ; 1 \right\}$$

of an MO located at this point.

Consider discrete time scale. Denote by τ time increment and by p the annihilation probability of a detected MO. For simplicity, we believe this probability is independent from the coordinates of a detection point, time and the speed of MOs; taking into account such dependencies seems a promising direction of further research. Next, let $e(x, y) = (x^* - x, y^* - y) / \sqrt{(x - x^*)^2 + (y - y^*)^2}$ be the unit direction vector to the TO at the point (x, y), $\rho((x, y); (q, w))$ designate the Euclidean distance between the points (x, y) and (q, w), and $s_\Delta(x, y)$ specify a circle with radius $\Delta \geq 0$ and center at the point (x, y).

Below we study several behavioral strategies (called patterns) of MOs.

Pattern I. Actually, the simplest one when each MO advances along a line connecting its initial position and the TO. A corresponding MO will be referred to as an *unintelligent object.*

Within the framework of Pattern I, at any instant each MO has to know just its current position and the position of the TO.

More "intelligent" MOs should consider the current and/or future probabilities of their detection. To describe their behavior, define a set of points such that 1) an MO may reach them in time τ by moving from the point (x, y) with the speed v_0; 2) the detection probability of an MO does not exceed a *threshold* δ; 3) the point (x, y) belongs to this set:

$$(3) \quad S_{v_0 \tau}^{\delta}(x, y) = \{(q, w) \mid \rho((x, y); (q, w)) = v_0 \tau; r(q, w) \leq \delta\}.$$

Assume that $\text{Proj}_{S_{v_0 \tau}^{\delta}(x, y)}(x^*, y^*)$ represents the projection of TO's position on the set $S_{v_0 \tau}^{\delta}(x, y)$ (the closest point of this set to the TO in the sense of Euclidean distance; choose randomly any projection if there are many of them).

Pattern II. Adopt the following path planning rules for MOs (*the noncooperative behavior algorithm*).

Step 1. At each instant, an MO located at the point (x, y) is annihilated by a defense system with the probability $pr(x, y)$ (and continues its motion with the probability $1 - pr(x, y)$).

Step 2. Continuing its motion, by the next instant an MO reaches the point (u, v) defined by

$$(4) \quad (u, v) \in \begin{cases} \mathrm{Proj}_{S^\delta_{v_0 \tau}(x,y)}(x^*, y^*), \; if \; (x, y) \notin \mathrm{Proj}_{S^\delta_{v_0 \tau}(u,v)}(x^*, y^*) \\ (x, y) + \min\{v_0 \tau; \rho((x, y); (x^*, y^*))\}\mathbf{e}(x, y), \; \mathrm{otherwise} \end{cases}.$$

The first case in (4) means that the distance to the TO is not decreasing (provided that the detection risk does not exceed the threshold). The second case corresponds to a "breakthrough" along the line to the TO (when keeping the detection probability smaller than the threshold increases the distance to the TO).

In the noncooperative behavior algorithm, an MO located at a certain point must have detection risk estimates only for the $v_0 \tau$-neighborhood of this point. In other words, the behavior of an MO appears locally optimal and requires just local information. In this context, we emphasize the following. Under active mode of detection ($k = 4$), this local information can be obtained via extrapolation of current measurements of sensor signals. In the passive mode, for all MOs it suffices to know the coordinates and sensitivity levels of sensors. This allows evaluating the risk (2) for an arbitrary point in a two-dimensional space.

The noncooperative behavior of a group of MOs will be characterized as follows. For each instant and MO, perform sequentially Step 1 and Step 2 (until all MOs are annihilated or all surviving MOs reach the TO).

"Cooperative" model. The interaction among MOs will be accounted as follows. Suppose that the detection probability of a given object depends on the current distances to sensors and the distance to other MOs (an example consists in growing efficient scattering surface). In other words, for convenience one can believe that MOs serve as "sensors" for each other; as soon as MOs approach each other, their detection probability increases.

Denote by

$$(5) \quad R_j(x_j, y_j) = \min \left\{ r(x_j, y_j) + \sum_{l \neq j} \frac{\alpha}{1 + (\sqrt{(x_j - x_l)^2 + (y_j - y_l)^2})^k}; 1 \right\}$$

the detection risk of MO j located at the point (x_j, y_j) (taking into consideration its interaction with other MOs); here α indicates a nonnegative constant.

Pattern III. Moving objects try to reach the TO along the line (by ignoring and not forecasting their detection probabilities). This situation resembles Pattern I – just substitute the risk (2) with the risk (5). The awareness of MOs is totally the same as in Pattern I.

Pattern IV. The "cooperative" behavior algorithm is described by Steps 1′ and 2′ being similar to Steps 1–2 (again, replace the risk (2) by the risk (5)). In the expression (5), summation takes place over all MOs that have survived by the current instant. Pattern IV corresponds to Pattern II if one substitutes (2) with (5).

In this case, path planning requires that each MO knows current coordinates of all MOs (in addition to information necessary in Pattern II)[22].

Reflexive model. Imagine that a group comprises MOs of two types. The objects of type 1 (known as *non-reflexing* objects) act according to the "cooperative" behavior algorithm (Pattern IV). The objects of type 2 (called *reflexing* objects) involve a sophisticated strategy. Each object considers the rest ones as non-reflexing objects (see Section 3.4) and forecasts their behavior. Notably, a reflexing MO evaluates the location of other objects at a subsequent instant (these objects follow the behavior of Pattern IV). And a reflexing MO chooses its direction by analyzing the forecasted positions of other MOs.

We define Steps $1''$ and $2''$ by replacing the risk (5) with the forecasted risk in Steps $1'$ and $2'$.

Pattern V. The reflexive behavior algorithm of a group of MOs. At each instant, perform successively Steps $1'$ and $2'$ (for a non-reflexing MO) and Steps $1''$ and $2''$ (for a reflexing MO) until all survived MOs reach the TO.

In the reflexive behavior algorithm (Pattern V), the awareness of each MO must coincide with the case of cooperative behavior (Pattern IV).

Adaptive model. A specific feature of intelligent agents consists in the following. Each agent corrects his/her beliefs about uncertain parameters based on the results of observing external environment and the behavior of other agents. Thus, an agent attempts to "explain" the choice of observed actions. For the diffuse bomb problem, this means that, even being unable to measure directly the detection probabilities in different points (or possessing no a priori information on them), an adaptive MO observes changes in paths of other MOs and recovers information on the threshold line.

Suppose there are MOs of two types. Assume that at each instant all MOs know their current position and the position of a target. In addition, at each instant MOs of type 1 know certain estimates of the detection risk for the $v_0\tau$-neighborhood of their current position, whereas MOs of type 2 know (or can measure) the current coordinates of all MOs of type 1.

Objects of type 1 act according to Pattern II. At each step, MOs of type 2 start with estimating the location of the threshold line. Then these objects act according to Pattern II, substituting their current estimate of the threshold line into the analog of formula (3). In other words, MOs of type 2 demonstrate an *adaptive behavior*.

MOs of type 1 can be *de bene esse* called "scouts." Indeed, they have a better awareness (perhaps, a higher price) and perform offensive reconnaissance. This yields information on a defense system (more specifically, on the threshold line) for other MOs of type 2.

The "extreme" situation is when all MOs have type 1 or type 2 (we have Pattern II or Pattern I, respectively).

---

[22]The "cooperative" model can be generalized to the following situation. Each MO possesses a fixed "radius of vision" and plans its path by considering (in formula (5)) only those MOs located within this radius.

Thus, we obtain six possible patterns of MOs' behavior:

No. of pattern	Taking into account detection probability	Taking into account the positions of other MOs	Forecasting the behavior of other MOs	Awareness
Noncooperative model				
I	NO	NO	NO	At any instant, each MO must know only his/her current position and the position of the TO.
II	YES	NO	NO	In addition to Pattern I, at any instant each MO must know the estimates of the detection risk for the $v_0\tau$-neighborhood of his/her current position.
Cooperative model				
III	NO	YES	NO	Similarly to Pattern I.
IV	YES	YES	NO	In addition to Pattern II, at any instant each MO must know the current coordinates of all other MOs.
Reflexive model				
V	YES	YES	YES	Similarly to Pattern IV.
Adaptive model				
VI				
MOs of type 1	YES	NO	NO	Similarly to Pattern II.
MOs of type 2	NO	YES	YES	In addition to Pattern I, at any instant each MO of type 2 must know the current coordinates of all MOs of type 1.

We naturally arrive at the issue regarding the interrelation between the efficiencies of different strategies applied by MOs. Providing an answer in the general analytic form seems almost impossible. And so, let us create a simulation model.

Simulation results. Consider the following simulation model implemented in *AnyLogic* by V.O. Korepanov, Cand. Sci. (Eng.). Set $K_0 = 100, N = 7, c_i = 0.25$, $p = 0.5$, and $\delta = 0.03$. The initial positions of MOs, TO, sensors and contour curves of the total "signal" are demonstrated in Fig. 4.37 (thick line corresponds to the threshold line).

Fig. 4.38 illustrates an example of collective penetration through a defense system for Pattern II (black circles denote annihilated MOs).

For Patterns I and II, Fig. 4.39 presents the efficiencies K of actions by a group of MOs as a function of the annihilation probability p for a detected MO. Here and in the sequel, each point in the efficiency plot reflects the result of averaging over 200 trials. Of course, the efficiency goes down as annihilation probability increases.

Obviously, transition from Pattern I to Pattern II (i.e., high intelligence level of MOs due to their analysis of annihilation probability in the $v_0\tau$-neighborhood of the current position) essentially improves the efficiency of defense system penetration. For instance, under $p = 0.5$ the efficiency rises from 38 to 53 (approximately by 40%).

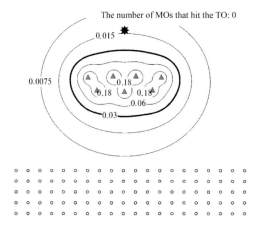

Figure 4.37 The initial position of MOs, TO (star), sensors (triangles) and the contour curves of the total "signal" (2).

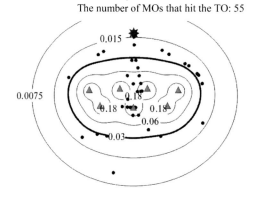

Figure 4.38 An example of collective penetration through a defense system for Pattern II.

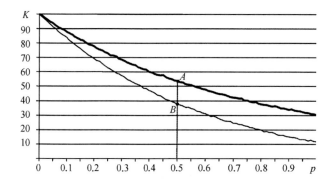

Figure 4.39 The efficiency K of actions by a group of MOs as a function of the annihilation probability p for a detected MO: Pattern I (thin line) and Pattern II (thick line).

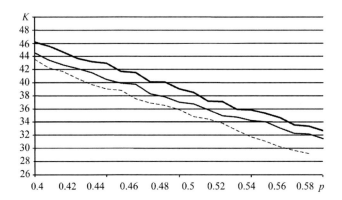

Figure 4.40 The efficiency K of actions by a group of MOs as a function of the annihilation probability p for a detected MO: Pattern III (dashed line), Pattern IV (thin line) and Pattern V (thick line).

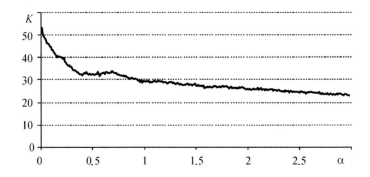

Figure 4.41 The efficiency K of actions by a group of MOs as a function of α.

Note that Patterns I–II and III–V appear incomparable, since the latter consider the interaction among MOs and the probabilities of their detection are greater. And so, we provide Fig. 4.40 with the efficiencies K of actions by a group of MOs as functions of the annihilation probability p of a detected MO under $\alpha = 0.03$ (in the case of Patterns III–IV). Here p varies between 0.4 and 0.6, whereas 50% of all MOs perform reflexion in Pattern V.

Again, growing intelligence level of MOs enhances the efficiency of defense system penetration (in the descending order of their efficiencies, the patterns form the following sequence: Pattern V, Pattern IV, and Pattern III).

Fig. 4.41 shows the efficiency K of actions by a group of MOs as a function of the parameter α reflecting the mutual influence of MOs.

Denote by $K^* \in \{0, 1, \ldots, K_0\}$ the number of reflexing MOs. Under $\alpha = 0.25$, the relationship $K(K^*)$ is given by Fig. 4.42.

Obviously, as the share of reflexing MOs increases, the efficiency of collective actions goes up. Moreover, the "*survival potential*" of reflexing agents appears higher – the average number of MOs that have reached the TO is higher among reflexing agents

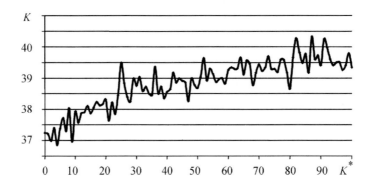

Figure 4.42 The efficiency K of actions by a group of MOs as a function of K*.

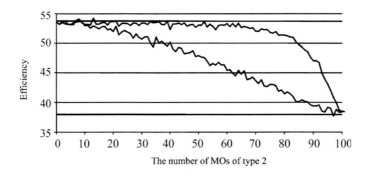

Figure 4.43 The efficiency K of actions by a group of MOs as a function of the number of MOs having type 2.

(as against non-reflexing ones). In the considered simulation model, this estimate is statistically significant, e.g., under 200 trials and identical numbers of reflexing and non-reflexing MOs.

In addition, we show simulation results for the adaptive model. Suppose that the annihilation probability of a detected MO equals 0.5. Fig. 4.43 demonstrates the relationship between the number of MOs that have reached the target and the number of MOs of type 2 in the following cases: (1) when all MOs advance simultaneously (the lower curve) and (2) when a defense system is first penetrated by MOs of type 1, with subsequent motion of MOs of type 2 (the upper curve). Here horizontal lines correspond to points A and B in Fig. 3, i.e., the efficiency levels of 53.74 and 37.97, respectively.

Clearly, 80% of MOs of type 2 ensure almost the same efficiency as all MOs of type 1 (possessing a higher price) – see Fig. 4.43.

Thus, in this subsection we have considered the diffuse bomb problem and performed simulation analysis by comparing six patterns with different "intelligence" levels in the behavior of MOs. It has been shown that the MO's capacity for accounting certain parameters of a defense system, as well as its capacity for forecasting and/or analyzing the behavior of other MOs improves the efficiency of solving the problem of collection penetration through a defense system. However, one naturally "pays for intelligence" by growing mass and dimensions of MOs, higher standards of power, computational and other resources of MOs. Thus, in a concrete problem, engineers have to optimize the balance between these criteria and the efficiency of penetration through a defense system.

A challenging issue consists in studying different modifications of the suggested models by varying the conditions of detection (annihilation) and path planning procedures adopted by MOs. For instance, the following pattern is possible. Denote by $R_0(x, y) = \max_{(q,w) \in s_\Delta(x,y)} r(q, w)$ the maximal risk of detection of an MO (over all MOs located within the distance of Δ from the point (x, y)). Let $N(x, y) = \#\{j \mid (x_j, y_j) \in s_\Delta(x, y)\}$ be the number of MOs lying in the Δ-neighborhood of the point (x, y); here the symbol # indicates the cardinal number of a set. Take into account "cooperation" as described below. Suppose that if $N(q, w) \geq K_{\max}$ (there exists a "critical mass"), then all MOs in the domain $s_\Delta(q, w)$ are detected with the probability $R_0(q, w)$ (and surely annihilated). This model is said to be the *critical mass model*. Introduce the following path planning rule for objects. At each instant, for an MO located at the point (x, y), verify the condition $N(u, v) \geq K_{\max}$, where the point (u, v) is defined by (4). If this condition fails, perform Step 1. Otherwise, all MOs in the domain $s_\Delta(x, y)$ are annihilated with the probability $R_0(x, y)$ (and continue their motion with the probability $1 - R_0(x, y)$). Within the framework of this model, a reflexing MO forecasts the behavior of other MOs. Imagine that, according to such a forecast, a reflexing MO evaluates that actions by Step 2 would lead to the domain with the critical mass. Then this object would strive for avoiding the domain in question.

In addition, the following directions of further research seem attractive:

1) performing synchronization and/or minimization of the hitting time of the TO by separate MOs;
2) making the detection probability of MOs dependent on their number and speeds;
3) making the detection probability of MOs dependent on their number, coordinates and/or speeds;
4) using (perhaps, as heuristics) the famous properties of optimal paths;
5) extending the "cooperative" model to the case when each MO has a specific fixed "radius of view" and plans its path by considering (in formula (5)) only those MOs located within this radius;
6) analyzing more sophisticated partitions of agents by reflexion ranks and taking into account their mutual awareness;
7) deriving analytic solutions for particular cases of the diffuse bomb problem.

4.26.2 The colonel Blotto game

The colonel Blotto game (CBG) is a two-person game, where players allocate their limited *resources* among a finite number of *objects* (battlefields or defense/attack objects, simultaneous contests/auctions, groups of voters, etc.) by one-time simultaneous independent choice of actions. This model is a classical pioneering example of game theory applications in military science.

Denote by $N = \{1, \ldots, n\}$ the set of objects, by $x = (x_1, \ldots, x_n)$ the action of agent 1, by $y = (y_1, \ldots, y_n)$ the action of agent 2; here $x_i \geq 0$ ($y_i \geq 0$) designate the quantity of resource allocated by player 1 (player 2, respectively) on object i, $i = \overline{1, n}$. The limitation of resources is specified by the conditions

$$(1) \quad \sum_{i \in N} x_i \leq R_x, \qquad \sum_{i \in N} y_i \leq R_y.$$

The auction model. Within the *auction model*, a victory on an object is gained by the player allocating the maximal quantity of resources (each player wins with the probability of 1/2 if the allocated quantities coincide). Let X_i (Y_i) be the value of object i for player 1 (player 2, respectively). Consequently, the players' payoffs in the auction model take the form

$$(2) \quad f_x(x, y) = \sum_{i \in N} X_i I(x_i > y_i) + \frac{1}{2} \sum_{i \in N} X_i I(x_i = y_i),$$

$$f_y(x, y) = \sum_{i \in N} Y_i I(y_i > x_i) + \frac{1}{2} \sum_{i \in N} Y_i I(x_i = y_i),$$

where $I(\cdot)$ stands for indicator function. In a more general model, the constraints (1) are absent, but the gain function (2) incorporates deducted costs being monotonic in the total quantity of the resources allocated by a player.

The cases of $n = 1$ and $n = 2$ seem trivial. Indeed, under $n = 1$, the winner is the player possessing the higher quantity of the resources (the available quantities being the same, each player wins equiprobably). If $n = 2$, the optimal strategy of each player consists in allocating the resources on most valuable object.

The *symmetrical* setting ($X_i = Y_i$, $i \in N$, $R_x = R_y$) of the *discrete* CBG (with discrete resources of players) representing a zero-sum matrix game is elementary.

The probabilistic model. In the *probabilistic model* of the CBG, the probability $p_x(x_i, y_i)$ of player 1 victory on object i turns out independent of other objects. Moreover, this probability is "directly proportional" to the quantity of resources allocated to the object and "inversely proportional" to the weighted sum of the resources allocated to this object by both players:

$$(3) \quad p_x(x_i, y_i) = \frac{\alpha_i (x_i)^{r_i}}{\alpha_i (x_i)^{r_i} + (y_i)^{r_i}}, \qquad p_y(x_i, y_i) = 1 - p_x(x_i, y_i).$$

Here $r_i \in (0; 1]$, $\alpha_i > 0$, $p_x(x_i = 0, y_i = 0) = \frac{\alpha_i}{\alpha_i + 1}$. In practical interpretation, the coefficients $\{\alpha_i\}$ serve for weighting the efficiencies of resources utilization by players on the same object.

The players' payoffs in the probabilistic model are defined through the expectation of the total payoff:

$$(4) \quad F_x(x, y) = \sum_{i \in N} X_i p_x(x_i, y_i), \qquad F_y(x, y) = \sum_{i \in N} Y_i p_y(x_i, y_i).$$

A pure strategies' Nash equilibrium (x^*, y^*) is a pair of vectors meeting the conditions (1) such that $\forall(x, y)$ satisfying (1) one has

(5) $\quad F_x(x^*, y^*) \geq F_x(x, y^*), \qquad F_y(x^*, y^*) \geq F_y(x^*, y).$

In some sense, the probabilistic model is "simpler" than its auction counterpart. The only Nash equilibrium in the case of $X_i = Y_i = \text{const}, r_i = 1, \alpha_i = 1, i \in N, R_x \neq R_y$ under arbitrary finite n (see (2)) lies in the following. Players adopt pure strategies of equal allocation of their available resources among all objects.

In the particular case of $X_i = Y_i = V_i, \alpha_i = r_i = 1, i \in N$, the expressions of the equilibrium actions and payoffs take the form

(6) $\quad x_i^* = \dfrac{V_i}{V} R_x, \quad y_i^* = \dfrac{V_i}{V} R_y, \quad i \in N,$

(7) $\quad F_x(x^*, y^*) = \dfrac{R_x}{R_x + R_y} V, \quad F_y(x^*, y^*) = \dfrac{R_y}{R_x + R_y} V,$

where $V = \sum_{i=1}^{n} V_i$. In other words, agents allocate their resources proportionally to the value of objects and gain payoffs being proportional to their total resources. Interestingly, the equilibrium actions of each player depend only on "their own" parameters. For instance, the actions of player 1 x^* are independent of the total quantity of the resources R_y available to player 2, and so on.

Strategic reflexion in the CBG. Consider the aspects of strategic reflexion in the CBG. Suppose that $X_i = Y_i = V_i, \alpha_i = r_i = 1, i \in N$; consequently, the equilibrium actions of players and their payoffs (within the probabilistic model) are defined by (6) and (7), respectively. The Nash equilibrium concept (as a forecasted stable outcome of a noncooperative game) implies that game parameters from a *common knowledge*. In the probabilistic model, the CBG is described by the following components. First, a tuple $(N, R_x, R_y, \{V_i\})$ comprising the set of objects, constraints on players' resources and the values of objects for players. Second, one should specify "the rules of play," namely, the payoff probabilities (3) and the goal functions (4) (whose maximization reflects the rationality of players' behavior). For convenience, imagine that informational reflexion (strategic reflexion) corresponds to the absence of common knowledge regarding the resource quantities of players and the values of objects for them (regarding the principles of decision-making by players).

It follows from (7) that, if $R_y > R_x$, then $\forall i \in N: x_i^* < y_i^*$. Thus, in the sense of the criterion (2), player 1 loses to the opponent on all objects. A rational player can analyze his/her behavior and modify the principles of decision-making.

Let $BR_x(y) = (u_1 y_1 + \varepsilon, \ldots, u_n y_n + \varepsilon)$ be the vector of the best response (in the sense of (2)) of player 1 to the choice of action vector y by player 2. Here the n-dimensional vector $u = (u_1, \ldots, u_n)$ represents a solution to the following knapsack problem:

(8) $\quad \begin{cases} \displaystyle\sum_{i=1}^{n} u_i V_i \rightarrow \max_{u_i \in \{0;1\}}, \\ \displaystyle\sum_{i=1}^{n} u_i y_i \leq R_x, \end{cases}$

with $\varepsilon = \frac{1}{n}(R_x - \sum_{i=1}^{n} u_i y_i)$. In other words, assume that the player seeks to win on the most valuable set of objects (under the resource constraints); the residual resources are equally shared by all objects.

By analogy, introduce $BR_y(x) = (v_1 x_1 + \delta, \ldots, v_n x_n + \delta)$ as the vector of the best response (in the sense of (2)) of player 2 to the choice of action vector x by player 1. Here the n-dimensional vector $v = (v_1, \ldots, v_n)$ solves the following knapsack problem:

$$
(9) \quad \begin{cases} \displaystyle\sum_{i=1}^{n} v_i V_i \to \max_{v_i \in \{0;1\}}, \\ \displaystyle\sum_{i=1}^{n} v_i x_i \le R_y, \end{cases}
$$

with $\delta = \frac{1}{n}(R_y - \sum_{i=1}^{n} u_i x_i)$.

In the auction model of the CBG, a Nash equilibrium can be constructed and studied by analyzing the properties of mappings of the best responses $BR_x(\cdot)$ and $BR_y(\cdot)$. However, we are concerned with strategic reflexion effects. Suppose that nonreflexing players choose a Nash equilibrium corresponding to the probabilistic model (6) of the CBG. A player possessing reflexion rank 1 chooses his/her actions as the best response (in the sense of (8) and (9)) to the actions by the nonreflexing opponent (believing that the latter acts within the probabilistic model).

Adhere to a tradition established in the theory of reflexive games. Assume that a player with a certain rank of strategic reflexion thinks that the opponent has the rank smaller by unity (see Section 3.4). In other words, the following "chain" takes place:

$$
(10) \quad x^1 = BR_x(y^*), \quad y^1 = BR_y(x^*),
$$

$$
x^2 = BR_x(y^1) = BR_x(BR_y(x^*)), \quad y^2 = BR_y(x^1) = BR_y(BR_x(y^*)), \ldots,
$$

$$
x^k = \underbrace{BR_x(BR_y(\ldots(\cdot)\ldots))}_{k}, \quad y^m = \underbrace{BR_y(BR_x(\ldots(\cdot)\ldots))}_{m},
$$

where x^k (y^m) is the action of player 1 (player 2) with reflexion rank k (reflexion rank m), $k, m = 1, 2, \ldots$.

Let us examine a game of *ranks* (see Section 3.2). Instead of resource quantities, players 1–2 choose their ranks. According to (10), these ranks determine resources' allocation, *ergo* the payoffs (2). In the game of ranks, player 1 chooses rank $k \in \{0, 1, 2, \ldots\}$, whereas player 2 chooses rank $m \in \{0, 1, 2, \ldots\}$. Each pair of ranks is assigned the pair of values $(f_x(k, m), f_y(k, m))$, i.e., the payoffs of players 1 and 2, respectively. The game of ranks under consideration is a (generally, infinite) bimatrix game (here, a constant-sum game). Interestingly, the games of ranks analyzed to-date (see references in Section 3.2) are finite but "superstructed" over matrix or bimatrix games.

For definiteness, suppose that $R_y > R_x$. Taking into account the expression (6) under $x = x^*$, $y = y^*$, rewrite the problems (8) and (9) as

$$
(11) \quad
\begin{cases}
\displaystyle\sum_{i=1}^{n} u_i V_i \to \max_{u_i \in \{0;1\}}, \\[2ex]
\displaystyle\sum_{i=1}^{n} u_i V_i \le V \frac{R_x}{R_y},
\end{cases}
$$

and

$$
(12) \quad
\begin{cases}
\displaystyle\sum_{i=1}^{n} v_i V_i \to \max_{v_i \in \{0;1\}}, \\[2ex]
\displaystyle\sum_{i=1}^{n} v_i V_i \le V \frac{R_y}{R_x}.
\end{cases}
$$

We have mentioned that, in the Nash equilibrium (6), player 2 dominates on all objects. By virtue of (12), the best response of player 2 to a fixed strategy of player 1 consists in the following. Allocate the same quantity to each object as player 1 does (e.g., distribute the residual quantity equally among all objects). In fact, any vector (even unit vector) answers the constraint in the problem (12). And player 2 has an advantage on all objects. In other words,

$$
(13) \quad \forall l = 0, 1, 2, \ldots \quad f_x(l, l+1) = 0, \quad f_y(l, l+1) = V.
$$

Now, analyze the behavior of player 1. Start with

Example 4.26.2.1. Set $n = 3$, $V_1 = 1$, $V_2 = 2$, $V_3 = 3$, $R_x = 3$, and $R_y = 4$. The Nash equilibrium (in the game with the criteria (4)) is $x^* = (1/2, 1, 3/2)$, $y^* = (2/3, 4/3, 2)$. In this equilibrium, the players' payoffs (2) constitute $f_x(x^*, y^*) = 0$, $f_x(x^*, y^*) = V = 6$.

Evaluation of players' best responses using (11)–(12) leads to the following payoff bimatrix in the game of ranks:

		Reflexion rank of player 2			
		0	1	2	3
Reflexion rank of player 1	0	(0; 6)	(0; 6)	(2; 4)	(2; 4)
	1	(4; 2)	(3; 3)	(0; 6)	(0; 6)
	2	(1; 5)	(4; 2)	(0; 6)	(0; 6)
	3	(3; 3)	(3; 3)	(5; 1)	(5; 1)

(We get confined with reflexion rank 3).

In the constant-sum game, the guaranteeing strategy of player 1 (player 2) is choosing reflexion rank 3 (reflexion rank 0 or 1, respectively).

We make an interesting observation. In this example, player 1 (having less resource) ensures 50% of the total payoff (3 of 6) in the ranks game. Compare this result with zero payoff in the Nash equilibrium. Furthermore, the combinations of ranks (3, 2) or (3, 3) provide even a higher payoff to player 1 (notably, 5 of 6). The described phenomenon takes place as we have "artificially" bounded the reflexion ranks of players. Imagine that the highest admissible rank is 4. Consequently, the maximal guaranteed payoff of player 1 will be attained at reflexion rank 4 (reflexion rank 3 becomes "dominatable" by reflexion rank 4 of player 2, and so on). Generally, formula (26) elucidates the following. For any reflexion rank of player 1, by choosing the rank greater by unity, player 2 ensures zero payoff of the opponent.

Example 4.26.2.1 brings us to an important conclusion. Under resources' shortage (in comparison with the opponent), a player can increase the payoff by incrementing the rank of his/her reflexion (under bounded reflexion ranks of the opponent). In other words, player 1 should always choose the maximal rank (if the latter exceeds the opponent's rank). •

To proceed, study the issue concerning the maximal rational reflexion rank of players (a rank being pointless to increase, even under potential unboundedness of their reflexion ranks).

We have conducted a numerical experiment to find repetitions of elements in the payoff matrices of players. This testifies to the boundedness of the maximal rational reflexion rank (a further increase in the rank gives no additional combinations of payoffs in the game of ranks).

During the experiment, 100 CBGs have been randomly generated with the following parameters: $n = 10$, $\{V_i\}$ indicate random real numbers from 2 to 100, $R_x = \Sigma V_i / 3$, and $R_y = 2R_x$.

For these CBGs, we have analyzed bimatrix games of ranks of with dimensions of 2000×2000 and evaluated the corresponding *maximal rational reflexion ranks* (MRRs). Let A be the payoff matrix of player 1 and A_i designate row i of the matrix A ($i = 1, \ldots, r, r = 2000$). Denote by m and d the minimal non-negative numbers such that the matrix A is representable as a sequence of rows $(A_1, \ldots, A_m, \{A_{m+1}, \ldots, A_{m+d}\}, A_{m+pd+1}, \ldots, A_{m+pd+q})$. Here the square brackets $\{\ldots\}$ mean, at least, double repetition, $p = \lfloor (r - m)/d \rfloor$, $q = r - m - pd$. Then the MRR of the player is $(m + d)$.

This experiment yielded the following outcomes. The maximal rational reflexion rank has reached 2000 only in 4 CBGs. On the average, the maximal rational reflexion rank of player 1 (player 2) equals 231.57 (230.16, respectively).

The MRR curve for player 1 is shown by Fig. 4.44. However, readers should acknowledge the formal nature of such an experiment. Indeed, one would hardly imagine a player possessing reflexion rank 100. In all 100 CBGs, the maximal guaranteed result of player 1 constitutes 0; on the other hand, the maximal guaranteed result of player 2 does not exceed 56.4% of the total payoff.

Informational reflexion in the CBG. Finally, consider informational reflexion; the mutual awareness of players is described by an awareness structure and their game results in an informational equilibrium. Due to (6), the equilibrium action of each player depends only on his/her estimated value of objects and the resource quantity available to him/her. Thus, the equilibrium action of a player turns out independent of the opponent's parameters.

Assume that the values of objects estimated by players and their beliefs about the opponent's resources may vary. Denote by V_{ij} (R_{il}) the estimated value of object

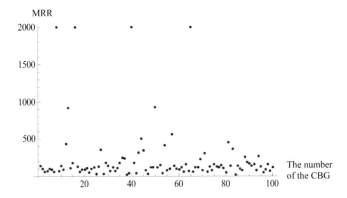

Figure 4.44 The MRR curve for a series of random CBGs.

j (resource quantity, respectively) by player i, $i, l = 1, 2, j \in N$. Naturally enough, suppose that a player surely knows the resource quantity available to him/her: $R_{11} = R_x$, $R_{22} = R_y$.

The informational equilibrium stability condition lies in the coincidence between the chosen actions of players and the actions expected by their opponents:

$$(14) \quad \frac{V_{ij}}{\sum_{l \in N} V_{il}} R_{ii} = \frac{V_{3-i,j}}{\sum_{l \in N} V_{3-i,l}} R_{3-i,i}, \quad i = 1, 2, \quad j \in N.$$

Speaking about informational control (an impact on players' beliefs about the values of objects, the beliefs about opponents and their available quantities of resources, the beliefs about beliefs, etc.), we emphasize the following. By an appropriate variation of V_{ij}, a principal can implement any combination of action vectors meeting (1) as an informational equilibrium.

4.26.3　The Cournot oligopoly: strategic reflexion

In the model of the Cournot oligopoly (also, see Section 4.16), agents make decisions regarding the amount of products manufactured by them. The market price represents a known decreasing function of the total offer (production output, the amount of production): $P(x) = a - bQ(x)$, where $Q(x) = \sum_{i \in N} x_i$, a and b are known nonnegative constants.

The goal function of agent i makes up the difference between the sales income (the unit price multiplied by the production output) and the quadratic manufacturing costs:

$$(1) \quad f_i(x_i, Q(x)) = (a - bQ(x))x_i - (x_i)^2/2, \quad i \in N.$$

Suppose that the agents' goal functions form a common knowledge. Accordingly, a Nash equilibrium in the game of agents is described by identical actions:

$$(2) \quad x_i^N = \frac{a}{1 + b + nb}, \quad i \in N,$$

leading to the equilibrium production output $Q(x^N) = \frac{na}{1+b+nb}$ and to the equilibrium price $P(x^N) = \frac{a(1+b)}{1+b+nb}$. Moreover, the Pareto point maximizing the sum of the agents' goal functions corresponds to the actions

$$(3) \quad x_i^P = \frac{a}{1+2nb}, \quad i \in N,$$

leading to the efficient production output $Q(x^P) = \frac{na}{1+2nb}$ and to the efficient price $P(x^P) = \frac{a(1+nb)}{1+2nb}$. In addition, $f(x^P) = \frac{a^2}{2(1+2nb)} \geq f(x^N) = \frac{a^2(1+2b)}{2(1+b+nb)^2}$, i.e., in the Pareto point the gain of each agent is not smaller than in the Nash point.

We pass to a numerical example. Set $n = 10, a = 2.1, b = 0.1$, and $\gamma_i^t = 0.5$. Then $x_i^N = 1$, $Q(x^N) = 10$, $P(x^N) = 1.1$, $x_i^P = 0.7$, $Q(x^P) = 7$, $P(x^P) = 1.4$, and $f(x^P) = 0.735 > f(x^N) = 0.6$.

Analyze the dynamics of collective behavior. Fix the vector x^0 of initial production outputs. Formula (4) implies that, as time evolves, the actions chosen by the agents satisfy the following expression:

$$(4) \quad x_i^t = x_i^{t-1} + \gamma_i^t \left[\frac{a - b \sum_{j \neq i} x_j^{t-1}}{1 + 2b} - x_i^{t-1} \right], \quad i \in N, \quad t = 1, 2, \ldots.$$

According to (4), the agents' actions converge to the Nash equilibrium.

Now, let us study the reflexive case. Under a given vector x^0 of initial actions, agents with reflexion rank 0 choose the actions

$$(5) \quad x_i = A + Bx_i^0, \quad i \in N_0,$$

where $A = \frac{a - bQ(x^0)}{1+2b}, B = \frac{b}{1+2b}$. Agents with reflexion rank 1 choose the actions

$$(6) \quad x1_j = A_1 + B^2 x_j^0, \quad j \in N_1,$$

where $A_1 = \dfrac{a(1 + 3b) - bna + b^2 Q(x^0)(n-2)}{(1+2b)^2}$.

Suppose that in the numerical example all initial actions coincide: $x_i^0 = 0.5, i \in N$. Consequently, $x_i = 31/24 = 1.291(6), x1_j = 103.5/144 = 0.71875$, and this result appears appreciably closer to the Pareto-efficient actions. By modifying the number of agents with reflexion rank 1, it is possible to vary the sum of agents' actions between ~ 7.2 and ~ 12.9. This range includes Nash equilibrium actions, but not the Pareto point. Notably, under the initial action vector $x_i^0 = 0.5, i \in N$, the number of agents with reflexion rank 1 is insufficient for realizing the Pareto-efficient point via reflexive control. However, one may realize the corresponding Nash equilibrium (production output); for this, the share of reflexing agents with rank 1 must be approximately 49%.

The feasibility of realizing the Pareto point depends on the vector of initial actions. For instance, reflexion rank 1 is maximum reasonable for Pareto point realization under the initial action vector $x_i^0 = 0.2, i \in N$. In this case, $x_i = 1.5, x1_j = 0.55$, and if the share of agents with reflexion rank 1 reaches 84%, the efficient price $P(x^P)$

stabilizes on the market. Nevertheless, such outcome appears unstable (see the stability conditions in Section 3.4).

Imagine that all initial actions of the agents coincide. Then the reflexive partition is defined only by the number of agents with the corresponding reflexion rank. Therefore, omitting the indexes being responsible for numbers of the agents, one obtains that agents with reflexion rank 2 choose the actions

(7) $x2 = \dfrac{1}{1 + 2b}[a - bn_0 x - b(n - n_0 - 1)x1]$.

And so, depending on reflexive partition, the following total action is realized:

(28) $Q(n_1, n_2) = (n - n_1 - n_2)x + n_1 x1 + n_2 x2$

$$= \frac{1}{1 + 2b}[(a - b(n - 1)x^0)(n - n_1 - n_2) + \frac{1}{1 + 2b}[a(1 + 3b)$$

$$- nab + b^2 x^0 (n - 1)^2]n_1 + \frac{1}{1 + 2b}[a - bn_0(n - n_1 - n_2)x]n_2].$$

Next, analyze the relationship between the production output $Q(n_1, n_2)$ and the number of reflexing agents with reflexion ranks 1 and 2. Under which values (n_1, n_2) does the total production output correspond to the Nash equilibrium production output? Reformulating the question, when does the condition $Q(n_1, n_2) = Q(x^N)$ take place (depending on the initial agents' actions x^0)?

For the example considered, the intersectional curve AB for the plotted function $Q(n_1, n_2)$ and the "Nash" plane $Q = 10$ is shown by Fig. 4.45. This curve turns out to be independent of x^0; indeed, the formula in the plane $Q = 10$ is given by

$$n_1 = 1 - n_2 - \frac{n - 1}{\left(1 + \frac{b}{b+1}n\right)\left(\frac{b}{2b+1}n_2 - 1\right)}.$$

Clearly, in Fig. 4.45 the total production output increases even if all agents possess reflexion rank 1.

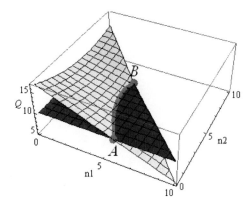

Figure 4.45 The curve *AB* implements the Nash equilibrium.

Certain dynamics being observed, the following aspect can be emphasized regarding "stability." If at step 1 the agents reach the Nash point, then nobody (neither nonreflexing, nor reflexing agents) has reasons to modify the actions.

Suppose that we are interested in a number of reflexing agents such that production output differs from $Q(x^N)$ (e.g., equals a value corresponding to the Pareto-optimal outcome). In this case, the curve AB changes depending on x^0. In the current example, the intersectional curve for the plotted function $Q(n_1, n_2)$ and any plane forms a second-order curve.

Now, let us pose the problem in the following way. By a proper choice of reflexive partition, realize the required total production output, e.g., 12 (greater than $Q(x^N)$). Assume that at the initial time instant the agents did not manufacture the products $(x^0 = 0)$. The required output is reachable – see Fig. 4.46.

If $x^0 \approx 0.305$, then the curve AB touches the plane $n_1 = 0$ (the point C in Fig. 4.47). This means that the required result is reachable provided that only agents with reflexion ranks 0 and 2 exist.

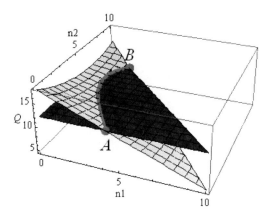

Figure 4.46 "Implementation" of the required total production output.

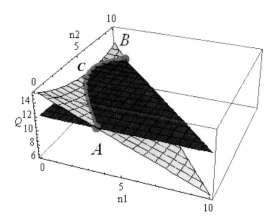

Figure 4.47 "Realization" of the required total production output in the absence of rank I agents (the point C).

Therefore, in the Cournot oligopoly, introducing reflexing agents allows for increasing the total production output and (or) realizing the corresponding Pareto-efficient value.

4.26.4 The consensus problem

In practice, the "consensus problem" (e.g., see [157]) possesses the following interpretation. Actions of the agents represent certain positions on a straight line (e.g., coordinates in a space, opinions), while the aggregated outcome means the average value of the agents' coordinates: $Q(x) = \frac{1}{n}\sum_{i \in N} x_i$. The goal function of an agent is defined as his/her "deviation" from the aggregated outcome:

$$(1) \quad f(x_i, Q(x)) = -(x_i - Q(x))^2, \quad i \in N.$$

Next, let the efficiency criterion be the "variance" of the agents' positions:

$$(2) \quad F(x) = -\frac{1}{n}\sum_{i \in N}(Q(x) - x_i)^2.$$

Note in the present example the principal's goal function depends not only on the aggregated outcome of the game, but also on the action vector of the agents.

From the game-theoretic viewpoint, such situations seem trivial. The goal functions of the agents being a common knowledge, the agents easily evaluate that any vector of identical actions makes up a Nash equilibrium. In this case, there is no conflict of interests among the agents, and any Nash equilibrium in the one-step game also maximizes the efficiency criterion (2). Nevertheless, even in the case of one-step collective behavior of the agents, the things get more sophisticated under incomplete awareness.

Reflexion rank 0. For given initial positions x^0 of the agents, according to the expression (3.4.2) agent i chooses the action

$$(3) \quad x_i = \frac{1}{n-1}\sum_{j \neq i} x_j^0 = \frac{1}{n-1}(nQ(x^0) - x_i^0), \quad i \in N.$$

In fact, this is the average position of the rest agents. The stated conclusion remains in force when the goal functions of the agents depend not on the aggregated outcome, but on the *aggregated opponents' action profile*: $g(x_i, Q_i(x_{-i})) = -(x_i - Q_i(x_{-i}))^2$, where $Q_i(x_{-i}) = \frac{1}{n-1}\sum_{j \neq i} x_j, i \in N$.

It follows from (3) that $Q(x) = Q(x^0)$, i.e., the average value of agents' coordinates is the same, while the value of the efficiency criterion increases by $(n-1)^2$ times: $F(x) = \frac{1}{(n-1)^2}F(x^0)$.

Reflexion rank 1. Suppose there exist n_1 agents with reflexion rank 1 (the rest $n_0 = n - n_1$ agents have reflexion rank 0). The latter choose the actions defined by (3), while the choice of the former makes up

$$(4) \quad x1_j = \frac{nQ(x) - x_j}{n-1} = \frac{n^2(n-2)Q(x^0) + x_j^0}{(n-1)^2}, \quad j \in N_1.$$

If all agents possess reflexion rank 1, then $Q(x1_{j\in N}) = Q(x) = Q(x^0)$. In other words, the average value of agents' coordinates is not modified (the described case seems ideal in the sense of stability of reflexive partition – all agents observe the expected values). As the result, the efficiency criterion again increases by $(n-1)^2$ times: $F(x1_{j\in N}) = \frac{1}{(n-1)^2} F(x) = \frac{1}{(n-1)^4} F(x^0)$.

Consider the example of $n = 2$. Consequently, depending on their reflexion ranks, the agents choose the actions presented in the table:

		Agent 1	Agent 2
Initial actions		x_1^0	x_2^0
Reflexion rank	0	x_2^0	x_1^0
	1	x_1^0	x_2^0
	2	x_2^0	x_1^0

First, the action vectors of both agents with reflexion rank 2 coincide with the action vectors of nonreflexing agents. Second, both agents possessing identical reflexion ranks, the value of the efficiency criterion is independent of the rank. Third, all four feasible combinations of the agents' actions are completely covered by reflexion ranks 0 and 1. Fourth, the maximal (actually, zero) value of the efficiency criterion (2) is attained when a certain agent has reflexion rank 0, while the other possesses reflexion rank 1.

Therefore, in the current example the maximal reasonable reflexion rank constitutes 1.

4.26.5 Active expertise: strategic reflexion

The example in this subsection indicates that the presence of reflexing agents may generate negative effects (so to say, from the whole collective viewpoint); also, see the models of team formation in [135].

The model of active expertise admits the following interpretation (see Section 4.15). There are n experts, $viz.$, agents reporting some information to the expertise organizer (a principal). The latter makes the decision $Q(x) = \frac{1}{n} \sum_{i \in N} x_i$ representing the arithmetical mean of agents' opinions.

Assume that messages of the agents are nonnegative. The agent's goal function is the "deviation" of the final opinion from his/her initial (true) opinion [133]:

(1) $\quad f(x_i, Q(x)) = -(x_i^0 - Q(x))^2, \quad i \in N.$

Let the agents be sorted in the ascending order of their initial opinions: $x_1^0 < x_2^0 < \cdots < x_n^0$.

From the game-theoretic viewpoint, all the goal functions representing a common knowledge among the agents, the latter easily evaluate the Nash equilibrium $x_i^N =$, $i = \overline{1, n-1}$, $x_n^N = nx_n^0$.

Define the set of agents $M(x^0) = \left\{ i \in N \mid x_i^0 \geq \dfrac{1}{n} \sum_{l \neq i} x_l^0 \right\}$.

Reflexion rank 0. Under given initial opinions x^0 of the agents, they choose the actions according to formula (10):

$$(2) \quad x_i = \max\{nx_i^0 - \sum_{l \neq i} x_l^0, 0\}, \quad i \in N.$$

Compute $Q(x) = \sum_{i \in M(x^0)} x_i - \dfrac{1}{n} \sum_{i \in M(x^0)} \sum_{l \neq i} x_l^0.$

Reflexion rank 1. Suppose there exist n_1 agents with reflexion rank 1, while the rest $n_0 = n - n_1$ agents possess reflexion rank 0. The agents with reflexion rank 0 choose the actions determined by the expression (2); at the same time, their first-rank opponents choose the actions

$$(3) \quad x1_j = \max\{nx_j^0 - \sum_{l \in M(x^0) \backslash \{j\}} x_l; 0\}, \quad j \in N_1.$$

Study a numerical example with 10 agents, whose initial opinions coincide with indexes of the agents. Actions of the agents are combined in the table below.

No.	1	2	3	4	5	6	7	8	9	10
x^0	1	2	3	4	5	6	7	8	9	10
$Q(x^0)$						5.5				
x	0	0	0	0	0	11	22	33	44	55
$Q(x)$						16.5				
$x1$	0	0	0	0	0	0	0	0	0	0

By changing the number of reflexing agents with reflexion rank 1 (between 0 and 10), the principal may vary the expertise results in the range from 0 to 16.5 (eleven feasible points totally). Note the above range appears wider than the interval of true opinions of the experts (see the analysis results for informational reflexion in expertise problems in Section 4.15). In other words, performing reflexive control, the principal has ample opportunities to manipulate the expertise results.

Reflexion rank 2. The agents with reflexion rank 2 choose the actions

$$(4) \quad x2_j = \max\left\{ nx_j^0 - \sum_{l \in N_1 \cup N_2 \cap M(x^0) \backslash \{j\}} x1_l - \sum_{l \in N_0 \cap M(x^0) \backslash \{j\}} x_l; 0 \right\}, \quad j \in N_2.$$

Assume that the principal involves the following reflexive partition: $N_0 = \{1, 2, 3, 4, 5\}, N_1 = \{6, 7, 8, 9\}, N_2 = \{10\}$. Then formulas (2)–(4) imply that all agents (except agent 10) choose zero actions; accordingly, agent 10 chooses the action 100.

Thus, in the example considered reflexion rank 2 is sufficient to obtain the outcome being identical to the Nash equilibrium one.

4.26.6 Transport flows and evacuation

Consider a room with n agents. The room has two exits; for convenience, we will call them the "left" exit (L) and the "right" exit (R). For a given exit, the evacuation time is defined by the instant when the last agent (going to this exit) leaves the room through it. Each agent chooses the exit for leaving the room (makes the decision) one-time. In the absence of "holdups," the agents' speeds are equivalent. Denote by $n_L(n_R)$ the number of agents going to the left exit (to the right exit, respectively), $n_L + n_R = n$.

Suppose that the relationship $T(k)$ between the evacuation time and the number of agents $k \geq 0$ is known. We will believe the relationship is continuous, convex (this requirement reflects the existing "holdups") and vanishing in the origin (i.e., a single agent faces no "holdups" and leaves the room without delays). For an agent, let $T_L(T_R)$ be the moving time to the left exit (to the right exit, respectively), where $T_L > T_R$ (the distance to the right exit is smaller than to the left one). Hence, the total evacuation time to the left (to the right) constitutes $T(n_L) = T_L + T(n_L)$ $(T(n_R) = T_R + T(n_R)$, respectively).

In models of evacuation, a common criterion used consists in the *evacuation time* T^* (the instant of leaving the room by the last agent). In the sense of this criterion, optimal distribution of the agents by moving directions $(n_L^*; n_R^*)$ satisfies the following system of equations (see Fig. 4.48):

(1)
$$\begin{cases} T(n_L^*) + T_L = T(n_R^*) + T_R, \\ n_L^* + n_R^* = n. \end{cases}$$

The minimal evacuation time makes up

(2) $T^* = T(n_L^*) + T_L = T(n_R^*) + T_R.$

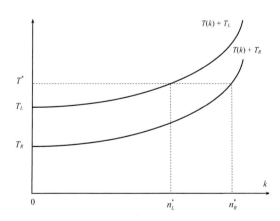

Figure 4.48 The relationship between evacuation time and the number of agents choosing the right (left) exit.

Now, consider collective behavior of the agents under the assumption that each agent strives for leaving the room as soon as possible. The agents with reflexion rank 0 choose the right exit (under the introduced assumptions, they would reach it faster). Forecasting the holdup in the right exit (to-be-generated by the agents with reflexion rank 0), their first-rank opponents definitely choose the left exit.

Depending on the number of agents with reflexion rank 1, the evacuation time is defined by

(3) $\quad T1(n_1) = \max\{T(n_1) + T_L; \; T(n - n_1) + T_R\}$,

see Fig. 4.49.

Apparently, both a small and great number of reflexing agents leads to negative consequences – increases the evacuation time (see Fig. 4.49). In other words, there exist an optimal number of reflexing agents corresponding to the minimal evacuation time.

Recall the properties of the function $T(\cdot)$ and the assumption $T_L > T_R$. The minimum of the expression (3) is attained for the number of agents n_1^* with reflexion rank 1, satisfying the equation

(4) $\quad T(n_1^*) + T_L = T(n - n_1^*) + T_R$.

In fact, the previous formula coincides with the condition (2), i.e., $n_1^* = n_L^*$, $T1(n_1^*) \equiv T^*$. Therefore, reflexion rank 1 appears maximal reasonable in the model considered.

We emphasize that agents with reflexion ranks 2, 3 (or even higher ranks) can be added to the suggested model. However, this is unreasonable, since the evacuation time (2) ensured by introducing agents with reflexion rank 1 would not be improved.

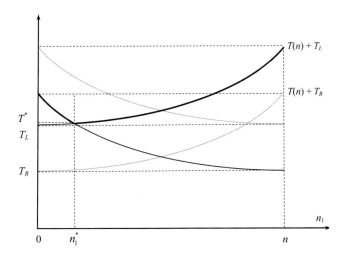

Figure 4.49 The relationship between evacuation time and the number of reflexing agents of rank 1.

4.26.7 A stock exchange

Let us discuss another possible extension for the described method of reflexive partitions. Using the example of the special model of a *stock exchange*, consider strategic
reflexion of the agents "over" their equilibrium strategies in the sense of Nash. Consider the following game-theoretic model of a stock exchange; at any time instant, each
agent possesses a certain quantity of money and an asset which can be purchased or
sold at a market price. The stated quantity satisfies dynamic balance constraints. The
market price depends on the trend θ (an external factor being a common knowledge)
and on the relationship between the demand and offer. Notably, growing demand
increases the market price of the asset (accordingly, growing offer reduces the price).
It is easy to make sure that, under the conditions of the agents' common knowledge
regarding all parameters of the game, the Nash equilibrium has the following structure.
Either all the agents purchase the asset by spending the available financial resources
(thus, "increasing" the relative price of the asset), or all the agents sell the existing
asset ("reducing" the relative price of the asset).

Consider the following model. At the initial time, each agent has the sum of money
$u_0 \geq 0$ and the asset quantity $x_0 \geq 0$. The agent chooses between two alternatives –
purchasing the asset for the whole sum u_0 or selling all the asset quantity x_0 (the
market is infinite).

The following price establishes depending on the agent's actions x. If all agents
purchase the asset, the price p equals $p^+ = p_0 + \theta + \alpha n x_0$; all agents selling the asset,
the resulting price p makes up $p^- = p_0 + \theta - \alpha n x_0$, where α is a certain price coefficient
defined by the demand-offer interrelation.

The initial value of the agent's goal function $u_0 + x_0 p_0$ is finite and constitutes

- $(x_0 + u_0/p_0)p^+ - u_0$, if the asset is purchased with the intention of being sold later;
- $u_0 + x_0 p_0$, if the asset is sold;
- $u_0 + x_0(p_0 + \theta)$, if the agent takes no actions.

To find out what action (purchasing, selling, or doing nothing) would a rational
agent choose, it is necessary to compare the above three quantities. In the case of
a positive trend ($\theta \geq 0$) or no trend ($\theta = 0$), the asset should be purchased. Under a
negative trend ($\theta < 0$), the things are somewhat complicated. Notably, the asset should
be purchased if

$$(1) \quad \theta \geq \frac{p_0 u_0}{p_0 x_0 + u_0} - \alpha n x_0.$$

Actually, this condition means that the agents (by purchasing the asset and
increasing the price in the subsequent period) can "overcome" the negative trend.
Otherwise, the asset should be sold.

A detailed analysis being necessary, we have to study all possible interrelations
among the parameters, i.e., for each of the three actions find certain conditions ensuring
optimality of the action. Proceeding in this way, we obtain that a rational agent should
apply the following algorithm: purchase the asset (if the condition (1) holds true) and
sell it (otherwise). Interestingly, passive behavior (no actions undertaken) turns out
unbeneficial for any combination of the model parameters.

The performed analysis yields the following conclusion. The existence of a constant trend for the price of an asset (in terms of "value" of money) implies that, if the trend is positive, one should invest all money in the asset. Yet, the trend being negative, it seems reasonable to get rid of the asset. The feasibility of agents' influence on the price of the asset (via certain actions – purchasing or selling) results in the following. In the case of a negative trend, purchasing the asset is reasonable only if such action enables "overcoming" the trend.

Therefore, we have described a Nash equilibrium in the game of agents. Now, let us discuss the reasoning of a reflexing agent with rank 1. Under the condition (1), he/she can forecast that all agents with reflexion rank 0 are going to purchase the asset. The condition (1) being violated, this agent forecasts that all agents with reflexion rank 0 would choose selling the asset (the price goes down); thus, he/she benefits from following their example. Hence, the actions of reflexing agents coincide with that of nonreflexing ones, i.e., in the model suggested adding reflexing agents with any reflexion rank does not change the market price.

This conclusion is valid for sufficiently "intelligent" nonreflexing agents. Indeed, they have been assumed able to forecast variations of the market price (depending on their actions).

Now, consider another model admitting less "intelligent" agents with reflexion rank 0. In particular, suppose that the agents pattern their behavior on the trend's sign only. Then under a positive trend, the agents with reflexion rank 0 purchase the asset (thus, increasing its price), and reflexing agents have to follow their example. The situation changes for a negative trend – the agents with reflexion rank 0 sell the asset, "reducing" the price. However, reflexing agents may attempt to "overcome the trend" by purchasing the asset. As a matter of fact, for this they should be confident that (a) the share q of reflexing agents forms a common knowledge and (b) the share is enough to increase the price. By analogy to (1), the last requirement can be rewritten as

$$(2) \qquad \theta \geq \frac{p_0 u_0}{p_0 x_0 + u_0} + \alpha n (1 - 2q) x_0,$$

or, equivalently,

$$(3) \qquad q \geq q^* = \frac{1}{2} + \frac{1}{2 \alpha n x_0} \left[\frac{p_0 u_0}{p_0 x_0 + u_0} - \theta \right].$$

We emphasize that the critical share q^* of reflexing agents makes up, at least, half of the total number of the agents (the condition $q^* \leq 1$ is equivalent to the condition (1)). Consider an illustrative example. Set $n = 100, u_0 = 1000, p_0 = 10, x_0 = 100, \alpha = 0.001$, and $\theta = -1$. Consequently, the condition (1) is satisfied. From the expression (3) we evaluate $q^* = 53\%$.

Note the assumption that the share of reflexing agents is a common knowledge among them contradicts the earlier assumption regarding the structure of subjective reflexive partitions (see Section 3.4). Indeed, the latter implies that reflexing agents "know nothing about the existence" of other agents with the same reflexion rank (and higher reflexion ranks). Under a negative trend, the market price grows as the result of any reflexive partition, where the shares of reflexing agents (with any ranks, except 0) in the sum exceed q^*, and this information is a common knowledge among reflexing agents of the corresponding levels. The stated assertion (allowing for clear interpretations in practice) indicates that the structure of subjective reflexive partitions (introduced in Section 3.4 and adopted in subsections 4.26.3–4.26.6) is not the only feasible and adequate one to all models being relevant in practice. Therefore, a promising direction of further research lies in analyzing other structures of subjective reflexive partitions.

Therefore, the method of reflexive partitions for the set of rational agents into subsets of agents possessing different ranks of strategic reflexion allows the following:

- from the decision theory viewpoint, extending the class of models of collective behavior for intelligent agents performing a joint activity under conditions of incomplete awareness and missed common knowledge;
- from the descriptive viewpoint, enlarging the set of outcomes that can be "explained" (within the framework of the model) as stable results of agents' interaction; accordingly, extending the domain of controllability (for control problems);
- from the normative viewpoint, posing and solving the problems of collective behavior by choosing a proper structure of agents' awareness.

The analysis of the considered examples enables claiming that the presence of reflexing agents may radically change their collective behavior.

In the example "The Cournot Oligopoly" for a specific range of initial actions of the agents it is possible to realize Pareto-efficient or Nash equilibrium levels of production (by introducing agents with reflexion ranks 1 and 2).

In the example "The Consensus Problem" introducing reflexing agents enlarges the set of action vectors chosen by the agents, as well as increases the value of the efficiency criterion.

In the example "Active Expertise" the presence of reflexing agents even with rank 1 appreciably extends the range of feasible expertise results. In addition, reflexion rank 2 makes it possible to realize a Nash equilibrium.

In the example "Transport Flows and Evacuation" the presence of reflexing agents with rank 1 ensures the minimal (optimal from the "centralized" viewpoint) evacuation time.

In the example "A Stock Exchange" it is demonstrated that only a certain "critical mass" of reflexing agents can change the outcome (in comparison with nonreflexive decision-making).

Finally, we should emphasize the following aspects. First, models in Section 4.26 almost do not consider agents with reflexion rank 2 and higher ranks (either they exceed the maximal reasonable rank, or the corresponding models turn out too complicated for analytical derivations).

Second, these models proceed from the assumption that agents with any reflexion rank are sufficiently "intelligent" (they choose actions striving to maximize their goal functions). It is possible to study less intelligent agents (known as *imitating agents*) possessing *de bene esse* reflexion rank -1. Their actions are defined by a known function of current and previous actions of the rest agents (e.g., choosing the action as the arithmetical mean of the actions chosen by the other agents or by the agents related to a given one; copying the action of a specific agent). Perhaps, such models provide an adequate description for some phenomena (such as innovation diffusion, etc.).

Third, models in Section 4.26 pay insufficient attention to the stability conditions.

Fourth, it seems interesting to establish a correspondence between the method of reflexive partitions and the theory of cognitive hierarchies, as well as to develop both approaches jointly. Recall that in the theory of cognitive hierarchies reflexion ranks are related to cognitive levels and a probabilistic model is involved (at a certain level, a player considers the rest players to be distributed over lower levels according to a Poisson distribution). This direction of research is intensively studied in experimental economics and the behavioral game theory (e.g., see [15, 30] and the overview in Section 3.4).

Moreover, control problems, the problem of the maximum reasonable reflexion rank and others may and should be posed within the framework of alternative (different from the suggested one) modifications of reflexive models of collective behavior. This represents a subject for future-oriented investigations. First of all, we mean the problems of *active forecast* (see [136] and Section 4.23), where agents involve the principal's information about the future state of a system to "retrieve" the current state and make decisions using this new information. Here introducing reflexive partitions appears to have definite prospects.

Conclusion

The present book has studied **reflexive games** that describe the interaction among agents making decisions based on hierarchies of their beliefs about essential parameters, the beliefs of other agents, and so on.

The key notions include (for technicalities and correct definitions, see the book):

a phantom agent is an agent existing in the beliefs of a real or phantom agent and possessing some awareness (within these beliefs);

an awareness structure is a graph displaying the mutual awareness of (real and phantom) agents;

a reflexive structure is a set of subjective reflexive partitions of agents (a *subjective reflexive partition* is agent's beliefs about the partition of all agents by the ranks of their strategic reflexion);

an informational equilibrium is an equilibrium of a reflexive game (i.e., the generalization of a Nash equilibrium to the case of a noncooperative game of real and phantom agents under a given awareness structure);

a reflexive equilibrium is the set of agents' actions representing the best responses to opponents' actions (within an existing reflexive structure);

informational control is searching for an admissible informational structure, which corresponds to the best informational equilibrium (in the view of a principal).

reflexive control is searching for an admissible reflexive structure, which corresponds to the best reflexive equilibrium (in the view of a principal).

In the context of game theory, the discussed concept of an informational equilibrium covers situations of agents' decision-making based on an hierarchy of (generally, uncoordinated) beliefs about the mutual awareness. It represents an extension of a Nash equilibrium.

In the context of the theory of collective behavior, the discussed concept of a reflexive equilibrium covers situations of agents' decision-making based on an hierarchy of (generally, uncoordinated) beliefs about the opponents' decision principles. It generalizes the models of cognitive hierarchies.

Furthermore, the suggested approaches to modeling of informational and strategic reflexion enable the following. First, they cover (provide a uniform description to) many applied problems as special cases (implicit control, informational reflexion by the mass media, reflexion in psychology and bélles-léttres – see Chapter 4).

Second, the framework of the constructed model allows studying the relationship between an informational and/or reflexive equilibrium and the payoffs of agents (on the one part) and awareness structures and/or reflexive structures (including reflexion ranks – on the other part). In particular, this framework assists in defining the maximal rational reflexion rank in specific situations (see Sections 2.6 and 3.2).

Third, the relationship between an informational (reflexive) equilibrium and an awareness structure (a reflexive structure) being available, one can formulate and solve the problems of informational and reflexive control. They lie in choosing a certain awareness structure (a reflexive structure) rendering a system to a required informational (reflexive) equilibrium.

Numerous applied models from different fields indicate of the following. The proposed approach appears efficient in description of **reflexion effects in control**.

As strategic objectives of future investigations, we mention integration of informational reflexion models with strategic reflexion ones. In other words, it seems promising to construct a language for uniform joint description of informational and reflexive structures.

References

[1]   Issacs R. (1965) Differential Game: A Mathematical Theory with Applications to Warfare and Pursuit, Control and Optimization. New York. Courier Dover Publications. 408 p.

[2]   Alaoui L. and Penta A. (2012) Level-k Reasoning and Incentives. Barcelona GSE Working Papers Series Working Papers #653. Barcelona GSE. 49 p.

[3]   Algorithmic Game Theory. (2009) (Ed. by N. Nisan, T. Roughgarden, E. Tardos and V. Vazirani). New York. Cambridge University Press. 776 p.

[4]   Ambroszkiewicz S. (2000) On the Concepts of Rationalizability in Games. Annals of Operations Research. No. 97. pp. 55–68.

[5]   Arad A. and Rubinstein A. The 11-20 Money Request Game: Evaluating the Upper Bound of k-Level Reasoning. Working Paper. http://ideas.repec.org/p/cla/levarc/661465000000000073.html.

[6]   Ashimov A.A., Novikov D.A., et al. (2011) Macroeconomic Analysis and Economic Policy Based on Parametric Control. Springer. 278 p.

[7]   Atkinson R., Bower G. and Crothers E. (1964) Introduction to Mathematical Learning Theory. New York. Wiley. 429 p.

[8]   Aumann R.J. (1976) Agreeing to Disagree. The Annals of Statistics. Vol. 4. No. 6. pp. 1236–1239.

[9]   Aumann R.J. and Brandenbunger A. (1995) Epistemic Conditions for Nash Equilibrium. Econometrica. Vol. 63. No. 5. pp. 1161–1180.

[10]   Aumann R.J. and Heifetz A. (2002) Incomplete Information. Handbook of Game Theory. Vol. III. Elsevier. pp. 1665–1686.

[11]   Aumann R.J. (1999) Interactive Epistemology I: Knowledge. International Journal of Game Theory. No. 28. pp. 263–300.

[12]   Bandler R. and Grinder D. (1998) Frogs into Prices: Neurolinguistic Programming. Boulder. Real People Press. 193 p.

[13]   Bandler R. and Grinder D. (1975) The Structure of Magic. Palo Alto. Science and Behavior Books. 225 p.

[14]   Barabanov I.N., Korgin N.A., Novikov D.A. and Chkhartishvili A.G. (2011) Dynamic Models of Informational Control in Social Networks. Automation and Remote Control. Vol. 71. No. 11. pp. 2417–2426.

[15]   Bardsley N., Mehta J., Starmer C. and Sugden R. (2008) Explaining Focal Points: Cognitive Hierarchy Theory versus Team Reasoning. Nottingham. The University of Nottingham. CeDEx Discussion Paper No. 17. 56 p.

[16]   Baron R. and Greenberg J. (2008) Behavior in Organizations (9th ed.). New Jersey. Pearson Education Inc. 775 p.

[17]   Berne E. (1964) Games People Play: The Psychology of Human Relationships. New York. Grove Press. 192 p.

[18] Bernheim D. (1984) Rationalizable Strategic Behavior. Econometrica. No. 5. pp. 1007–1028.

[19] Bottero M. (2010) Cognitive Hierarchies and the Centipede Game. Stockholm. SSE/EFI Working Paper Series in Economics and Finance No. 723. 38 p.

[20] Brams S.J. (1995) Theory of Moves. Cambridge. University of Cambridge. 248 p.

[21] Brandenburger A. and Dekel E. (1993) Hierarchies of Beliefs and Common Knowledge. Journal of Economic Theory. Vol. 59. pp. 189–198.

[22] Breer V.V. (2012) A Game-theoretic Model of Non-anonymous Threshold Conformity Behavior. Automation and Remote Control. Vol. 73. No. 7. pp. 1256–1264.

[23] Breer V.V. and Novikov D.A. (2013) Models of Mob Control. Automation and Remote Control. Vol. 74 (in press).

[24] Burchardi K. and Penczynski S. (2010) Out of Your Mind: Estimating the Level-k Model. London. London School of Economics. Working Paper. 44 p.

[25] Burkov V. and Enaleev A.K. (1994) Stimulation and Decision-making in the Active Systems Theory: Review of Problems and New Results. Mathematical Social Sciences. Vol. 27. pp. 271–291.

[26] Burkov V.N., Enaleev A.K. and Novikov D.A. (1996) Operation Mechanisms of Social Economic Systems with Information Generalization. Automation and Remote Control. Vol. 57. No. 3. pp. 305–321.

[27] Burkov V.N., Enaleev A.K. and Novikov D.A. (1993) Stimulation Mechanisms in Probability Models of Socioeconomic Systems. Automation and Remote Control. Vol. 54. No. 11. pp. 1575–1598.

[28] Burkov V.N., Goubko M.V., Korgin N.A. and Novikov D.A. Theory of Control in Organizations: An Introductory Course (Ed. by D. Novikov) (in press).

[29] Burkov V.N., Zaloshnev A. Yu. and Novikov D.A. (2001) Risk Management: Mechanisms of Mutual and Endowment Insurance. Automation and Remote Control. Vol. 62. No. 10. pp. 1651–1657.

[30] Camerer C., Ho T. and Chong J. (2004) A Cognitive Hierarchy Model of Games. The Quarterly Journal of Economics. No. 8. pp. 861–898.

[31] Camerer C., Ho T. and Chong J. (2001) Behavioral Game Theory: Thinking, Learning, and Teaching. Working Paper. 70 p.

[32] Camerer C., Ho T. and Chong J. (2005) Cognitive Hierarchy: A Limited Thinking Theory in Games. Experimental Business Research. Vol. 3. pp. 203–228.

[33] Camerer C., Ho T. and Chong J. (2003) Models of Thinking, Learning and Teaching in Games. AEA Papers and Proceedings. Vol. 92. No. 2. pp. 192–195.

[34] Camerer C. and Weigelt K. (1991) Information Mirages in Experimental Asset Markets. Journal of Business. Vol. 64. pp. 463–493.

[35] Carnegie D. (2004) How to Feed Friends and Influence People. London. Wiley. 192 p.

[36] Cialdini R. (2001). Influence: Science and Practice (4th ed.). Boston: Allyn & Bacon.

[37] Chkhartishvili A.G. (2003) Bayes–Nash Equilibrium: Infinite-Depth Point Information Structures. Automation and Remote Control. Vol. 64. No. 12. pp. 1922–1927.

[38] Chkhartishvili A.G. (2012) Concordant Informational Control. Automation and Remote Control. Vol. 73. No. 8. pp. 1401–1409.

[39] Chkhartishvili A.G. (2004) Game-theoretical Models of Informational Control. Moscow. PMSOFT. 227 p.

[40] Chkhartishvili A.G. (2003) Informational Equilibrium. Large-Scale Systems Control. No. 3. pp. 100–119.

[41] Chkhartishvili A.G. (2010) Reflexive Games: Transformation of Awareness Structure. Automation and Remote Control. Vol. 71. No. 6. pp. 1208–1216.

[42] Chkhartishvili A.G. and Novikov D.A. (2004) Models of Reflexive Decision-Making. Systems Science. Vol. 30. No. 2. pp. 45–59.

[43]   Chkhartishvili A.G. and Shikin E.V. (2002) Geometry of Dynamical Search for Objects. Journal of Mathematical Sciences. Vol. 110. No. 2. pp. 2508–2527.

[44]   Chkhartishvili A.G. and Shikin E.V. (1998) Geometry of Search Problems with Informational Discrimination. Journal of Mathematical Sciences. Vol. 90. No. 3. pp. 2192–2213.

[45]   Choi S. (2006) A Cognitive Hierarchy Model of Learning in Networks. London. Centre for Economic Learning and Social Evolution. Working Papers 238. http://eprints.ucl.ac.uk/1446.

[46]   Choi S., Gale D. and Kariv S. (2006) A Quantal Response Equilibrium Analysis of Social Learning in Networks. Working paper. UCL.

[47]   Choi S., Gale D. and Kariv S. (2005) Behavioral Aspects of Learning in Social Networks: An Experimental Study. Advances in Behavioral and Experimental Economics (Ed. by John Morgan). JAI Press.

[48]   Clark H.H. and Marshall C.R. (1981) Definite Reference and Mutual Knowledge. Elements of Discourse Understanding (Ed. by A.K. Joshi, B.L. Webber, I.A. Sag). Cambridge. Cambridge University Press. pp. 10–63.

[49]   Cooper D. and Van Huyck J. (2003) Evidence on the Equivalence of the Strategic and Extensive Form Representation of Games. Journal of Economic Theory. Vol. 110. No. 2. pp. 290–308.

[50]   Copic J. and Galeotti A. (2007) Awareness Equilibrium. Essex. University of Essex. Mimeo. 34 p.

[51]   Costa-Gomes M. and Broseta B. (2001) Cognition and Behavior in Normal-Form Games: An Experimental Study. Econometrica. Vol. 69. No. 5. pp. 1193–1235.

[52]   Costa-Gomes M. and Crawford V. (2006) Cognition and Behavior in Two-Person Guessing Games: An Experimental Study. AER. Vol. 96. pp. 1737–1768.

[53]   Crawford V., Costa-Gomes M. and Iriberri N. (2013) Structural Models of Nonequilibrium Strategic Thinking: Theory, Evidence and Applications. Journal of Economic Literature. Vol. 51 (in press).

[54]   Dasgupta P., Hammond P. and Maskin E. (1979) The Implementation of Social Choice Rules: Some General Results on Incentive Compatibility. Review of Economic Studies. Vol. 46. No. 2. pp. 185–216.

[55]   Ereshko F.I. (2001) Modeling of Reflexive Strategies in Control Systems. Moscow. CC RAS. 37 p. (in Russian).

[56]   Fagin R., Geanakoplos J., Halpern J. and Vardi M. (1999) The Hierarchical Approach to Modeling Knowledge and Common Knowledge. International Journal of Game Theory. Vol. 28. pp. 331–365.

[57]   Fagin R., Halpern J., Moses Y. and Vardi M. (1999) Common Knowledge Revisited. Annals of Pure and Applied Logic. Vol. 96. pp. 89–105.

[58]   Fagin R., Halpern J., Moses Y. and Vardi M. (1995) Reasoning about knowledge. Cambridge. MIT Press.

[59]   Fagin R., Halpern J. and Vardi M. (1991) A Model-Theoretic Analysis of Knowledge. Journal of the Association for Computing Machinery. Vol. 38. No. 2. pp. 382–428.

[60]   Fedyanin D.N. and Chkhartishvili A.G. (2011) On a Model of Informational Control in Social Networks. Automation and Remote Control. Vol. 72. No. 10. pp. 2181–2187.

[61]   Feinberg Y. (2004) Subjective Reasoning – Games with Unawareness. Research Paper No. 1875. Stanford. Graduate School of Business. 38 p.

[62]   Fudenberg D. and Levine D. (1998) The Theory of Learning in Games. Massachusetts. MIT Press. 292 p.

[63]   Fudenberg D. and Tirole J. (1995) Game Theory. Cambridge. MIT Press. 579 p.

[64]   Gale D. and Kariv S. (2003) Bayesian Learning in Social Networks. Games and Economic Behavior. Vol. 45. No. 2. pp. 329–346.

[65]   Gamov G. and Stern M. (1958) Puzzle Math. New York. Viking Press. 128 p.

[66]   Geanakoplos J. (1994) Common Knowledge. Handbook of Game Theory. Vol. 2. Amsterdam. Elsevier. pp. 1438–1496.

[67]   Germeier Yu. (1986) Non-Antagonistic Games, 1976. Dordrecht. D. Reidel Publishing Company. 331 p.

[68]   Gontarev A.V. and Chkhartishvili A.G. (2009) Implicit and Explicit Coalitions in Reflexive Games. Large-Scale Systems Control. No. 26. pp. 47–63 (in Russian).

[69]   Gorelik V.A. and Kononenko A.F. (1982) Game-Theoretical Models of Decision-Making in Ecological and Economic Systems. Moscow. Radio and Communication. 144 p. (in Russian).

[70]   Gray J. (1978) Notes on Database Operating System. Operating Systems: An Advanced Course. Lecture Notes in Computer Science. Vol. 66. Berlin. Springer. pp. 393–481.

[71]   Gubanov D.A. and Chkhartishvili A.G. (2008) On Strategic Reflexion in Bimatrix Games. Large-Scale Systems Control. No. 21. pp. 49–57 (in Russian).

[72]   Gubanov D.A., Kalashnikov A.O. and Novikov D.A. (2011) Game-Theoretic Models of Informational Confrontation in Social Networks. Automation and Remote Control. Vol. 72. No. 9. pp. 2001–2008.

[73]   Gubanov D.A., Novikov D.A. and Chkhartishvili A.G. (2011) Informational Influence and Informational Control Models in Social Networks. Automation and Remote Control. Vol. 72. No. 7. pp. 1557–1567.

[74]   Gubanov D.A., Novikov D.A. and Chkhartishvili A.G. (2010) Social Networks: Models of Informational Influence, Control, and Confrontation (Ed. by D. Novikov). Moscow. Fizmatlit. 228 p. (in Russian).

[75]   Gubko M.V. and Novikov D.A. (2002) Game Theory in Control of the Organization Systems. Moscow. Sinteg. 148 p. (in Russian).

[76]   Halpern J. and Moses Y. (1990) Knowledge and Common Knowledge in a Distributed Environment. Journal of the Association for Computing Machinery. Vol. 37. No. 3. pp. 549–587.

[77]   Harris R. (2009) A Cognitive Psychology of Mass Communication (5th ed.). London. Routledge. 480 p.

[78]   Harsanyi J. (1967) Games with Incomplete Information Played by "Bayesian" Players. Management Science. Part I: 1967. Vol. 14. No. 3. pp. 159–182. Part II: 1968. Vol. 14. No. 5. pp. 320–334. Part III: 1968. Vol. 14. No. 7. pp. 486–502.

[79]   Harsanyi J. and Selten R. (1988) A General Theory of Equilibrium Selection in Games. Massachusetts. MIT Press. 365 p.

[80]   Heifetz A. (1999) Iterative and Fixed Point Belief. Journal of Philosophical Logic. Vol. 28. pp. 61–79.

[81]   Heifetz A., Meier M. and Schipper B. (2008) A Canonical Model of Interactive Unawareness. Games and Economic Behavior. No. 62. pp. 304–324.

[82]   Heinrich T. and Wolff I. (2012) Strategic Reasoning in Hide-and-Seek Games: A Note. Research Paper Series No. 74. Konstanz. Thurgau Institute of Economics. 15 p.

[83]   Hill B. (2010) Awareness Dynamics. Journal of Philosophical Logic. No. 39. pp. 113–137.

[84]   Hintikka J. (1962) Knowledge and Belief. Ithaca. Cornell University Press. 148 p.

[85]   Ho T. and Su X. (2010) A Dynamic Level-k Model in Games. Berkeley. University of California. Working Paper. 45 p.

[86]   Howard N. (1974) General Metagames: An Extension of the Metagame Concept. Game Theory as a Theory of Conflict Resolution. Dordrecht. Reidel. pp. 258–280.

[87]   Howard N. (1966) Theory of Meta-Games. General Systems. No. 11. pp. 187–200.

[88]   Huizinga J. (2008) Homo Ludens. London. Routledge. 272 p.

[89]   Jackson M. (2008) Social and Economic Networks. Princeton. Princeton University Press. 520 p.

[90]   Kahneman D., Slovic O. and Tversky A. (1982) Judgment under Uncertainty: Heuristics and Biases. Cambridge. Cambridge University Press. 544 p.

[91]   Korepanov V.O. and Novikov D.A. (2013) A Diffuse Bomb Problem. Automation and Remote Control. Vol. 74. No. 5. pp. 863–874.

[92]   Korepanov V.O. and Novikov D.A. (2012) Method of Reflexive Partitions in the Problems of Group Behavior and Control. Automation and Remote Control. Vol. 73. No. 8. pp. 1424–1441.

[93]   Kozielecki J. (1982) Psychological Decision Theory. London. Springer. 424 p.

[94]   Kripke S. (1959) A Completeness Theorem in Modal Logic. Journal of Symbolic Logic. No. 24. pp. 1–14.

[95]   Laing R. (1972) Self and Others. Amsterdam. Penguin Books. 192 p.

[96]   Lefebvre V.A. (2010) Algebra of Conscience (2nd ed.). Berlin. Springer. 372 p.

[97]   Lefebvre V.A. (1965) Basic Ideas of the Reflexive Games Logic. Proceedings of "Problems of Systems and Structures Researches". Moscow. USSR Academy of Science (in Russian).

[98]   Lefebvre V.A. (1973) Conflicting Structures. Moscow. Soviet Radio, 1973. 158 p. (in Russian). The Structure of Awareness: Toward a Symbolic Language of Human Reflexion. New York. Sage Publications, 1977. 199 p.

[99]   Lefebvre V. (2010) Lectures on the Reflexive Games Theory. New York. Leaf & Oaks Publishers. 220 p.

[100]   Lewis D. (1969) Convention: A Philosophical Study. Cambridge. Harvard University Press. 232 p.

[101]   Lezina Z.M. (1985) Choice Manipulation: Agenda Theory. Automation and Remote Control. No. 4. pp. 5–22.

[102]   Li J. (2009) Information Structures with Unawareness. Journal of Economic Theory. Vol. 144. No. 3. pp. 977–993.

[103]   Luft J. (1969) On Human Interaction. Palo Alto. National Press. 177 p.

[104]   Luft J. and Ingham H. (1955) The Johari Window: A Graphic Model for Interpersonal Relations. University of California, Western Training Lab. 55 p.

[105]   Luce R. and Raiffa H. (1989) Games and Decisions. New York. Dover Publications. 509 p.

[106]   Mansour Y. (2003) Computational Game Theory. Tel Aviv. Tel Aviv University. 150 p.

[107]   Martino J. (1973) An Introduction to Technological Forecasting. New York. Routledge. 118 p.

[108]   Mas-Collel A., Whinston M. and Green J. (1995) Microeconomic Theory. New York. Oxford University Press. 981 p.

[109]   McCain R. (2010) Learning Level-k Play in Noncooperative Games. Philadelphia. Drexel University. Working Paper. 23 p.

[110]   McCarthy J., Sato M., Hayashi T. and Igarishi S. (1979) On the Model Theory of Knowledge. Technical Report STAN-CS-78-657. Stanford. Stanford University. 78 p.

[111]   McKelvey R. and Palfrey T. (1992) An Experimental Study of the Centipede Game. Econometrica. Vol. 60. No. 3. pp. 803–836.

[112]   McKelvey R. and Palfrey T. (1995) Quantal Response Equilibria for Normal Form Games. Games and Economic Behavior. Vol. 10. No. 1. pp. 6–38.

[113]   Mechanisms Design and Management (Ed. by D. Novikov). New York. Nova Science Publishing, 2013. 163 p.

[114]   Mertens J.F. and Zamir S. (1985) Formulation of Bayesian Analysis for Games with Incomplete Information. International Journal of Game Theory. No. 14. pp. 1–29.

[115]   Miller G. (1956) The Magical Number Seven Plus or Minus Two: Some Limits on Capacity for Information Processing. Psychological Review. Vol. 63. No. 1. pp. 81–92.

[116]   Morris S. (1999) Approximate Common Knowledge Revisited. International Journal of Game Theory. Vol. 28. pp. 385–408.

[117]   Morris S. and Shin S.S. (1997) Approximate Common Knowledge and Coordination: Recent Lessons from Game Theory. Journal of Logic, Language and Information. Vol. 6. pp. 171–190.

[118]   Moulin H. (1995) Cooperative Microeconomics: A Game-Theoretic Introduction. Princeton. Princeton University Press. 440 p.

[119]   Moulin H. (1986) Game Theory for Social Sciences. New York. New York Press. 228 p.

[120]   Myers D. (2012) Social Psychology (12th ed.). Columbus. McGraw-Hill. 768 p.

[121]   Myerson R. (1991) Game Theory: Analysis of Conflict. London. Harvard University Press. 568 p.

[122]   Nagel R. (1995) Experimental Results on Interactive Competitive Guessing. American Economic Review. Vol. 85. No. 6. pp. 1313–1326.

[123]   Nagel R. (1995) Unraveling in Guessing Games: An Experimental Study. AER. Vol. 85. pp. 1313–1326.

[124]   Nash J.F. (1951) Non-cooperative Games. Annals of Mathematics. Vol. 54. pp. 286–295.

[125]   Neumann J. and Morgenstern O. (1944) Theory of Games and Economic Behavior. Princeton. Princeton University Press. 776 p.

[126]   Nisan N., Roughgarden T., Tardos E., et al. (2007) Algorithmic Game Theory. Cambridge. Cambridge University Press. 776 p.

[127]   Novikov A.M. and Novikov D.A. (2007) Methodology. Moscow. Sinteg. 668 p. (in Russian).

[128]   Novikov A.M. and Novikov D.A. (2013) Research Methodology: from Philosophy of Science to Research Design. Amsterdam. CRC Press. 200 p.

[129]   Novikov D.A. (2010) "Cognitive Games": A Linear Impulse Model. Automation and Remote Control. Vol. 71. No. 4. pp. 718–730.

[130]   Novikov D.A. (2013) Control Methodology. New York. Nova Science Publishing. 76 p.

[131]   Novikov D.A. (2001) Management of Active Systems: Stability or Efficiency. Systems Science. Vol. 26. No. 2. pp. 85–93.

[132]   Novikov D.A. (2012) Models of Strategic Behavior. Automation and Remote Control. Vol. 73. No. 1. pp. 1–19.

[133]   Novikov D.A. (2011) Reflexive Models of Collective Behavior. Proceedings of 18 IFAC World Congress. Milan. pp. 1971–1976.

[134]   Novikov D.A. (1997) Stimulation Mechanisms in Dynamic and Composite Social Economic Systems. Automation and Remote Control. Vol. 58. No. 6. pp. 887–904.

[135]   Novikov D.A. (2013) Theory of Control in Organizations. New York. Nova Science Publishing. 276 p.

[136]   Novikov D.A. and Chkhartishvili A.G. (2002) Active Forecast. Moscow. Institute of Control Sciences. 101 p. (in Russian).

[137]   Novikov D.A. and Chkhartishvili A.G. (2003) Information Equilibrium: Punctual Structures of Information Distribution. Automation and Remote Control. Vol. 64. No. 10. pp. 1609–1619.

[138]   Novikov D.A. and Ckhartishvili A.G. (2003) Reflexive Games. Moscow. Sinteg. 160 p. (in Russian).

[139]   Novikov D.A. and Chkhartishvili A.G. (2005) Stability of Information Equilibrium in Reflexive Games. Automation and Remote Control. Vol. 66. No. 3. pp. 441–448.

[140]   Orlovski S.A. (1987) Optimization Models Using Fuzzy Sets and Possibility Theory. Berlin. Springer. 452 p.

[141] Owen G. (1969) Game Theory. Philadelphia. W.B. Saunders Company. 228 p.

[142] Pearce D.G. (1984) Rationalizable Strategic Behavior and the Problem of Perfection. Econometrica. No. 5. pp. 1029–1050.

[143] Pease A. and Pease B. (2006) The Definitive Book of Body Language. New York. Bantam Books. 400 p.

[144] Penta A. (2012) Higher Order Uncertainty and Information: Static and Dynamic Games. Econometrica. Vol. 80. No. 2. pp. 631–660.

[145] Petrosyan L.A. (1993) Differential Games of Pursuit. London. World Scientific Publishing. 325 p.

[146] Petrosyan L.A. and Yeung D.W.K. (2005) Cooperative Stochastic Differential Games. Berlin. Springer. 242 p.

[147] Pyndick R. and Rubinfeld D. (2001) Microeconomics (7th ed.). Upper Saddle River. Prentice Hall. 768 p.

[148] Rapoport A. and Guyer M. (1966) A Taxonomy of 2×2 Games. General Systems: Yearbook of the Society for General Systems Research. No. 11. pp. 203–214.

[149] Rêgo L. and Halpern J. (2012) Generalized Solution Concepts in Games with Possibly Unaware Players. International Journal of Game Theory. No. 41. pp. 131–155.

[150] Ross L., Greene D. and House P. (1977) The "False Consensus" Effect: An Egocentric Bias in Social Perception and Attribution. Journal of Experimental Social Psychology. Vol. 13. pp. 279–301.

[151] Rubinstein A. (1989) The Electronic Mail Game: Strategic Behavior Under "Almost Common Knowledge". American Economic Review. Vol. 79. pp. 385–391.

[152] Sakovics J. (2001) Games of Incomplete Information without Common Knowledge Priors. Theory and Decision. No. 50. pp. 347–366.

[153] Sandage C., Fryburger V. and Rotzoll K. (1989) Advertizing: Theory and Practice. Edinburgh. Longman Group. 483 p.

[154] Schelling T. (1960) The Strategy of Conflict. Cambridge. Harvard University Press. 328 p.

[155] Senger H. (1993) The Book of Stratagems: Tactics for Triumph and Survival. Amsterdam. Penguin Books. 397 p.

[156] Shibutani T. (1991) Society and Personality. Piscataway. Transaction Publishers. 648 p.

[157] Shoham Y. and Leyton-Brown K. (2009) Multiagent Systems: Algorithmic, Game-Theoretical and Logical Foundations. Cambridge. Cambridge University Press. 504 p.

[158] Simon H. (1996) The Sciences of the Artificial (3rd ed.). Massachusetts. The MIT Press. 247 p.

[159] Simon R. (1999) The Difference of Common Knowledge of Formulas as Sets. International Journal of Game Theory. Vol. 28. pp. 367–384.

[160] Sobel J. (2011) Giving and Receiving Advice. Working paper. URL: http://econweb.ucsd.edu/~jsobel/Papers/Advice.pdf.

[161] Soros G. (1994) The Alchemy of Finance: Reading the Mind of the Market. New York. Wiley. 391 p.

[162] Stahl D. (1993) Evolution of Smart$_n$ Players. Games and Economic Behavior. No. 5. pp. 604–617.

[163] Stahl D. and Wilson P. (1994) Experimental Evidence on Players' Models of Other Players. Journal of Economic Behavior and Organization. Vol. 25. pp. 309–327.

[164] Stahl D. and Wilson P. (1995) On Players Models of Other Players: Theory and Experimental Evidence. Games and Economic Behavior. Vol. 10. pp. 213–254.

[165] Strogats S. (2001) Nonlinear Dynamics and Chaos: With Applications to Physics, Biology, Chemistry, and Engineering (Studies in Nonlinearity). Boulder. Westview Press. 512 p.

[166] Strzalecki T. (2010) Depth of Reasoning and Higher Order Beliefs. Harvard. Harvard University. Working Paper. 32 p.

[167] Sutter M., Czermak S. and Feri F. Strategic sophistication of individuals and teams in experimental normal-form games. University of Innsbruck. Working Paper. 55 p.

[168] Taha H. (2011) Operations Research: An Introduction (9th ed.). New York. Prentice Hall. 813 p.

[169] Taran T.A. and Shemaev V.N. (2004) Boolean Reflexive Control Models and their Application to Describe the Information Struggle in Socio-Economic Systems. Automation and Remote Control. Vol. 65. No. 11. pp. 1834–1846.

[170] The Cambridge Handbook of Expertise and Expert Performance. (2006) (Ed. by K. Ericsson). Cambridge. Cambridge University Press. 918 p.

[171] The Handbook of Experimental Economics. (1995) (Ed. by J. Kagel and A. Roth). Princeton. Princeton University Press. 740 p.

[172] Van Huyck J., Cook J. and Battalio R. (1997) Adaptive Behavior and Coordination Failure. Journal of Economic Behavior and Organization. Vol. 32. pp. 483–503.

[173] Vanderschraaf P. (1998) Knowledge, Equilibrium and Conventions. Erkenntnis. Vol. 49. pp. 337–369.

[174] Weber R. (2001) Behavior and Learning in the "Dirty Face" Game. Experimental Economics. Vol. 4. pp. 229–242.

[175] Weibull J. (1995) Evolutionary Game Theory. Cambridge. MIT Press. 265 p.

[176] Wolter F. (2000) First Order Common Knowledge Logics. Studia Logica. Vol. 65. pp. 249–271.

[177] Wright J. and Leyton-Brown K. (2010) Beyond Equilibrium: Predicting Human Behavior in Normal Form Games. Proceedings of the Conference on Association for the Advancement of Artificial Intelligence (AAAI-10). pp. 461–473.

[178] Zimbardo P. and Leippe M. (1991) Psychology of Attitude Change and Social Influence. Columbus. McGraw-Hill. 431 p.

Subject index

action 8, 17
active forecast 81, 143, 231–234
activity 5
agent 6
  active 146
  passive 146
  phantom 12, 32
  real 12, 32
anonymity 200
artificial intelligence 10
awareness 8, 32, 89
awareness structure 32, 89
axiom of self-awareness 33

basis 36
beliefs 1, 4, 29
  about beliefs 4, 29
  second order 29
  third order 29
best response 19
  quantal 139

coalition 221
cognitive hierarchy 138
collective behavior 19, 25
  threshold 145
common knowledge 8, 24
consensus problem 263
control 4
  implicit 145
  informational 75
  institutional 74
  methodology 5
  motivational 74
  object 6
  optimal 6
  reflexive 6, 27

subject 5
theory 5
convention 9
coordinated attack problem 3
Cournot oligopoly 205, 259

decision making 17, 19
dictator 200
diffuse bomb problem 244

equilibrium 8, 19
  Bayes-Nash 23, 68
  dominant strategy 20
  focal 232
  informational 37, 94
    false 64
    stable 61
    true 63
    unstable 61
  maximin 20
  Nash 10, 20
  quantal 139
  reflexive 136
  subjective 21
ethics 160
evacuation 266
experimental economics 26

feedback 5
focal point 137, 233

game 8
  Battle of Sexes 128
  Bayesian 23
  Beauty Contest 139
  bimatrix 60, 118
  Centipede 140
  Chess 155

game (*Continued*)
    Dirty Face 3
    Electronic Mail 3
    hierarchical 59
    normal-form 19
    of ranks 130
    of search 167
    Prisoners' Dilemma 127
    reflexive 10, 32
    Stackelberg 148
game theory 7, 10
    algorithmic 7
    computational 7
    evolutionary 7
goal function 17, 19
    single-peaked 106
graph of worlds 87
graph of reflexive game 40

hierarchy of beliefs 9, 24
hypothesis
    of benevolence 77
    of determinism 18
    of rational behavior 19

indicator behavior 26, 134
informational impact 75

Johari Window 159

maximal guaranteed result 18, 31
maximal rational rank of reflexion 124
mechanism 29, 106
    expertise 199, 264
    insurance 210
    planning 106
    resource allocation 207
    strategy-proof 106
message 3, 82
    simple 82
mob control 240
model 6

modeling 6
multiagent system 143

observation function 60, 95
opponents 8

Pareto efficiency 128
plan 106
player 8
principal 4, 23

rank of reflexion 14
reflexion 1, 11, 24
    informational 11, 24, 29
    rank 2, 30, 118
    strategic 11, 24, 113
reflexive game 1, 10, 32
reflexive mapping 53
reflexive partition 134
reflexive structure 136

self-appraisal 2
self-reflexion 1, 160
sequence of moves 8, 74
set
    of agents 8, 19
    of feasible actions 8, 17
social network 234
state of nature 9, 17
stock exchange 268
strategy
    guaranteeing 113
    rationalizable 23

transaction 157

unanimity condition 200
uncertainty 18
universal beliefs space 69

world 85
    possible 90
    real 91

Communications in Cybernetics, Systems Science and Engineering

Book Series Editor: Jeffrey 'Yi-Lin' Forrest

ISSN: 2164-9693

Publisher: CRC Press/Balkema, Taylor & Francis Group

1. A Systemic Perspective on Cognition and Mathematics
    Jeffrey Yi-Lin Forrest
    ISBN: 978-1-138-00016-2 (Hb)

2. Control of Fluid-Containing Rotating Rigid Bodies
    Anatoly A. Gurchenkov, Mikhail V. Nosov & Vladimir I. Tsurkov
    ISBN: 978-1-138-00021-6 (Hb)

3. Research Methodology: From Philosophy of Science to Research Design
    Alexander M. Novikov & Dmitry A. Novikov
    ISBN: 978-1-138-00030-8 (Hb)

4. Fast Fashion Systems: Theories and Applications
    Tsan-Ming Choi
    ISBN: 978-1-138-00029-2 (Hb)

5. Reflexion and Control: Mathematical Models
    Dmitry A. Novikov & Alexander G. Chkhartishvili
    ISBN: 978-1-138-02473-1 (Hb)